ASTER 20th Anniversary

ASTER 20th Anniversary

Special Issue Editors

Yasushi Yamaguchi
Michael J. Abrams

MDPI • Basel • Beijing • Wuhan • Barcelona • Belgrade

Special Issue Editors

Yasushi Yamaguchi
Nagoya University
Japan

Michael J. Abrams
Jet Propulsion Laboratory,
California Institute of Technology
USA

Editorial Office
MDPI
St. Alban-Anlage 66
4052 Basel, Switzerland

This is a reprint of articles from the Special Issue published online in the open access journal *Remote Sensing* (ISSN 2072-4292) from 2018 to 2020 (available at: https://www.mdpi.com/journal/remotesensing/special_issues/ASTER).

For citation purposes, cite each article independently as indicated on the article page online and as indicated below:

LastName, A.A.; LastName, B.B.; LastName, C.C. Article Title. *Journal Name* **Year**, *Article Number*, Page Range.

ISBN 978-3-03928-684-3 (Pbk)
ISBN 978-3-03928-685-0 (PDF)

Cover image courtesy of Michael J. Abrams.

Contents

About the Special Issue Editors vii

Preface to "ASTER 20th Anniversary" ix

Yasushi Yamaguchi and Michael Abrams
Editorial for the Special Issue "ASTER 20th Anniversary"
Reprinted from: *Remote Sens.* **2020**, *12*, 884, doi:10.3390/rs12050884 1

Michael Abrams and Yasushi Yamaguchi
Twenty Years of ASTER Contributions to Lithologic Mapping and Mineral Exploration
Reprinted from: *Remote Sens.* **2019**, *11*, 1394, doi:10.3390/rs11111394 3

Michael S. Ramsey and Ian T.W. Flynn
The Spatial and Spectral Resolution of ASTER Infrared Image Data: A Paradigm Shift in
Volcanological Remote Sensing
Reprinted from: *Remote Sens.* **2020**, *12*, 738, doi:10.3390/rs12040738 31

Kana Kurata and Yasushi Yamaguchi
Integration and Visualization of Mineralogical and Topographical Information Derived from
ASTER and DEM Data
Reprinted from: *Remote Sens.* **2019**, *11*, 162, doi:10.3390/rs11020162 71

Louis Gonzalez, Valérie Vallet and Hirokazu Yamamoto
Global 15-Meter Mosaic Derived from Simulated True-Color ASTER Imagery
Reprinted from: *Remote Sens.* **2019**, *11*, 441, doi:10.3390/rs11040441 91

Han Fu, Bihong Fu, Yoshiki Ninomiya and Pilong Shi
New Insights of Geomorphologic and Lithologic Features on Wudalianchi Volcanoes in the
Northeastern China from the ASTER Multispectral Data
Reprinted from: *Remote Sens.* **2019**, *11*, 2663, doi:10.3390/rs11222663 106

**Toru Kouyama, Soushi Kato, Masakuni Kikuchi, Fumihiro Sakuma, Akira Miura, Tetsushi
Tachikawa, Satoshi Tsuchida, Kenta Obata and Ryosuke Nakamura**
Lunar Calibration for ASTER VNIR and TIR with Observations of the Moon in 2003 and 2017
Reprinted from: *Remote Sens.* **2019**, *11*, 2712, doi:10.3390/rs11222712 122

Hideyuki Tonooka and Tetsushi Tachikawa
ASTER Cloud Coverage Assessment and Mission Operations Analysis Using Terra/MODIS
Cloud Mask Products
Reprinted from: *Remote Sens.* **2019**, *11*, 2798, doi:10.3390/rs11232798 140

Tom Cudahy, Pilong Shi, Yulia Novikova and Bihong Fu
Satellite ASTER Mineral Mapping the Provenance of the Loess Used by the Ming to Build their
Earthen Great Wall
Reprinted from: *Remote Sens.* **2020**, *12*, 270, doi:10.3390/rs12020270 158

**Satoshi Tsuchida, Hirokazu Yamamoto, Toru Kouyama, Kenta Obata, Fumihiro Sakuma,
Tetsushi Tachikawa, Akihide Kamei, Kohei Arai, Jeffrey S. Czapla-Myers, Stuart F. Biggar
and Kurtis J. Thome**
Radiometric Degradation Curves for the ASTER VNIR Processing Using Vicarious and
Lunar Calibrations
Reprinted from: *Remote Sens.* **2020**, *12*, 427, doi:10.3390/rs12030427 193

Jigjidsurengiin Batbaatar, Alan R. Gillespie, Ronald S. Sletten, Amit Mushkin, Rivka Amit, Darío Trombotto Liaudat, Lu Liu and Gregg Petrie
Toward the Detection of Permafrost Using Land-Surface Temperature Mapping
Reprinted from: *Remote Sens.* **2020**, *12*, 695, doi:10.3390/rs12040695 . . . . **210**

Amit Mushkin, Alan R. Gillespie, Elsa A. Abbott, Jigjidsurengiin Batbaatar, Glynn Hulley, Howard Tan, David M. Tratt and Kerry N. Buckland
Validation of ASTER Emissivity Retrieval Using the Mako Airborne TIR Imaging Spectrometer at the Algodones Dune Field in Southern California, USA
Reprinted from: *Remote Sens.* **2020**, *12*, 815, doi:10.3390/rs12050815 . . . . **232**

Hiroyuki Fujisada, Minoru Urai and Akira Iwasaki
Technical Methodology for ASTER Global Water Body Data Base
Reprinted from: *Remote Sens.* **2018**, *10*, 1860, doi:10.3390/rs10121860 . . . . **255**

About the Special Issue Editors

Yasushi Yamaguchi is the Japan ASTER Science Team Leader and a professor at Graduate School of Environmental Studies, Nagoya University, Japan. He graduated Tohoku University, Japan, and obtained Doctor of Science degree in geologic remote sensing from the same university in 1989. Prior to joining Nagoya University in 1996, he spent 16 years with the Geological Survey of Japan. He was a visiting scientist at Department of Applied Earth Sciences, Stanford University, USA, from 1984 to 1986. He served as the Dean of the Graduate School of Environmental Studies at Nagoya University from 2009 to 2011, as president of the Remote Sensing Society of Japan from 2012 to 2014, and as all Japan Chapter Chair of the IEEE Geoscience and Remote Sensing Society from 2016 to 2017. His research is focused on remote sensing using data from ASTER and other optical sensor images for geological mapping, terrestrial vegetation monitoring, and heat balance analysis in urban centers. In addition, he uses data from a lunar radar sounder on-board the Kaguya lunar probe to analyze the subsurface structure of the moon.

Michael Abrams has been on the U.S./Japan ASTER Science Team since 1988, and became the U.S. ASTER Science Team Leader in 2003. Since 1973, he has worked at NASA's Jet Propulsion Laboratory in geologic remote sensing. He has been on the science team for many instruments, including Skylab, HCMM, Landsat, EO-1, HyspIRI, and ASTER. Combining geologic research and technology validation, he has helped design field spectrometers and aircraft scanners in the optical reflected and emissive parts of the electromagnetic spectrum. His areas of geologic specialization are mineral exploration, ophiolites, natural hazards, and volcanology. He is the author of over 100 journal publications, several book chapters, and serves on the Editorial Board for earth science journals and books.

Preface to "ASTER 20th Anniversary"

The Advanced Speceborne Thermal Emission and Reflection Radiometer (ASTER) is a research facility instrument on NASA's Terra spacecraft. We celebrated the 20th anniversary of ASTER's launch in December 1999. ASTER has been providing high spatial resolution multispectral data in the VNIR, SWIR, and TIR regions, and along-track stereo data. Starting April 2016, ASTER data have been distributed to the public at no cost. The most popular data set is the ASTER Global DEM, which covers almost the entire land surface with a 30 m grid size. ASTER data have been widely used in a variety of applications such as land surface mapping and change detection, volcano and other natural hazard monitoring, mineral exploration, and urban heat island monitoring. This Special Issue consists of 12 papers (2 reviews, 9 articles, and 1 technical note) and covers topics including development of new techniques to process ASTER data, calibration activities to ensure long-term consistency of ASTER data, validation of the ASTER data products, and scientific achievements using ASTER data. Abrams and Yamaguchi (2019) provide a comprehensive review of ASTER contributions to lithological mapping and mineral exploration. Ramsey and Flynn (2020) present the history of ASTER's contribution to volcanology, highlighting unique aspects of the instrument and its data. Kurata and Yamaguchi (2019) propose a method for combining and visualizing multiple lithological indices derived from ASTER data and topographical information derived from DEM data. Gonzales et al. (2019) propose a new methodology to build an Earth-wide mosaic using ASTER images in pseudo-true color. Fu et al. (2019) analyzed the geomorphologic and lithologic features of Wudalianchi volcanoes in Northeastern China using ASTER multispectral and DEM data. Kouyama et al. (2019) assess the sensitivity degradations of the ASTER bands based upon the lunar and deep-space observation data obtained in 2003 and 2017. Tonooka and Tachikawa (2019) developed a method for ASTER cloud coverage assessment using the MODIS cloud mask product and also evaluated performance of the cloud avoidance function implemented in the ASTER observation scheduler. Cudahy et al. (2020) show that ASTER mineral maps revealed both the compositional heterogeneity of loess as well as the complexity of the sediment transport pathways of individual loess components around the Great Wall of China, built during the Ming Dynasty. Tsuchida et al. (2020) discuss the sensor degradation curves of the ASTER VNIR bands based on the results of the onboard calibrator, the vicarious calibration, and the cross calibration since February 2014. Batbaatar et al. (2020) propose a method to map the "zero curtain" as a precursor for delineating permafrost boundaries, determined from ASTER and MODIS land-surface temperature data. Mushkin et al. (2020) provide validation of the ASTER emissivity product using data from the airborne TIR hyperspectral Mako sensor. Fujisada et al. (2018) describe the technical methodology for improving the initial tile-based waterbody data that are created during production of the ASTER GDEM.

Yasushi Yamaguchi, Michael J. Abrams
Special Issue Editors

Editorial

Editorial for the Special Issue "ASTER 20th Anniversary"

Yasushi Yamaguchi [1,*] **and Michael Abrams** [2]

1   Graduate School of Environmental Studies, Nagoya University, Nagoya 464-8601, Japan
2   Jet Propulsion Laboratory, California Institute of Technology, Pasadena, CA 91104, USA;
    mjabrams@jpl.nasa.gov
*   Correspondence: yasushi@nagoya-u.jp

Received: 5 March 2020; Accepted: 9 March 2020; Published: 10 March 2020

The Advanced Spaceborne Thermal Emission and Reflection Radiometer (ASTER) is a research facility instrument on NASA's Terra spacecraft. We celebrated the 20th anniversary of ASTER since its launch in December 1999. ASTER has been providing high spatial resolution multispectral data in the visible to near infrared (VNIR), short wave infrared (SWIR) and thermal infrared (TIR) regions, and along-track stereo data. Starting April 2016, ASTER data have been distributed to the public at no cost. Another important, and the most popular data set, is the ASTER Global digital elevation model (DEM), which covers almost the entire land surface at 30 m grid size. ASTER data have been widely used in a variety of application areas such as land surface mapping and change detection, volcano and other natural hazard monitoring, mineral exploration, and urban heat island monitoring.

This special issue consists of 12 papers (2 reviews, 9 articles and 1 technical note), and covers topics including the development of new techniques to process ASTER data, calibration activities to ensure long-term consistency of ASTER data, validation of the ASTER data products, and scientific achievements using ASTER data. Abrams and Yamaguchi [1] provide a comprehensive review on ASTER contribution to lithological mapping and mineral exploration. Ramsey and Flynn [2] present the history of ASTER's contribution to volcanology, highlighting unique aspects of the instrument and its data. Kurata and Yamaguchi [3] propose a method of combining and visualizing multiple lithological indices derived from ASTER data and topographical information derived from DEM data. Gonzalez et al. [4] propose a new methodology to build an Earth-wide mosaic using ASTER images in pseudo-true color. Fu et al. [5] analyze the geomorphologic and lithologic features of Wudalianchi volcanoes in northeastern China by using the ASTER multispectral and DEM data. Kouyama et al. [6] assess sensitivity degradations of the ASTER bands based upon lunar and deep-space observation data obtained in 2003 and 2017. Tonooka and Tachikawa [7] develop a method for ASTER cloud coverage assessment using the Moderate Resolution Imaging Spectroradiometer (MODIS) cloud mask product, and also evaluated performance of the cloud avoidance function implemented in the ASTER observation scheduler. Cudahy et al. [8] show that ASTER mineral maps revealed both the compositional heterogeneity of loess, as well as the complexity of the sediment transport pathways of individual loess components around the Great Wall of China, built during the Ming Dynasty. Tsuchida et al. [9] discuss the sensor degradation curves of the ASTER VNIR bands based on the results of the onboard calibrator, the vicarious calibration, and the cross calibration since February 2014. Batbaatar et al. [10] propose a method to map the "zero curtain" as a precursor for delineating permafrost boundaries, determined from ASTER and MODIS land-surface temperature data. Mushkin et al. [11] provide validation of the ASTER emissivity product by using data from the airborne TIR hyperspectral Mako sensor. Fujisada et al. [12] describe the technical methodology for improving the initial tile-based waterbody data that are created during production of the ASTER GDEM.

Author Contributions: The two guest editors contributed equally to all aspects of this editorial. All authors have read and agreed to the published version of the manuscript.

Acknowledgments: The guest editors would like to thank the authors who contributed to this Special Issue and to the reviewers who dedicated their time to provide the authors with valuable and constructive recommendations.

Conflicts of Interest: The guest editors declare no conflict of interest.

References

1. Abrams, M.; Yamaguchi, Y. Twenty Years of ASTER Contributions to Lithologic Mapping and Mineral Exploration. *Remote Sens.* **2019**, *11*, 1394. [CrossRef]
2. Ramsey, M.S.; Flynn, I.T.W. The Spatial and Spectral Resolution of ASTER Infrared Image Data: A Paradigm Shift in Volcanological Remote Sensing. *Remote Sens.* **2020**, *12*, 738. [CrossRef]
3. Kurata, K.; Yamaguchi, Y. Integration and Visualization of Mineralogical and Topographical Information Derived from ASTER and DEM Data. *Remote Sens.* **2019**, *11*, 162. [CrossRef]
4. Gonzalez, L.; Vallet, V.; Yamamoto, H. Global 15-Meter Mosaic Derived from Simulated True-Color ASTER Imagery. *Remote Sens.* **2019**, *11*, 441. [CrossRef]
5. Fu, H.; Fu, B.; Ninomiya, Y.; Shi, P. New Insights of Geomorphologic and Lithologic Features on Wudalianchi Volcanoes in the Northeastern China from the ASTER Multispectral Data. *Remote Sens.* **2019**, *11*, 2663. [CrossRef]
6. Kouyama, T.; Kato, S.; Kikuchi, M.; Sakuma, F.; Miura, A.; Tachikawa, T.; Tsuchida, S.; Obata, K.; Nakamura, R. Lunar Calibration for ASTER VNIR and TIR with Observations of the Moon in 2003 and 2017. *Remote Sens.* **2019**, *11*, 2712. [CrossRef]
7. Tonooka, H.; Tachikawa, T. ASTER Cloud Coverage Assessment and Mission Operations Analysis Using Terra/MODIS Cloud Mask Products. *Remote Sens.* **2019**, *11*, 2798. [CrossRef]
8. Cudahy, T.; Shi, P.; Novikova, Y.; Fu, B. Satellite ASTER Mineral Mapping the Provenance of the Loess Used by the Ming to Build their Earthen Great Wall. *Remote Sens.* **2020**, *12*, 270. [CrossRef]
9. Tsuchida, S.; Yamamoto, H.; Kouyama, T.; Obata, K.; Sakuma, F.; Tachikawa, T.; Kamei, A.; Arai, K.; Czapla-Myers, J.S.; Biggar, S.F.; et al. Radiometric Degradation Curves for the ASTER VNIR Processing Using Vicarious and Lunar Calibrations. *Remote Sens.* **2020**, *12*, 427. [CrossRef]
10. Batbaatar, J.; Gillespie, A.R.; Sletten, R.S.; Mushkin, A.; Amit, R.; Trombotto Liaudat, D.; Liu, L.; Petrie, G. Toward the Detection of Permafrost Using Land-Surface Temperature Mapping. *Remote Sens.* **2020**, *12*, 695. [CrossRef]
11. Mushkin, A.; Gillespie, A.R.; Abbott, E.A.; Batbaatar, J.; Hulley, G.; Tan, H.; Tratt, D.M.; Buckland, K.N. Validation of ASTER Emissivity Retrieval Using the Mako Airborne TIR Imaging Spectrometer at the Algodones Dune Field in Southern California, USA. *Remote Sens.* **2020**, *12*, 815. [CrossRef]
12. Fujisada, H.; Urai, M.; Iwasaki, A. Technical Methodology for ASTER Global Water Body Data Base. *Remote Sens.* **2018**, *10*, 1860. [CrossRef]

 *remote sensing*

Review

Twenty Years of ASTER Contributions to Lithologic Mapping and Mineral Exploration

Michael Abrams [1,*] **and Yasushi Yamaguchi** [2]

[1] Jet Propulsion Laboratory, California Institute of Technology, Pasadena, CA 91104, USA
[2] Graduate School of Environmental Studies, Nagoya University, Nagoya 464-8601, Japan; yasushi@nagoya-u.jp
* Correspondence: mjabrams@jpl.nasa.gov; Tel.: +1-626-375-5922

Received: 30 April 2019; Accepted: 8 June 2019; Published: 11 June 2019

Abstract: The Advanced Spaceborne Thermal Emission and Reflection Radiometer is one of five instruments operating on the National Aeronautics and Space Administration (NASA) Terra platform. Launched in 1999, the Advanced Spaceborne Thermal Emission and Reflection Radiometer (ASTER) has been acquiring optical data for 20 years. ASTER is a joint project between Japan's Ministry of Economy, Trade and Industry; and U.S. National Aeronautics and Space Administration. Numerous reports of geologic mapping and mineral exploration applications of ASTER data attest to the unique capabilities of the instrument. Until 2000, Landsat was the instrument of choice to provide surface composition information. Its scanners had two broadband short wave infrared (SWIR) bands and a single thermal infrared band. A single SWIR band amalgamated all diagnostic absorption features in the 2–2.5 micron wavelength region into a single band, providing no information on mineral composition. Clays, carbonates, and sulfates could only be detected as a single group. The single thermal infrared (TIR) band provided no information on silicate composition (felsic vs. mafic igneous rocks; quartz content of sedimentary rocks). Since 2000, all of these mineralogical distinctions, and more, could be accomplished due to ASTER's unique, high spatial resolution multispectral bands: six in the SWIR and five in the TIR. The data have sufficient information to provide good results using the simplest techniques, like band ratios, or more sophisticated analyses, like machine learning. A robust archive of images facilitated use of the data for global exploration and mapping.

Keywords: ASTER; mineral exploration; geologic mapping

1. Introduction

The Advanced Spaceborne Thermal Emission and Reflection Radiometer (ASTER) is one of five instruments on the U.S. Terra spaceborne platform (the other instruments are the Moderate Resolution Imaging Spectroradiometer (MODIS), Clouds and the Earth's Radiant Energy System (CERES), Multi-angle Imaging SpectroRadiometer (MISR), and Measurement of Pollution in the Troposphere (MOPITT)). Launched in December 1999, ASTER has been continuously acquiring image data for 20 years. ASTER is a joint project between Japan's Ministry of International Trade and Industry (MITI) (later changed to Ministry of economy, Trade and Industry (METI)) and the U.S. National Aeronautics and Space Administration (NASA). Japanese aerospace companies built the ASTER subsystems for METI; NASA provided the Terra platform and the Atlas 2AS launch vehicle. Both organizations are responsible for instrument calibration, scheduling, data archiving, processing, and distribution.

ASTER was conceived as a geologic mapping instrument. It was designed to provide several improvements over instruments existing at the time, like Landsat. The science team pushed for better spatial resolution, high spectral resolution short wave infrared (SWIR) bands, multispectral thermal infrared (TIR) bands, and along-track stereo capability. Until April 2008, the ASTER subsystems performed nominally. At that time, the SWIR subsystem had a failure of the detector cooling apparatus (not the Stirling cycle cooler), so no further SWIR data could be captured. Since April 2016, all of the ASTER scenes in the data archive, and all of their derived products, were made available to all users at no cost.

As a general-purpose imaging instrument, ASTER-acquired data are used in numerous scientific disciplines, including land use/land cover, urban monitoring, urban heat island studies, wetlands studies, agriculture monitoring, forestry, etc. [1]. Significant resources are devoted to monitoring 1500 active volcanoes and 3000 valley glaciers. However, of special importance is the use of ASTER data for geologic applications: lithologic mapping and mineral exploration.

This article reviews the geologic applications of early spaceborne optical instruments, discusses the history of the ASTER instrument, describes the instrument, and reviews applications of ASTER data for lithologic mapping and mineral exploration

2. Early Geologic Applications of Spaceborne Instruments

The first optical satellite data applied for geologic mapping were acquired by NASA's Landsat 1 (also known as ERTS (Earth Resources Technology Satellite)) Multispectral Scanner (MSS) with four bands in the visible to near infrared wavelengths (VNIR), and about 80 m spatial resolution, launched in 1972. Several studies demonstrated the usefulness of these data for geologic mapping. An early publication by Goetz and his team in 1975 [2] described mapping the geology of the Coconino Plateau on the south rim of the Grand Canyon in northern Arizona. They were able to distinguish lithologic units and faults, applying the results for ground water exploration. Identification of the composition of sedimentary geologic units was not possible, given the scanner's broad bands and restricted wavelength position; different mappable units could, however, be separated and mapped. Baker [3] reviewed similar applications in presentations delivered at one of the first geology applications symposia, focusing on additional results using Landsat MSS data.

In 1978 and 1980, two very influential textbooks describing the usage of Landsat 1 for geologic applications alerted the general applications community to the value of remote sensing data. The first of many textbooks describing geologic applications of satellite data was *Remote Sensing: Principles and Interpretations* written by Floyd Sabins in 1978 [4]. He introduced remote sensing to Chevron, leading to oil discoveries in Sudan and Papua New Guinea. His programs for digitally processing Landsat images discovered world-class Chile copper deposits. A second influential book was *Remote Sensing in Geology*, edited by Siegal and Gillespie [5] published in 1980. These two books alerted the general geologic community to the possible applications of satellite remote sensing data, and led several resource exploration companies to form their own in-house technical divisions.

In 1982, NASA launched the first in a series of Landsat Thematic Mapper (TM) instruments, on Landsat 4. This groundbreaking scanner had 30 m spatial resolution, repeat global coverage, and seven spectral bands: four in the VNIR like the MSS, two in the SWIR, and one in the TIR [6]. The late addition of SWIR band 7 in the 2–2.5 micron region, was directly the result of studies using aircraft instrument data, demonstrating the application of data from the SWIR region for detecting hydrothermal alteration minerals [7] (a change in mineralogy as a result of interaction of the rock with hot fluids). The Cuprite, Nevada test site, used for this study, became a spectral validation site for most future optical instruments.

A seminal study, detailing a 4-year NASA sponsored project, was the joint NASA and Geosat test case report, published in 1985 [8]. The project was a joint collaboration between NASA and the non-renewable resources exploration industry to test and document the applications of remote sensing data for porphyry copper, uranium, and petroleum exploration. Methods used for data processing were adopted by later researchers and exploration geologists.

For the next 18 years after 1982, Landsat TM data were the workhorse of the geology remote sensing community. A sampling of relevant publications includes geologic mapping and mineral targeting in Precambrian terrain in India [9]; mapping mafic and ultramafic rocks in the Oman Ophiolite, and targeting hydrothermal alteration zones related to possible massive sulfide copper deposits [10] (band 7 data were crucial for detecting hydrothermal alteration); mapping ophiolite lithologies in Cyprus, and delineation of a critical lithologic boundary to target possible massive sulfide deposits (all TM bands were needed to distinguish two types of lava flows [11]). Several geologic mapping studies in the Arabian Shield and Middle East focused on mineral exploration for uranium, gold, copper, and other metal deposits [12,13]; again, the availability of SWIR data was critical to detect hydrothermal alteration. In 1999, Sabins [14] published a comprehensive review of the application of remote sensing data for mineral exploration. Following on the heels of his earlier book on remote sensing applications to geology, this publication was key in influencing exploration companies to further embrace remote sensing technology in their reconnaissance exploration strategies.

A fascinating application of Landsat 5 TM data for geobotanical mapping was reported by Almeida et al. [15]. Their goal was to detect geobotanical anomalies associated with hydrothermal alteration in epithermal high-sulfidation gold deposits in the Amazon region in an area of virgin tropical rain forest, Brazil. Their method was to concentrate information and reduce data dimensionality by applying spectral indices, principal component analyses to the indices, another principal component analysis to the original VNIR and SWIR bands, and convolutional filtering. Field information showed a near-perfect spatial correlation between color classes highlighted in this Landsat image product and hydrothermal alteration facies identified in outcrops.

In 1992, NASDA/MITI/STA (National Space Development Agency/Science and Technology Agency) launched the Japanese Earth Resources Satellite (JERS-1) with a synthetic aperture radar instrument, and a high resolution optical scanner (OPS) with eight bands. In addition to three VNIR bands, OPS had four SWIR bands. The instrument was designed to provide remote sensing data to Japan's resource exploration industry to search globally for non-renewable resources. The instrument had a 75 km swath width, and about 25 m spatial resolution [16]. The SWIR band performance was fairly poor, with excessive noise (the VNIR and SAR bands performed well). In one report [17], advanced image processing techniques were applied to the SWIR data. When combined with Landsat TM data for a test site in Eritrea, JERS-1 data permitted routine identification of marbles, and allowed distinction of rocks bearing either Al-OH or Mg-OH phyllosilicates. In 1998, JERS-1 ceased operations, with very few geologic studies reported. However, it did provide the foundation for the next generation Japanese instrument, by demonstrating the value of multispectral SWIR bands for geologic applications. As a result, ASTER came into being, as described in the following sections.

3. ASTER History

ASTER has its roots in several moderate resolution imaging sensors [18]. The Landsat instruments (Multispectral Scanner, and Thematic Mapper) had developed a large and devoted user community adapted to analyzing multispectral data. The second most used data was provided by the French SPOT (Satellite Pour l'Observation de la Terre) instruments, with 10–20 m VNIR wavelength data, and cross-track stereo. Between 1992 and 1998, the JERS-1 OPS acquired three bands of VNIR data, four bands of SWIR data, and along-track stereo data.

The Earth Observing System (EOS) ASTER program began as two separate instruments proposed separately by the U.S. and Japan [19] in the 1980s. The US had a proposal for the Thermal Infrared Ground Emission Radiometer (TIGER), a 14 channel imager plus a profiling spectrometer. At the same time Japan's MITI was designing and proposed the Intermediate Thermal Infrared Radiometer (ITIR) with five SWIR bands and four TIR bands [20,21] as a follow-on to JERS-1. Once again, the design of the Japanese instrument focused on geologic applications. Starting in 1989, the joint U.S. and Japan science team worked jointly to come up with a compromise design for a VNIR-SWIR-TIR instrument to go on NASA's EOS AM-1 platform (re-named Terra after launch). The number of TIR bands was increased to five; the number of SWIR bands was increased to six; spatial resolution was decided; VNIR band 3 was selected for the along-track stereo; and bandpasses of all the channels were determined. In December 1999, ASTER was launched on NASA's Terra spacecraft, along with four other Earth observing instruments.

4. ASTER Instrument

The ASTER instrument comprises three separate scanners, located on different sites on the Terra platform (Figure 1). Each was built by a different Japanese aerospace company.

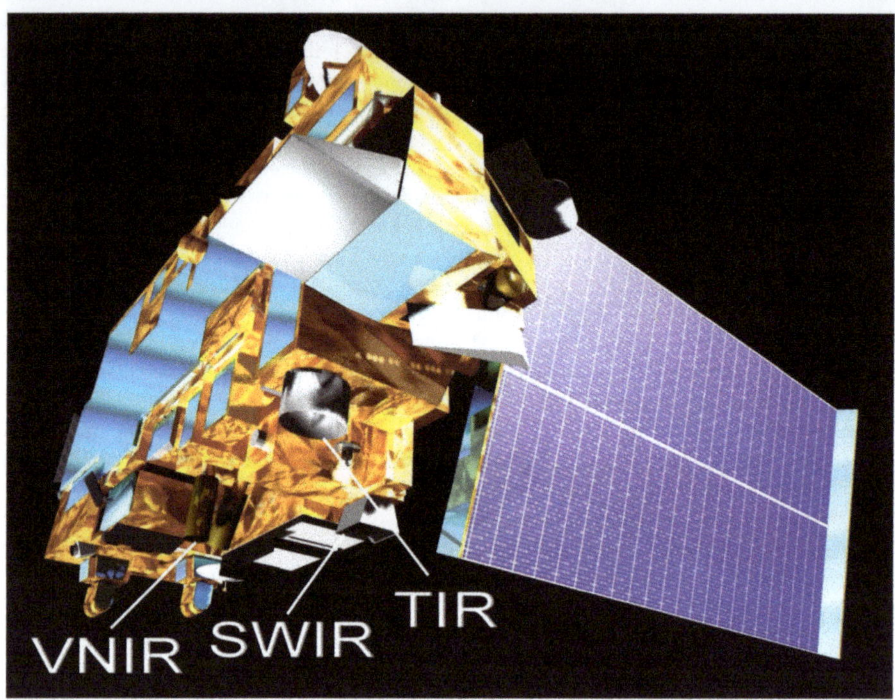

Figure 1. Artist's rendition of Terra platform and position of Advanced Spaceborne Thermal Emission and Reflection Radiometer (ASTER)'s visible to near infrared wavelengths (VNIR), short wave infrared (SWIR) and thermal infrared (TIR) imaging sensors.

In accordance with the scientific objectives of the mission, the ASTER instrument was designed to meet certain baseline performance requirements. In addition, several specific improvements were included to better ASTER's performance compared to existing optical sensors such as Landsat TM, SPOT HRV and JERS OPS:

- increased number of SWIR bands to six to improve mapping of surface composition;
- increased number of TIR bands to five to derive accurate surface temperature and emissivity measurements [22];
- improved radiometric accuracy and resolution [23].
- increased base-to-height (b/h) ratio of the stereo data, from 0.3 to 0.6, to improve surface elevation determination

ASTER acquires swaths of images that are 60 km wide, while orbiting the earth at 705 km altitude, in a sun-synchronous near-polar descending orbit. The equatorial crossing time is 10:30 am, a few minutes behind Landsat 7. ASTER (on the Terra platform) flew in formation with NASA's EO-1 satellite and Argentina's SAC-C satellite to form the morning constellation. ASTER must be tasked to acquire data; however, because the instrument has been in operation for almost 20 years, near-global coverage of the land has been achieved (Appendix A).

The three VNIR bands, with 15 m spatial resolution, have bandpasses similar to Landsat TM bands 2, 3, and 4 and the optical sensor of the JERS-1 OPS (Figure 2). In addition, the VNIR has along-track stereo coverage in band 3, with nadir and backward-looking telescopes. The base-to-height ratio of 0.6 allows calculation of Digital Elevation Models with 30 m postings, and about 15 m vertical accuracy [24]. The SWIR bands, with 30 m spatial resolution, were chosen mainly for the purpose of surface soil and mineral mapping. Band 4 is similar to TM band 5 located at 1.6 μm; bands 5–9 are narrow SWIR bands, replacing TM's single band 7, positioned in the 2–2.5 μm region to detect the presence of mineral absorption features, such as occur in clays, carbonates, and sulfates (Figure 2). The TIR bands, with 90 m spatial resolution, provide two major improvements over TM's single TIR band: derivation of emissivity values allows estimation of silica content, which is important in characterizing silicate rocks, the most prevalent rocks at the earth's surface [25]; and by correcting for emissivity, accurate surface kinetic temperature can be determined for energy flux modeling and climate modeling. Thus, ASTER data have greater mineral and lithologic mapping capability than Landsat data due to more SWIR and TIR bands with corresponding higher spectral resolution (Figure 2).

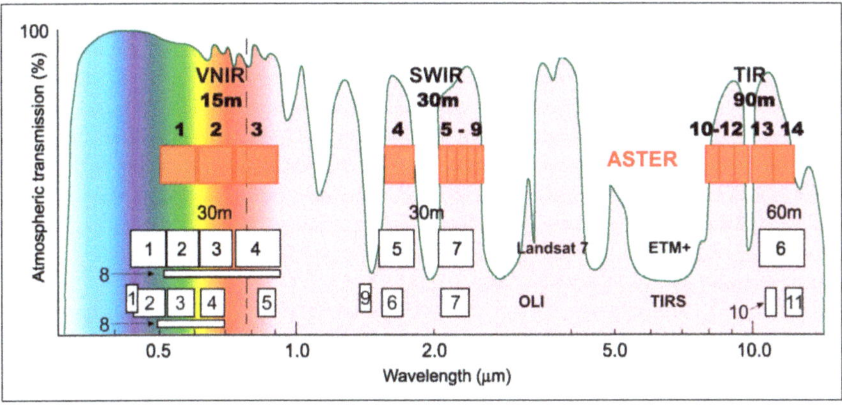

Figure 2. Spectral bandpasses of ASTER, Landsat Thematic Mapper, and the Operational Land Imager (OLI), the newest Landsat 8. Background is atmospheric transmission (Testa, F. et al. 2018).

5. Lithologic Mapping with ASTER Data

5.1. Mapping Using Only ASTER Data

In 2003, Rowan and Mars were one of the first researchers to report on lithologic mapping using ASTER data, over the rare earth mineral deposit at Mountain Pass, California [26]. Using all 14 ASTER bands, they were able to distinguish calcite from dolomite, mapped skarn deposits and marble in the contact metamorphic zones, distinguished Fe-muscovite from Al-muscovite in the granites and gneisses, and discriminated quartzose rocks. None of these discriminations could be accomplished with Landsat TM data, due to the lack of multispectral SWIR and TIR spectral bands. Watts and Harris [27] applied this method to map granite and gneiss in domes in the Himalayas. Yamaguchi and Naito [28] developed spectral indices using orthogonal transformation with ASTER SWIR bands for lithologic mapping. Their method relied on band ratios and thresholding. Wherever there were good rock exposures, ASTER data produced good results. Analyzing all of the ASTER spectral bands, Byrnes et al. [29] mapped volcanic lava flows from the Maunu Ulu eruption, Island of Hawaii. The TIR data highlighted variations in the silica coatings, the VNIR and SWIR bands indicated relative ages of the flows as they developed surface weathering products. Gomez et al. [30] published a paper describing lithologic mapping in arid Namibia using ASTER's VNIR and SWIR bands. They first converted the data to apparent reflectance, then processed the data with principal components analysis and supervised classification. Comparison with an existing 1:250,000 scale geologic map indicated that the ASTER data could be used to discriminate most of the geologic units. Several additional units were found, based on compositional differences, though the geologic map defined them as belonging to the same stratigraphic unit. Kargel et al. [31] reviewed applications of ASTER data for glaciological studies. The data were used to distinguish differences in surface cover on top of glaciers, as well as the extent of glacial lakes.

One of the first applications of ASTER TIR data for lithologic mapping was published by Ninomiya et al. [32]. They developed mineralogical indices using blended ratios of the TIR bands to create a Quartz Index, Carbonate Index, and Mafic Index; then tested their technique over arid parts of northwest China, eastern central Australia, and southern Tibet. Their results demonstrated the stability of the mineral indices to temperature and atmospheric changes. In succeeding years, many projects used Ninomiya "indices" for lithologic mapping with ASTER TIR data. More quantitative analyses of ASTER TIR data were reported by Hook et al. [33], who determined weight percent silica in igneous rocks, and validated the findings with laboratory measurements of field samples.

About the same time, Rowan et al. [34] reported on lithologic mapping of ultramafic rocks with ASTER data in Mordor Pound, NT, Australia. Analysis of the data, coupled with lab measurements, showed dominantly Al-OH and ferric-iron VNIR–SWIR absorption features in felsic rock spectra, and ferrous-iron and Fe,Mg-OH features in the mafic–ultramafic rock spectra. ASTER ratio images, matched-filter, and spectral-angle mapper processing were evaluated for mapping the lithologies. Combining analyses of VNIR, SWIR and TIR data resulted in discrimination of four mafic–ultramafic categories; three categories of alluvial–colluvial deposits; and a significantly more completely mapped quartzite unit than could be accomplished by using either data set alone. Hewson et al. [35] described a method to seamlessly mosaic 35 ASTER scenes to produce a regional mosaic for analysis. They then mapped Al-OH and carbonate from SWIR data, and quartz content from TIR data. SWIR bands were also used to map Al-OH composition. Comparison with large-scale maps and airborne hyperspectral data, supplemented with field sampling, constrained the ASTER map accuracy.

Qiu et al. [36] compared several spectral classification techniques, using a laboratory spectral library, ground spectral measurements, or selecting endmembers from the image. In the Allaqi-Heiani suture, Egypt, they found all three methods fairly similar, and allowed successful mapping of the well-exposed lithologies. Mapping carbonates and associated rocks in Oman was published by Rajendran et al. [37] and Rajendran and Nasir [38]. They were able to separate the ophiolitic rocks, carbonates, quartz-rich silicates, and surficial deposits. The remote sensing maps were very similar to the published geologic maps, and they recommended using ASTER data for mapping in other, similar environments. Guha et al. [39] also mapped carbonates, using ASTER data for the Kolkhan limestone in India. They applied different spectral mapping techniques; the results were similar with each of them. In the region near Askja volcano, Iceland, Grattinger et al. [40] used ASTER data to map glaciovolcanic deposits, including glaciovolcanic tuffs and subaerial pumice. The results were applied to paleo-ice reconstruction in a relatively inaccessible area. Tayebi et al. [41] mapped salt diapirs and surrounding areas using neural network models in the Zagros fold belt, Iran. Field observations and X-ray diffraction analysis of field samples confirmed the minerals identified remotely.

Yajima and Yamaguchi [42] used simple color composites of TIR data to separate mafic-ultramafic rocks (such as gabbro, dolerite and dunite) from various quartz-rich felsic rocks (such as granite and alluvium). This is a simple method to display lithological information from the TIR bands. Ninomiya and Fu [43] applied Ninomiya indices (Quartz Index, Mafic Index, and Carbonate Index) for ASTER TIR data to do lithological mapping in Tibet. Mapping relied on classification of quartzose rocks based on variations on the carbonate and mafic indices, and the granitic rocks based on the feldspar content. Ozyavas [44] applied standard image processing techniques to ASTER data over the study area around the Salt Lake Fault, Turkey. They were able to map gypsum and carbonate rocks, primarily based on spectral differences in the SWIR bands. A variation of the band ratio method was described by

Askari et al. [45] for lithologic mapping of sedimentary rocks, north Iran. Using VNIR and SWIR bands, they mapped quartz, carbonate, Al, Fe, Mg-OH bearing minerals, and created lithologic maps that matched well with existing geology maps. Hubbard et al. [46] used ASTER TIR emissivity images to characterize the physiochemical characteristics of sand dunes at seven different sites. They found that " … less dense minerals typically have higher abundances near the center of the active and most evolved dunes in the field, while more dense minerals and glasses appear to be more abundant along the margins of the active dune fields." ASTER data were applied to characterize limestones for industrial rock resource assessment in Oman by Rajendran et al. [47]. They were able to separate dolomite from calcite-bearing marble; this has a direct application in the industrial rock business.

Several studies focused on mapping granitoid rocks. Massironi et al. [48], working in the Saghro massif, Morocco, used all the ASTER bands, applying simple processing techniques. They were able to distinguish different granitoid rocks with similar silica content based on secondary minerals, and separate plutons with varying silica content using TIR data. Bertoldi et al. [49] used field and laboratory data to guide processing of ASTER data to map lichen-covered granitic rocks in the western Himalayas. Their maps were based on characterizing spectral differences from various lichens, with spectral characteristics of muscovite. In the Dahab Basin, Egypt, Omran et al. [50] used band ratios of ASTER VNIR, SWIR, and TIR data, combined with field investigations, to map and separate granitoid rocks of Cambrian and Cretaceous ages. They revised and updated existing geologic maps by adding rock units and re-interpreting the geologic history. Zheng and Fu [51] used band ratios of ASTER SWIR and TIR data to separate alkali-feldspar granite, granite, granodiorite, and monzogranite. These distinctions relied on determination of silica difference expressed in the TIR bands. In the Anti-Atlas Mountains, Morocco, El Janati et al. [52] used classification algorithms to map the spatial distribution of porphyritic granites, granodiorites, and peraluminous leucogranites. They also were able to map different kinds of metamorphic rocks and carbonate cover rocks. Guha and Kumar [53] developed a variation of Ninomiya's thermal indices to map granitoids in Dharwar Craton, India. They found that their mafic index was comparable with Ninomiya's index, but their quartz index was better, for their study area. Asran et al. [54] used band ratios of ASTER data to separate granodiorite, monzogranite, syenogranite, and alkali-feldspar granite. Mapping their distribution, with structural information and microfabric data, led to revised interpretation of their deformation history.

Another group of reports described applications of ASTER data for mapping ophiolite complexes. These unusual terranes expose dominantly mafic and ultramafic rocks. ASTER data have been shown to be particularly effective in separating the dark rocks. Li et al. [55] mapped the Derni ophiolite complex using spectral matching methods with spectra from a spectral library. Their results were of mixed accuracy, dependent on the quality of their spectral library. Using similar methods, Huang and Zhang [56] were more successful in mapping various rock types, in the Yarlung-Zangpo suture zone, Tibet. They used both the VNIR-SWIR data, and TIR data to achieve their results. In the Neyruz ophiolite, Iran, Tangestani et al. [57] used field and laboratory spectral measurements to train supervised classification of ASTER data with spectral feature fitting algorithm. Results suggested that this method could be applied to map other, more poorly mapped, ophiolite complexes. Ozkan et al. [58] mapped an accretionary complex in Turkey using hybrid color composite images combining ratio images and principal components images. They were able to delineate peridotite, gabbro, basalt, epi-ophiolitic sedimentary rocks, siliceous and carbonate rocks, and degrees of serpentinization.

5.2. Lithologic Mapping with ASTER and Other Remote Sensing Data

Deller and Andrews [59] combined Landsat, ASTER, and Advanced Land Imager (ALI) data to discriminate three laterite facies in Eritrea and Arabia, based on differences of iron and clay minerals. The results can be used to assess ground water quality, agricultural land, building resources, and potential mineralization sites. Qari et al. [60] used Landsat data to identify lineaments, and ASTER data to map lithology of the basement rocks in an area of Saudi Arabia. The resulting 1:100,000 geologic map was validated by fieldwork. Lithologic mapping in the Sighan ophiolite complex, Iran, was the subject of work by Pournamdari et al. [61]. They used band ratios and principal components analysis to separate mafic and ultramafic rocks. In the Neyriz ophiolite, Iran, Eslami et al. [62] also applied ASTER and Landsat data for lithologic mapping, using similar techniques. In North Africa, Adiri et al. [63] combined Landsat and ASTER data to map sedimentary rocks in Morocco. Lamri et al. [64] combined airborne magnetic and gamma-ray spectrometry data with ASTER data to map geology in a hardly accessible area in Saharan Africa. Integration of the two data sets allowed mapping of the Paleozoic and aeolian sand sedimentary cover, and the underlying granitoids. In West Africa, Metelka et al. [65] analyzed ASTER, Landsat, Radarsat, and Japanese Phased Array type L-band Synthetic Aperture Radar (PALSAR) radar data, combined with airborne gamma-ray spectrometry, to map geomorphological landform units. The resulting maps, over an area of long-term lateritic weathering history, were more accurate than existing maps. Yang et al. [66] combined ASTER data with high spatial resolution Chinese GaoFen-1 data to map lithologic units in the Tien Shan mountains, China. In comparison with lithologic mapping results using ASTER data alone, the fused data set was more accurate. Ge at al. [67] combined ASTER data with Sentinel-2A and digital elevation data to map lithologies in the Shinbanjing Ophiolite Complex, Mongolia. They found their method yielded high classification accuracy.

Hassan et al. [68] combined analyses of Landsat and ASTER data to map the basement rocks associated with the Meatiq dime, Egypt. ASTER data analysis revealed four granitic varieties, and Landsat data analysis allowed regional geologic mapping. Ali-Bik et al. [69] used data from ASTER, Landsat, and Sentinel-2 to map gneiss complex, low-grade ophiolitic and island-arc assemblages in the Gebel Zabarra area, Egypt. They used ASTER mineral indices, principal components analysis, and color composites to map the different rock types. They were able to propose revised metamorphic and tectonic history for the area based on the remote sensing results. Hadigeh and Ranjbar [70] combined ASTER data with panchromatic Indian Remote Sensing (IRS) data (moderate to high spatial resolution VNIR data) to map lithologies in Iran. Their classification maps closely matched published geologic maps. By combining ASTER data with SPOT-5 data, Lohrer et al. [71] mapped weathered wadi deposits in Jordan. They found that the initial transformation from hematite to goethite is the dominant process, and it is possible to predict new archaeological areas using remote sensing techniques. Over the Newer Volcanic Province, Australia, Boyce et al. [72] used ASTER data, airborne magnetic data, digital elevation models, and Google Earth images to identify eruptive centers. Seven previously identified eruptive centers were brought into question, and three new ones were identified.

Soltaninejad et al. [73] compared ASTER data and Landsat data for mapping evaporite minerals of Sirjan Playa, Iran. They used both spectra from field samples, and endmember spectra to classify both sets of images. Classification accuracy was about 92% for both data sets; better accuracy was achieved using image derived spectra. Chen et al. [74] mapped variations in metamorphic rocks in the Wuliangshan Mountains, China. They looked for spectral differences from minerals such as actinolite, chlorite, epidote, biotite, muscovite, hornblende, and sillimanite. Identification of metamorphic rocks in five mapped areas were consistent with existing data. Two new areas of unmapped metamorphic rocks were identified for further field study. In a vegetated area in the Yanshan Mountains, China, Wang et al. [75] used ASTER data to classify quartz sandstone, carbonate rocks, gneiss and andesite. Compared to the outcrop geologic map, accuracies were 90%, 87%, 77%, and 52% respectively. This is excellent, considering the presence of vegetation contamination and cover. A final example, published in this 20th Anniversary ASTER Special Issue, by Kurata and Yamaguchi [76] proposed a method of combining and visualizing multiple lithological indices derived from ASTER data, and topographical information derived from digital elevation model data, in a single color image that can be easily interpreted from a geological point of view. Indices highlighted silicate rocks, carbonate rocks, and amounts and types of clay minerals. Results were verified by field survey and comparison with previous studies in the test area.

6. Mineral Exploration with ASTER Data

6.1. Technique Development for Alteration Mapping

Many studies focused on developing techniques to extract mineralogical information from ASTER data. Crosta et al. [77] applied principal components analysis to data over epithermal deposits in Patagonia. The method was adopted by many later users and is referred as the "Crosta" method. Galvao et al. [78] demonstrated that ASTER data could be useful for alteration detection even in tropical savannah environments where there was some surface material showing between the vegetation cover. More image processing technique development by Moore et al. [79] applied principal components analysis and matched filter processing to identify unknown targets based on training sets over known deposits. Data from ASTER's thermal infrared bands were used by Rockwell and Hofstra [80] to identify quartz and carbonate minerals in Nevada based on detection of emissivity features in the TIR region. ASTER's unique multispectral TIR bands particularly allow mapping of SiO_2 variation, often a key characteristic of alteration associated with mineralization. Hosseinjani and Tangestani [81] used sub-pixel unmixing to determine the relative proportions of different minerals within each pixel. This technique relies on having a laboratory of mineral spectral responses, or end-member spectra extracted from the data. Another application of principal components analysis by Honarmand et al. [82] combined this with spectral angle mapper [83] to determine the probability of a mineral being present in a pixel.

In Birjand, Iran, Abdi and Karimpour [84] also applied spectral angle mapper to ASTER data to discriminate hydrothermal alteration. Spectral feature fitting method was used in Rabor area, Iran, to enhance hydrothermal alteration by Abbaszadeh and Hezarkhani [85]. This method compares the fit of image spectra to reference spectra using a least-squares technique. Additionally, in Iran, in the Dehaj-Sarduiyeh Copper Belt, Zadeh et al. [86] used mixture tuned matched filtering approach to map alteration. This method estimates the relative degree of match to each reference spectrum, and estimates the sub-pixel abundance [87]. More sophisticated (complicated) image processing methods were reported by Tayebi et al. [88]. They integrated coded spectral ratio images with SOM neural network models. Results were acceptable, though the method is transportable to other areas, and able to be used by other researchers, with great difficulty. Continued interest in application of sub-pixel unmixing was published by Modaberri et al. [89] for areas in Iran. They sought to map the usual assortment of alteration minerals, including alunite and jarosite.

6.2. General Alteration Mapping

In one of the first studies to appear, Rowan et al. [90] used ASTER data to map hydrothermal alteration minerals and zones at the Cuprite, Nevada test site. The work verified earlier results obtained with airborne scanners, and validated the mineralogical information extractable from the ASTER VNIR, SWIR, and TIR data. Numerous subsequent studies used newly-developed image processing techniques (described in the previous section) for alteration mapping. Bhadra et al. [91] reported on their analysis of ASTER data for mineral potential mapping in central Rajasthan, India, using standard image processing classification methods. In Jiafushaersu Area, China, Liu et al. [92] used color composites and principal components to map alteration associated with molybdenum mineralization adjacent to granitoid intrusions. Popov and Bakardjiev [93] used band ratios and spectral angle mapper analyses to identify alteration minerals in a humid and vegetated area. Known deposits with pits and tailings ponds were found. Several other possible, unknown targets were identified. In the Gobi Desert area, China, Son et al. [94] used band ratios to characterize a known Cu-Au mineral deposit. Using the indicator mineral assemblage, they identified a new area with pervasive argillic alteration. Two test sites in India, exposing intrusive and volcanic rocks, were sampled by Canbaz et al. [95] for material to analyze in the laboratory, using XRD (X-ray diffraction) and spectrometer measurements. They used the lab measurements to help interpret ASTER ratio and principal components images, and found that the combination of lab and remote sensing data were effective in mapping argillic alteration zones. In a highly vegetated area in India, Mahanta and Maiti [96] used band ratios and principal components analysis of ASTER data to map alteration assemblages: iron-rich gossans, sericitization, ferruginization, and chloritization. They identified two potential mineral prospects, one for hydrothermal polymetallic sulfides, and secondary iron and manganese mineralization. In Egypt, band ratios and principal components analysis were used by Abdelkareem and El-Baz [97] to identify hydrothermal alteration zones. Spectral analyses allowed characterization of chlorite, kaolinite, muscovite, and hematite. Validation by field investigations and XRD analyses contributed to the delineation of important prospects for gold and massive sulfide mineralization.

In another example, ASTER data were even used for lithological and alteration mapping in Antarctica. Pour et al. [98] reported work in the Oscar II coast area, northeastern Graham Land, Antarctic Peninsula. They applied special band ratios and band combinations with all 14 ASTER bands to detect muscovite, kaolinite, illite, montmorillonite, epidote, chlorite, and biotite. In their three-scene strip area, good geologic maps existed for the central scene, providing some control. However, poor or no maps existed for the northern and southern areas. Despite shadows, snow, and glaciers, they concluded that their approach for lithological and alteration mapping was highly successful.

In 2012, Australia's Commonwealth Scientific and Industrial Research Organization (CSIRO) released a series of Geoscience mineral maps for the continent of Australia, described by Cudahy [99]. These maps were created from thousands of ASTER scenes, acquired between 2000 and 2008. The CSIRO maps are the first (and only) continental-scale mineral maps generated from an imaging satellite, designed to measure clays, quartz, and other minerals. There are 17 Australia Geoscience products, such as kaolinite abundance, iron oxide species, and quartz content. Each product is a calibrated index of the geoscience product it represents. Color-coded displays are a visualization of the product's values. Their 100 m spatial resolution allows them to be used at the deposit as well as the regional scale. An example of the iron oxide abundance map is shown in Figure 3.

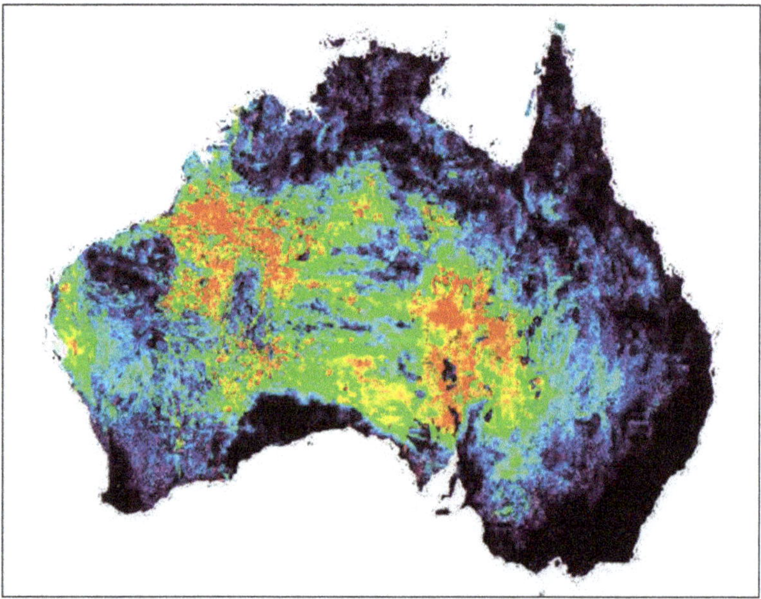

Figure 3. Australian ASTER Geoscience map of iron oxide abundance. Color code is red for high, to black for low [99].

More recent publications attest to the continuing use of ASTER data for mineral exploration. Testa et al. [100] described their work on the eastern flank of the Andean Cordillera, Argentina to do lithologic and hydrothermal alteration mapping of epithermal, porphyry, and tourmaline breccia districts. Argillic, phyllic, propylitic, and silicic alteration mineral assemblages were identified and mapped from the ASTER images. The results from field control areas confirmed the presence of the targeted minerals. They concluded "ASTER image processing of large areas has the ability to effectively discriminate smaller targets where it is possible to find mineral deposits. We believe it is critical to understand that interpreted hydrothermal alteration zones may not be real, so field verification is essential."

6.3. Alteration Mapping with ASTER Data and Other Data Sources

A few years after launch, geologists starting using ASTER data in combination with other remote sensing data and geophysical data for alteration mapping. An early study by Hubbard et al. [101] compared alteration mapping with ASTER, Hyperion, and ALI data in the VNIR and SWIR regions. Hyperion's hyperspectral data provided more information about mineralogy than either ASTER or ALI. However, the 7.5 km swath width was a distinct limitation. In Greenland, Bedini [102] used aircraft HyMap data with ASTER data to detect alteration minerals (HyMap has 128 bands in the 0.4–2.45 μm region and 5 m spatial resolution). Results and conclusions were similar to the previous study. In the Kerman magmatic arc in Iran, Honarmand et al. [103] used ASTER and ALI data to map alteration. Since ALI has broad spectral bands, similar to Landsat, very little additional information was added to the ASTER data. In another project Pour and Hashim [104] used ASTER, ALI, and Hyperion data for both lithologic and alteration mapping. Hyperion added additional mineralogical separation information, but ALI added very little to the ASTER data. Similarly, in a project reported by Ramos et al. [105] in the Andes, Hyperion data supplemented ASTER data for alteration mineral mapping, but was limited by its spatial coverage. In Gabal Dara, Egypt, Gemail et al. [106] used airborne magnetic geophysical data to complement mineralogical information from ASTER data for mineral exploration. The geophysical data added lithologic characteristics not available from optical remote sensing data. In Australia, most of the country is covered with weathered regolith, making exploration challenging. Lampinen et al. [107] used ASTER SWIR data with surface geochemistry analysis and gamma-ray spectrometry over a known base-metal deposit. The geophysical data allowed discrimination of lithologic units underneath the regolith. This guided interpretation of the geochemical and ASTER spectral information. Two recent studies over sites in China described combined use of ASTER data with other, seldom-reported, remote sensing instruments. Liu et al. [108] combined information from the Chinese hyperspectral scanner on the Tiangong-1 space station, with ASTER data for regional alteration mapping in the Jintanzi-Malianquan area. The hyperspectral instrument allowed detection of muscovite, kaolinite, chlorite, epidote calcite and dolomite, while ASTER data allowed detection of the first five, not dolomite. Hu et al. [109] combined ASTER data with Sentinel-2A multispectral data and Hyperion data. The image processing results were validated by field investigations. Identified hydrothermally altered rocks corresponded with five porphyry copper deposits. By extrapolation, three new prospects were discovered as a result.

6.4. ASTER Data Applied to Porphyry Copper Exploration

As early as 2003, a report was published by Volesky et al. [110] describing use of ASTER data to characterize massive sulfide copper deposits in Saudi Arabia. ASTER's SWIR data were crucial to characterize hydrothermal alteration minerals associated with these deposits.

Starting about the same time, and continuing to the present, studies in Iran dominate the literature on porphyry copper deposits, characterized by analysis of ASTER data. There are several reasons for this: (1) Iran hosts the largest number of porphyry copper deposits yet found; (2) the deposits are located in arid, desert landscapes: vegetation cover is minimal, exposures of the surface are almost 100%; (3) a wide range of alteration minerals are present, as erosion has cut into different depths of individual porphyry copper systems, exposing alteration zones from potassic and phyllic to propylitic and argillic.

One type of report concentrated on characterizing existing, known deposits in Iran. Karimpour et al. [111] described discrimination of erosion levels in the Maherabad, Shadan, and Chah Shaljami areas. Mohebi et al. [112] reported on delineation of structural controls on alteration and mineralization around Hanza Mountain. Sojdehee et al. [113] described discriminating hydrothermal alteration zones using SWIR data at the Daralu copper deposit. Farahbanksh [114] combined ASTER data with QuickBird data to characterize the Naysian porphyry copper deposit. Yousefi et al. [115,116] discriminated alteration zones using SWIR and TIR data to map sericitic, phyllic, and quartz-rich alteration zones in the Kerman Magmatic Arc. Safari et al. [117] combined Landsat and ASTER data to characterize the Shar-e-Babak deposit. In every one of these studies, the unique mineralogical information contained in the ASTER SWIR bands was critical to detect and map hydrothermal alteration zones.

A second type of report on the Iranian copper deposits focused on using ASTER data as an exploration tool to identify promising targets, usually extracting information gleaned by analyzing known deposits, then extrapolating to poorly explored or unexplored areas. The use of ASTER SWIR data featured strongly in Pour and Hashim's [118,119] in the Urumieh-Dokhtar Volcanic Belt. Applying a porphyry copper formation model, that postulated concentric alteration zones with characteristic mineral assemblages, promising targets were identified. Honarpazhouh [120] combined stream sediment geochemistry with ASTER data for reconnaissance mapping in the Khatun Abad area. This was a more effective exploration strategy than using the remote sensing data alone. Pazand et al. [121] applied ASTER data for reconnaissance exploration for porphyry copper mineralization in the Ahar area. In the Daraloo-Sarmeshk area, Alimohammadi et al. [122] used ASTER data to explore for undetected copper deposits using ASTER SWIR data to highlight alteration zones. Similar projects were reported by Yazdi et al. [123] in the Chahargonbad area, by Saadat [124] in the Feyz-Abad area, and by Zadeh and Honarmand [125] in the Dehaj-Sarduiyeh copper belt.

One of the best applications of ASTER data for regional mineral exploration in a copper belt was published by Mars and Rowan in 2006 [126]. They mosaicked 62 ASTER scenes covering a 900 km-wide belt in the Zagros magmatic arc, Iran. They first developed a series of logical operators involving band ratios and thresholds of ASTER data to highlight the presence of spectral absorption features associated with phyllic and argillic alteration. The operators were tested over the Cuprite, Nevada calibration and validation site [7], before being applied to the Iran data. Based on the alteration patterns, ~50 potential porphyry coper deposits were mapped northwest of the Zagros-Makran transform zone, and 11 potential deposits were mapped southeast of the transform. A small part of the mapped area is shown in Figure 4, around the Meiduk copper mine. Note the two large alteration centers northwest and southeast of the mine.

Studies focused on other areas of the world include Carrino et al.'s [127] project in the Chapi Chiara area of southern Peru. They used ASTER data to map the geology and alteration mineralogy of the region to define possible copper targets. Ibrahim et al. [128] applied ASTER and Landsat data, with field data in the North Hamisana shear zone, Egypt, to detect structural and lithologic controls for base metal sulfide deposits. By combining information extracted from ASTER and Landsat data, Zhang et al. [129] mapped hydrothermal alteration minerals around the Duolong copper deposit in Tibet. Rajendran and Nasir [130] characterized the spectral response of ASTER bands to map alteration zones of volcanogenic massive sulfide deposits in several known deposits. Additional work in China, by Zhang and Zhou [131] over the Baogutu porphyry copper deposit, used ASTER data to identify the associated alteration zones. This information could be used to explore other nearby areas with similar geology. In the Bangonghu-Nujiang metallogenic belt, Tibet, Dai et al. [132] used alteration detection from ASTER data to define new target areas with characteristic spectral features related to desired mineralogical assemblages.

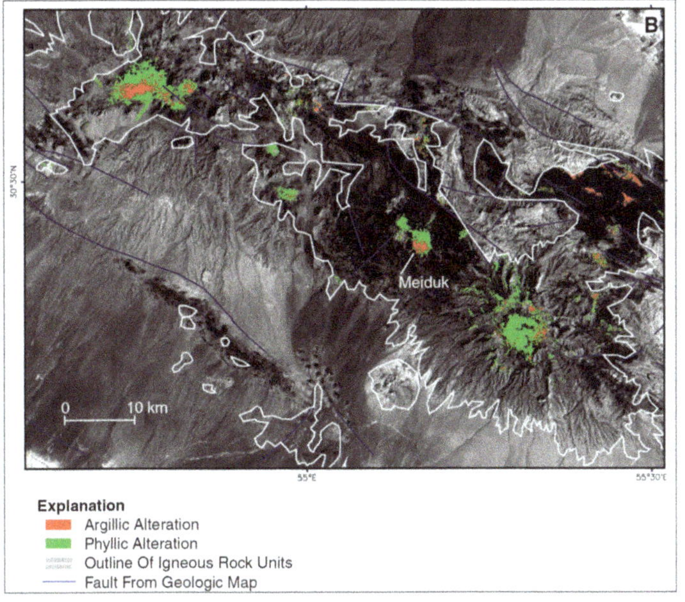

Figure 4. Landsat TM band 7 image with ASTER-derived argillic and phyllic alteration around the Meiduk copper mine, Iran. (From [126], Figure 17).

6.5. ASTER Data Applied to Gold Exploration

Second in numbers to reports on applications of ASTER data for porphyry copper exploration, are publications describing exploration for gold. The largest number of studies focused on describing alteration associated with gold deposits, and developing methods to identify new targets in Egypt. The mineralogical information provided in the ASTER SWIR bands provided the unique tool to successfully detect and map different intensities of hydrothermal alteration, in the same way as was done for porphyry copper exploration. Egypt is an arid, desert environment, with near 100% surface rock and soil exposures; ideal for application of optical remote sensing data.

Amer et al. [133] used ASTER data to detect gold-related alteration in the Um Rus area. Salem et al. [134] combined geologic mapping and alteration mapping with ASTER data to identify new exploration targets at the Barramiya District. Mapping alteration associated with potential gold deposits at Wadi Allaqi was published by Salem and Soliman [135]. Hasan et al. [136] combined spectral analysis of ASTER data with aeromagnetic data to identify promising gold exploration targets in the Eastern Desert. The two data sets provided complementary information on rock types and mineralogy, allowing a fuller picture to be created of promising areas. By combining ASTER spectral analysis with geochemical data from surface-collected samples, Salem et al. [137] were able to validate alteration zones detected on the remote sensing data, and strengthen the association of ASTER-defined targets with potential gold deposits. Abdelnasser [138] and Salem et al. [139], in separate studies, reported similar ASTER-plus-geochemistry studies at the Atud gold deposit and the Samut area, respectively. By establishing a physical tie between alteration and gold occurrence, an exploration strategy using only remote sensing data was formulated.

Orion Gold announced in 2013 [140] the discovery of new exploration gold target areas in Queensland, Australia, based on analysis of ASTER images. The area has known productive mines in low sulfidation epithermal gold systems and porphyry gold-copper systems. ASTER data were interpreted to locate occurrences of high temperature illite, crystalline kaolinite, dickite, and possible vegetation anomalies. Along with K/Th radiometric anomalies, five target areas were identified (Figure 5) for further ground- and laboratory-based analyses. Follow-up drilling in 2015 intersected multiple epithermal veins and stockwork zones.

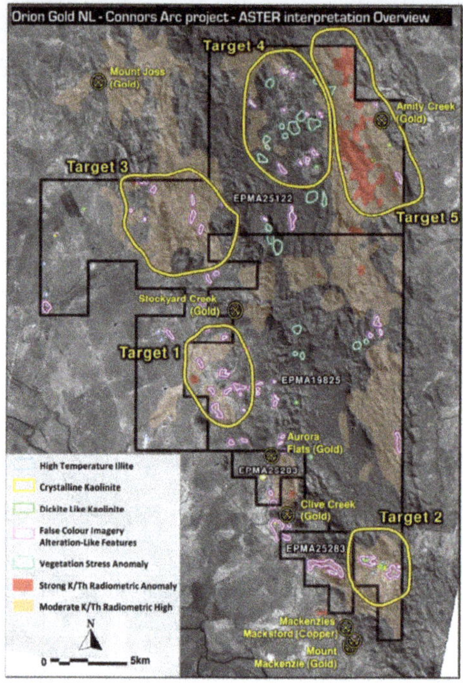

Figure 5. ASTER interpretation of clay anomalies and delineation of five target areas. Pick/shovel symbol represents known gold of copper prospects. ASX (Australian Stock Exchange) announcement [140].

The first gold discovery using the Australia ASTER Geoscience minerals maps (described previously) was announced in 2014 by Kentor Gold Limited on the Australian Stock Exchange soon after the public release of the satellite products [141]. Their discovery at Chukbo in the east Arunta of the Northern Territory, Australia was based on recognition in the ASTER geoscience maps of coincident phyllic and propyllitic alteration (Figure 6). Similar new additional targets are also apparent.

Articles have been published describing applications of ASTER data for gold exploration in other parts of the world. An early work by Zhang et al. [142] applied ASTER data for lithologic mapping and alteration detection in the Chocolate Mountains, California. They analyzed the geologic setting of a known gold mine, then used the extracted characteristics to search for similar environments. A similar study by de Palomera [143] was carried out in the Deseado Massif, Argentina to prospect for epithermal gold-silver deposits. At Mount Olympus, Australia, Wells et al. [144] used ASTER data to characterize alteration associated with sediment-hosted gold mineralization. Their results defined a different suite of alteration minerals compared with hydrothermal alteration found in epithermal or Carlin-type gold deposits. At the Gua Musang Goldfield, China, Yao et al. [145] applied ASTER SWIR and TIR data to map rock types and quartz content to search for promising host rocks for possible gold mineralization. Yousefi et al. [146] integrated ASTER and Landsat data to map geologic setting of the

Zarshuran Carlin-type deposit in Iran. ASTER data provided mineralogical and lithologic information, and Landsat provided regional structural information. Rani et al. [147] combined mineralogical information from ASTER data with ground magnetic data, ground spectroscopy and gravity data to identify potential targets for gold sulfide mineralization. This project was an improvement over many other, remote-sensing only projects: bringing in geophysical data provided a more complete geologic picture of the setting; ground spectroscopy provided validation of the mineralogical information derived from the ASTER data.

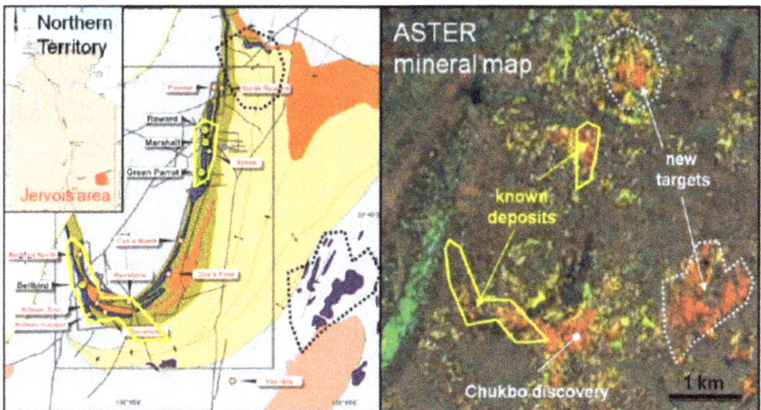

Figure 6. Published (from Kentor Gold) geology and mineral occurrences in the Jervois area, Northern Territory (**left**) and propyllitic alteration (warmer colors) evident in the ASTER "MgOH" product, which was critical in the discovery of Chukbo (**right**).

6.6. ASTER Data Applied to Exploration for Other Minerals

Several papers have appeared describing application of ASTER data in the search for iron ore deposits. Using ASTER data alone, or in combination with data from other satellites (Landsat and Hyperion) or airborne geophysics, methods were developed over targets in India, Iran, Brazil, and Australia by Rajendran et al. [148], Huang et al. [149], Duuring et al. [150], Mansouri et al. [151], and Mazhari et al. [152]. One report by Moghtaderi et al. [153] used ASTER and Landsat data to determine iron mineral contamination in an iron mine area in Iran. These studies mainly relied on the VNIR bands, as this spectral region covers the diagnostic spectral absorption features associated with ferric and ferrous iron minerals.

A few articles have appeared where ASTER data have been applied in the search for specific minerals. Cardoso-Fernandez [154] reported on the use of ASTER data in the search for lithium-bearing pegmatites. TIR data were one of the key inputs to their exploration model. Shawky et al. [155] processed ASTER data to detect the presence of known uranium localities in Egypt, hoping to develop a more general exploration tool. In Australia, Hewson et al. [156] analyzed ASTER data to map geology associated with manganese mineralization. In Kurdistan, Othman and Gloaguen [157] discovered a new chromite body in the Mawat Ophiolite Complex. They combined lithologic mapping to delineate possible host rocks, then identified targets using Support Vector Machine classification, searching for unknown deposits using known targets as training sites.

ASTER data have been applied in oil exploration to search for hydrocarbon seepage induced alteration by Fu et al. [158], Shi et al. [159], Siydon et al. [160], Pena and Abdelsalam [161], and Petrovic et al. [162]. Generally, surface effects are reduction/oxidation reactions, and alteration to produce clay minerals. In a few cases, over known oil fields, alteration has been successfully detected. Application as an exploration tool is still in the experimental stage; but ASTER data are considered an addition to an exploration protocol.

7. Discussion and Conclusions

All of these examples conclusively demonstrate the tremendous advances in lithologic mapping and mineral exploration provided by ASTER's multispectral SWIR and TIR bands. For the first time, global image data were acquired at sufficient spatial resolution to be applicable to deposit-scale mapping, as well as regional reconnaissance exploration. In addition, the systematic search for indicators of potential base and precious metal deposits was enabled by the ability to detect minerals associated with propylitic, argillic, potassic, phyllic, and silicic hydrothermal alteration. Clays, carbonates, sulfates, and other hydrous minerals were discriminated, not just lumped into a single category as with Landsat data.

Researchers described a variety of initial ASTER data products (radiance-at-the-sensor, surface reflectance, etc.) as their inputs for analysis. Further pre-processing steps varied, depending on the final analysis and information extraction algorithms used. This variety of data analysis and processing methods, then applied to the pre-processed ASTER data, shows that there is no "perfect workflow" to successfully extract mineralogical information. The data have sufficient information to provide satisfactory results using the simplest techniques like color composites and band ratios of radiance-at-the-sensor data, or more sophisticated analyses, like machine learning. Mineralogical indices seem to be a good middle ground (see Appendix B).

The aerospace commercial sector took note of the sizable exploration geology user community using ASTER and Landsat data, and gauged there was a sufficiently large market to add multispectral SWIR capability to a for-hire satellite scanner. Since 2015, DigitalGlobe's WorldView-3 instrument has provided data for sale to customers desiring high spatial resolution data. The instrument provides eight SWIR bands with 7.5 m pixel size, four in the 1.6–1.75 μm region and four in the 2.15–2.35 μm region; and eight VNIR bands with ~1.2 m pixel size, and has bands with similar bandpasses to ASTER VNIR-SWIR bands except for ASTER SWIR band 9. The images have a swath width of 13.1 km, so the data are not suitable for reconnaissance of medium to large areas, as can be done with ASTER data. However, the band positions allow better clay and carbonate mineral identification than with ASTER, as reported by Mars [163] over the Mountain Pass, California site. No funded, future instrument by any country or agency currently exists to provide high spatial resolution TIR data. Several hyperspectral VNIR-SWIR scanners are either operational or planned for launch in the next few years. None of these provide global coverage, as they are all sampling missions, with narrow (~30 km) swath widths.

Since the launch of the Landsat MSS scanner in 1972, geologists have increasingly turned to satellite-based remote sensing data as an integral tool in lithologic mapping and mineral exploration programs. The 1982 launch of the Landsat Thematic Mapper scanner, with its single 2–2.5 μm band, was a breakthrough for detecting hydrothermal alteration that could be associated with potential mineral deposits. Japan's 1992 launch of OPS with its four SWIR bands demonstrated the potential capability to not only detect hydrothermal alteration minerals, but to identify individual and classes of these minerals. The ASTER instrument, with its six SWIR bands, and five TIR bands, was the next logical step in developing a spaceborne instrument with greatly enhanced geological mapping capabilities. For the past 20 years, ASTER data have been shown again and again to be an important tool to map the surface of the Earth. Several announced mineral deposit discoveries (and undoubtedly many unannounced discoveries) attest to the success of ASTER's design.

Author Contributions: Conceptualization and writing by M.A. and Y.Y.

Funding: Yamaguchi was funded by JSPS KAKENHI (Grant No. 15H04225).

Acknowledgments: Work by Abrams was performed at the Jet Propulsion Laboratory, California Institute of Technology under contract with the National Aeronautics and Space Administration.

Conflicts of Interest: The authors declare no conflict of interest.

Acronyms

AIST	National Institute of Advanced Industrial Science and Technology
ALI	Advanced Land Imager
ASTER	Advanced Spaceborne Thermal Emission and Reflection Radiometer
CSIRO	Commonwealth Scientific and Industrial Research Organisation
EO-1	Earth Observer
EOS	Earth Observing System
ERTS	Earth Resources Technology Satellite
GDEM	Global digital elevation model
HRV	High resolution visible
IRS	Indian Remote Sensing
ITIR	Intermediate Thermal Infrared Radiometer
JERS	Japan Earth Resources Satellite
LPDAAC	Land Processes Distributed Active Archive Center
METI	Ministry of Economy, Trade and Industry
MITI	Ministry of International Trade and Industry
MSS	Multispectral scanner
NASA	National Aeronautics and Space Administration
NASDA	National Space Development Agency
OLI	Operational Land Imager
OPS	Optical sensor
PALSAR	Phased Array type L-band Synthetic Aperture Radar
SAC-C	Scientific Application Satellite-C
SPOT	Satellite pour l'Observation de la Terre
STA	Science and Technology Agency
SWIR	Short wave infrared
TIGER	Thermal Infrared Ground Emission Radiometer
TIR	Thermal infrared
TM	Thematic Mapper
VNIR	Visible and near infrared

Appendix A ASTER Operations

Appendix A.1 Data Acquisition

Due to its limited duty cycle, the ASTER instrument is scheduled each day for specific data collections. On any given day there are thousands of possible data acquisitions that could be collected. The ASTER science team developed an automatic scheduler to prioritize the possible acquisitions and produce a daily acquisition schedule. On average, about 500–550 scenes are collected daily, limited by the capacity of the onboard data recorders. To date, ASTER has acquired almost 4 million images (Figure A1).

Figure A1. Mosaic of ASTER browse images showing near-global coverage achieved over the lifetime of the mission.

Appendix A.2 Data Products

The ASTER project provides the user community with Standard Data Products throughout the life of the mission. Algorithms to create these products were created by the ASTER Science Team, and were rigorously peer reviewed by outside scientists [164]. This ensured that the products met NASA standards for accuracy and scientific soundness. Adhering to NASA procedures, products are labeled by level, from 1 to 3. "Level 1A are reconstructed, unprocessed instrument data at full resolution, time-referenced and annotated with ancillary information, including geometric and radiometric calibration coefficients and georeferencing parameters computed and appended but not applied to Level 0 data. Level 1B data are L1A data that have been processed to sensor units. Level 2 data are derived geophysical variables at the same resolution and location as Level 1 source data. Level 3 data are variables mapped on uniform space-time grid scales, usually with some completeness and accuracy." [165]. A list of the ASTER data products can be found here [166].

Appendix A.3 Data Archiving and Distribution

In the U.S., NASA's Land Processes Distributed Active Archive Center (LPDAAC; https://lpdaac.usgs.gov) is responsible for archiving, processing, and distribution of all ASTER data products. In Japan, Japan Space Systems (http://jspacesystems.or.jp) is responsible for archiving the entire ASTER Level 0 and Level 1 data. Distribution of a limited suite of ASTER products in Japan is implemented by the National Institute of Advanced Industrial Science and Technology (AIST) through the Geologic Survey of Japan (https://gbanks.gsj.jp/madas/). As of January 2019 ASTER had acquired over 3.8 million images. A complete catalog of links to ASTER data providers can be found on the ASTER website (https://asterweb.jpl.nasa.gov/data.asp).

Since April 2016, both METI and NASA agreed to distribute all ASTER data products to all users at no cost. For the period April 2016 to April 2019, over 32 million ASTER Level 1, 2, and 3 files have been distributed (excluding the Global Digital Elevation Model (GDEM)). For the past two years, the most popular product has been the ASTER GDEM. The number of these files ordered has exceeded 55 million.

Appendix B Spectral Indices

Appendix B.1 TIR Spectral Indices

In 2005, Ninomiya et al. [32] proposed mineralogical spectral indices using ASTER's TIR bands. The indices were based on spectral absorption features of silicates and carbonates, located in the 8–12 micron wavelength region.

CI = Carbonate index = B13/B14, where B13 is the value of radiance-at-sensor of ASTER TIR band 13; CI is high for calcite and dolomite
QI = Quartz Index = $(B11 \times B11)/(B10 + B12)$
MI = Mafic Index = $(B12 \times B14^3)/B13^4$

Appendix B.2 Logical Operators

In 2006, Mars and Rowan [126] developed logical operators to map argillic and phyllic alteration. Their algorithms were written using IDL (Interactive Data Language). Input data were the ASTER calibrated surface reflectance products, atmospherically corrected. In addition, the ASTER data were adjusted using Hyperion data, resampled to ASTER bandpasses.

Argillic alteration:

((float(B3/B2 le 1.35) and (B4 gt 260) and ((float (B4)/B5) gt 1.25) and ((float(B5/B6) le 1.05) and ((float(b7)/B6) ge 1.03)
The first term masks vegetation; the second term masks dark pixels; the third term maps the 2.165 micron feature; the fourth term delineates argillic from phyllic alteration; the last term maps the 2.20 micron feature.

Phyllic alteration:

((float(B3)/B2) le 1.35) and (B4 gt 260) and ((float(B4)/B6) gt 1.25) and ((float(B5)/(B6) gt 1.05) and ((float(B7)/B6) ge 1.03)
The first term masks vegetation; the second term masks dark pixels; the third term maps the 2.20 micron feature; the fourth term delineates argillic from phyllic alteration; the last term maps the 2.20 micron feature.

Appendix B.3 Australian Geoscience Maps

In 2012, CSIRO released "Satellite ASTER Geoscience Products for Australia" [99]. These consisted of 17 geoscience products for the entire continent of Australia, created from 35,000 ASTER images. Fourteen were from the VNIR and SWIR bands, three from the TIR bands. The geoscience maps started with ASTER radiance-at-sensor data products; further processing included geometric correction, cloud and vegetation masking, mosaicking images to make a seamless mosaic, and application of product masks/thresholds to generate the geoscience products.

The 17 products are false color (Bands 3,2,1 in RGB); Landsat TM Regolith ratios; green vegetation content; ferric oxide content; ferric oxide composition; ferrous iron index; opaque index; AlOH group content; AlOH group composition; Kaolin group index; FeOH group content; MgOH group content; MgOH group composition; ferrous iron content in MgOH/carbonate; silica index; quartz index; and gypsum index.

An example of the algortihms used is the MgOH group content, designed to highlight the abundance of calcite, dolomite, magnesite, chlorite, epidote, amphibole, talc and serpentinite.

Algorithm = (B6+B9)/(B7+B8)
Masks = green vegetation <1.4 and cloud, water, shadow and sun glint
Stretch = linear 1.05–1.2

References

1. Abrams, M.; Tsu, H.; Hulley, G.; Iwao, K.; Pieri, D.; Cudahy, T.; Kargel, J. The Advanced Spaceborne Thermal Emission and Reflection Radiometer (ASTER) after Fifteen Years: Review of Global Data Product. *Int. J. Appl. Earth Obs. Geoinform.* **2015**, *38*, 292–301. [CrossRef]
2. Goetz, A.; Billingsley, F.; Elston, D.; Lucchitta, I.; Shoemaker, E.; Abrams, M.; Gillespie, A.; Squires, R. *Applications of ERTS Images and Image Processing to Regional Problems and Geologic Mapping in Northern Arizona*; NASA/JPL Technical Reports 32-1597; NASA: Pasadena, CA, USA, 1975.
3. Baker, R. Landsat Data: A new perspective for geology. *Photogramm. Eng. Remote Sens.* **1975**, *41*, 1233–1239.
4. Sabins, F. *Remote Sensing: Principles and Interpretation*; W. H. Freeman: San Francisco, CA, USA, 1975; p. 426.
5. Siegal, B.; Gillespie, A. (Eds.) *Remote Sensing in Geology*; Wiley: New York, NY, USA, 1980; p. 702.
6. Landsat Science. Available online: https://landsat.gsfc.nasa.gov/the-thematic-mapper/ (accessed on 29 January 2019).
7. Abrams, M.; Ashley, R.; Rowan, L.; Goetz, A.; Kahle, A. Mapping hydrothermal alteration in the Cuprite mining district, Nev., using aircraft scanner imagery for the 0.46–2.36 micron spectral region. *Geology* **1977**, *5*, 713–718. [CrossRef]
8. Abrams, M. *The Joint NASA Geosat Test Case Project: Final Report*; American Association of Petroleum Geologists: Tulsa, OK, USA, 1985; 3 volumes.
9. Abdelhamid, G.; Rabba, I. An investigation of mineralized zones revealed during geologic mapping, Jabal-Hamra-Faddan Wadi area, Jordan, using Landsat-TM data. *Int. J. Remote Sens.* **1994**, *15*, 1495–1506. [CrossRef]
10. Rothery, D.; Abrams, M. Improved discrimination of rock units using Landsat Thematic Mapper imagery of the Oman Ophiolite. *J. Geol. Soc. Lond.* **1987**, *144*, 587–597. [CrossRef]
11. Van Der Meer, F.; Vazquez-Torres, M.; Van Dijk, P. Spectral characterization of ophiolite lithologies in the Troodos Ophiolite complex of Cyprus and its potential in prospecting for massive sulfide deposits. *Int. J. Remote Sens.* **1997**, *18*, 1245–1257. [CrossRef]
12. Elrakaiby, M. The use of enhanced Landsat-TM image in the characterization of uraniferous granitic rocks I the central eastern desert of Egypt. *Int. J. Remote Sens.* **1995**, *16*, 1063–1074. [CrossRef]
13. Qari, M. Application of Landsat TM data to geological studies, Al-Khabt area, southern Arabian shield. *Photogramm. Eng. Remote Sens.* **1991**, *57*, 421–429.
14. Sabins, F. Remote sensing for mineral exploration. *Ore Geol. Rev.* **1999**, *14*, 157–183. [CrossRef]
15. Almeida, T.; Filho, C.; Juliani, C.; Branco, F. Application of Remote Sensing to Geobotany to Detect Hydrothermal Alteration Facies in Epithermal High-Sulfidation Gold Deposits in the Amazon Region. *Rev. Econ. Geol.* **2006**, *16*, 135–142.
16. eoPortal Directory. Available online: https://earth.esa.int/web/eoportal/satellite-missions/j/jers-1 (accessed on 30 January 2019).
17. DeSouza, C.; Drury, S. Evaluation of JERS-1 (FUYO-1) OPS and Landsat TM images for mapping of gneissic rocks in arid areas. *Int. J. Remote Sens.* **1998**, *19*, 3569–3594. [CrossRef]
18. Salomonson, V.; Abrams, M.; Kahle, A.; Barnes, W.; Xiong, X.; Yamaguchi, Y. Evolution of NASA's earth Observing System and development of the Moderate Resolution Imaging Spectrometer and the Advanced Spaceborne Thermal Emission and Reflection radiometer instruments. In *Land Remote Sensing and Global Environmental Change*; Ramachandran, R., Justice, C., Abrams, M., Eds.; Springer: New York, NY, USA, 2011; Chapter 1; pp. 3–34.
19. Yamaguchi, Y.; Tsu, H.; Fujisada, H. Scientific basis of ASTER instrument design. *Proc. SPIE* **1998**, *1939*, 150–160.
20. Plafcan, D. Technoscientific Diplomacy: The practice of international policies in the ASTER collaboration. In *Land Remote Sensing and Global Environmental Change*; Ramachandran, R., Justice, C., Abrams, M., Eds.; Springer: New York, NY, USA, 2011; Chapter 4; pp. 483–508.
21. Yamaguchi, Y.; Sato, I.; Tsu, H. ITIR design concept and science mission. In *The Use of EOS for Studies of Atmospheric Physics*; Gille, J.C., Visconti, G., Eds.; Elsevier: New York, NY, USA, 1992; pp. 229–310.
22. Yamaguchi, Y.; Kahle, A.; Tsu, H.; Kawakami, T.; Pniel, M. Overview of Advanced Spaceborne Thermal Emission and Reflection Radiometer (ASTER). *IEEE Trans. Geosci. Remote Sens.* **1998**, *36*, 1062–1071. [CrossRef]

23. Fujisada, H. Terra ASTER Instrument Design and Geometry. In *Land Remote Sensing and Global Environmental Change*; Ramachandran, R., Justice, C., Abrams, M., Eds.; Springer: New York, NY, USA, 2011; Chapter 4; pp. 59–82.

24. Fujisada, H.; Bailey, G.; Kelly, G.; Hara, S.; Abrams, M. ASTER DEM Performance. *IEEE Trans. Geosci. Remote Sens.* **2005**, *43*, 2702–2714. [CrossRef]

25. Balick, L.; Gillespie, A.; French, A.; Danilina, I.; Allard, J.P.; Mushkin, A. Longwave thermal infrared spectral variability in individual rocks. *IEEE Geosci. Remote Sens. Lett.* **2009**, *6*, 52–56. [CrossRef]

26. Rowan, L.; Mars, J. Lithologic mapping in the Mountain Pass, California area using ASTER data. *Remote Sens. Environ.* **2003**, *84*, 350–366. [CrossRef]

27. Watts, D.; Harris, N. Mapping granite and gneiss in domes along the North Himalayan antiform with ASTER SWIR band ratios. *Geol. Soc. Am. Bull.* **2005**, *117*, 879–886. [CrossRef]

28. Yamaguchi, Y.; Naito, C. Spectral indices for lithologic discrimination and mapping by using the ASTER SWIR bands. *Int. J. Remote Sens.* **2003**, *24*, 4311–4323. [CrossRef]

29. Byrnes, J.; Ramsey, M.; Crown, D. Surface unit characterization of the Mauna Ulu flow field, Kilauea Volcano, Hawaii, using integrated field and remote sensing analyses. *J. Volcanol. Geotherm. Res.* **2004**, *135*, 169–193. [CrossRef]

30. Gomez, C.; Delacourt, C.; Allemand, P.; Ledru, P.; Wackerle, R. Using ASTER remote sensing data set for geological mapping, in Namibia. *Phys. Chem. Earth* **2005**, *30*, 97–108. [CrossRef]

31. Kargel, J.; Abrams, M.; Bishop, M.; Bush, A.; Hamilton, G.; Jiskoot, H.; Kaag, A.; Kieffer, H.; Lee, E.; Paul, F.; et al. Multispectral imaging contributions to global land ice measurements from space. *Remote Sens. Environ.* **2005**, *99*, 187–219. [CrossRef]

32. Ninomiya, Y.; Fu, B.; Cudahy, T. Detecting lithology with Advanced Spaceborne Thermal Emission and Reflection radiometer (ASTER) multispectral thermal infrared "radiance-at-sensor" data. *Remote Sens. Environ.* **2005**, *99*, 127–139. [CrossRef]

33. Hook, S.; Dmochowski, J.; Howard, K.; Rowan, L.; Karlstrom, K.; Stock, J. Mapping variations in weight percent silica measured from multispectral thermal infrared imagery; examples from the Hiller Mountains, Nevada, USA and Tres Virgenes-La Reforma, Baja California Sur, Mexico. *Remote Sens. Environ.* **2005**, *95*, 273–289. [CrossRef]

34. Rowan, L.; Mars, J.; Simpson, C. Lithologic mapping of the Mordor, NT, Australia ultramafic complex by using Advanced Spaceborne Thermal Emission and Reflection Radiometer (ASTER). *Remote Sens. Environ.* **2005**, *99*, 105–126. [CrossRef]

35. Hewson, R.; Cudahy, T.; Mizuhiko, S.; Ueda, K.; Mauger, A. Seamless geological map generation using ASTER in the Broken Hill-Curnamona province of Australia. *Remote Sens. Environ.* **2005**, *99*, 159–172. [CrossRef]

36. Qiu, F.; Abdelsalam, M.; Thakkar, P. Spectral analysis of ASTER data covering part of the Neoproterozoic Allaqi-Heiani suture, Southern Egypt. *J. Afr. Earth Sci.* **2006**, *44*, 169–180. [CrossRef]

37. Rajendran, S.; Hersi, O.; Al-Harthy, A.; Al-Wardi, M.; El-Ghali, M.; Al-Abri, A. Capability of Advanced Spaceborne Thermal Emission and Reflection radiometer (ASTER) on discrimination of carbonates and associated rocks and mineral identification of eastern mountain region (Saih Hatat window) of Sultanate of Oman. *Carbonates Evaporates* **2011**, *26*, 352–364.

38. Rajendran, S.; Nasir, S. ASTER spectral sensitivity of carbonate rocks—Study in Sultanate of Oman. *Adv. Space Res.* **2014**, *53*, 656–673. [CrossRef]

39. Guba, A.; Kumar, K.; Rao, E.; Parveen, R. An image processing approach for converting ASTER derived spectral maps for mapping Kolhan limestone, Jharkhand, Indian. *Curr. Sci.* **2014**, *106*, 40–49.

40. Graettinger, A.; Ellis, M.; Skilling, I.; Reata, K.; Ramsey, M.; Lee, R.; Hughes, C.; McGarvie, D. Remote sensing and geologic mapping of glaciovolcanic deposits in the region surrounding Askja (Dyngjufjoll) volcano, Iceland. *Int. J. Remote Sens.* **2013**, *34*, 7178–7198. [CrossRef]

41. Tayebi, M.; Tangestani, M.; Roosta, H. Mapping salt diapirs and salt diapir-affected areas using MLP neural network model and ASTER data. *Int. J. Dig. Earth* **2013**, *6*, 143–157. [CrossRef]

42. Yajima, T.; Yamaguchi, Y. Geological mapping of the Francistown area in northeastern Botswana by surface temperature and spectral emissivity information derived from ASTER thermal infrared data. *Ore Geol. Rev.* **2013**, *53*, 134–144. [CrossRef]

43. Ninomiya, Y.; Fu, B. Regional Lithological Mapping Using ASTER-TIR Data: Case Study for the Tibetan Plateau and the Surrounding Area. *Geosciences* **2016**, *6*, 39. [CrossRef]

44. Ozyavas, A. Assessment of image processing techniques and ASTER SWIR data for the delineation of evaporates and carbonate outcrops along the Salt Lake Fault, Turkey. *Int. J. Remote Sens.* **2016**, *37*, 770–781. [CrossRef]

45. Askari, G.; Pour, A.; Pradhan, B.; Sarfi, M.; Nazemnejad, F. Band Ratios Matrix Transformation (BRMT): A Sedimentary Lithology Mapping Approach Using ASTER Satellite Sensor. *Sensors* **2018**, *18*, 3213. [CrossRef] [PubMed]

46. Hubbard, B.; Hooper, D.; Solano, F.; Mars, J. Determining mineralogical variations of aeolian deposits using thermal infrared emissivity and linear deconvolution methods. *Aeolian Res.* **2018**, *30*, 54–96. [CrossRef]

47. Rajendran, S.; Nasir, S.; El-Ghali, M.; Alzebdeh, K. Spectral Signature Characterization and Remote Mapping of Oman Exotic Limestones for Industrial Rock Resource Assessment. *Geosciences* **2018**, *8*, 145. [CrossRef]

48. Massironi, M.; Bertoldi, L.; Calafa, P.; Visona, D. Interpretation and processing of ASTER data for geological mapping and granitoids detection in the Saghro massif (eastern Anti-Atlas, Morocco). *Geosphere* **2008**, *4*, 736–759. [CrossRef]

49. Bertoldi, L.; Massironi, M.; Visona, D.; Caros, R.; Montomoli, C.; Gubert, F.; Naletto, G.; Pelizzo, M. Mapping the Buraburi granite in the Himalaya of Western Nepal: Remote sensing analysis in a collisional belt with vegetation cover and extreme variation of topography. *Rem Sens. Environ.* **2011**, *115*, 1129–1144. [CrossRef]

50. Omran, A.; Hahn, M.; Hochschild, V.; El-Reyes, A. Lithological Mapping of Dahab Basin, South Sinai, Egypt, using ASTER Data. *Photogramm. Fernerkund. Geoinform.* **2012**, *6*, 711–726. [CrossRef]

51. Zheng, S.; Fu, B. Lithological mapping of granitiods in the western Junggar from ASTER SWIR-TIR multispectral data: Case study in Karamay pluton, Xinjiang. *Acta Petrol. Sin.* **2013**, *29*, 2936–2948.

52. El Janati, M.; Soulaimani, A.; Admou, H.; Youbi, M.; Hafid, A.; Hefferan, K. Application of ASTER remote sensing data to geological mapping of basement domains in arid regions: A case study from the Central Anti-Atlas, Iguerda inlier, Morocco. *Arab. J. Geosci.* **2014**, *7*, 2407–2422. [CrossRef]

53. Guha, A.; Kumar, K. New ASTER derived thermal indices to delineate mineralogy of different granitoids of an Archaean Craton and analysis of their potentials with reference to Ninomiya's indices for delineating quartz and mafic minerals of granitoids—An analysis in Dharwar Craton, Indian. *Ore Geol. Rev.* **2016**, *74*, 76–87.

54. Asran, A.; Emam, A.; E-Fakharani, A. Geology, structure, geochemistry and ASTER-based mapping of Neoproterozoic Gebel El-Delihimmi granites, Central Eastern Desert of Egypt. *Lithos* **2017**, *282*, 358–372. [CrossRef]

55. Li, P.; Long, X.; Liu, L. Ophiolite mapping using ASTER data: A case study of Derni ophiolite complex. *Acta Petrol. Sin.* **2007**, *23*, 1175–1180.

56. Huang, Z.; Zhang, X. Lithological mapping of ophiolite composition in Zedan-Luobusha, Yarlung Zangbo suture zone using Advanced Spaceborne Thermal Emission and Reflection Radiometer (ASTER) data. *Acta Petrol. Sin.* **2010**, *26*, 3589–3596.

57. Tangestani, M.; Jaffari, L.; Vincent, R.; Sridhar, B. Spectral characterization and ASTER-based lithological mapping of an ophiolite complex: A case study from Neyriz ophiolite, SW Iran. *Remote Sens. Environ.* **2011**, *115*, 2243–2254. [CrossRef]

58. Ozkan, M.; Celik, O.; Ozyavas, A. Lithological discrimination of accretionary complex (Sivas, northern Turkey) using novel hybrid color composites and field data. *J. Afr. Earth Sci.* **2018**, *138*, 75–85. [CrossRef]

59. Deller, M.; Andrews, S. Facies discrimination in laterites using Landsat Thematic Mapper, ASTER and ALI data—Examples from Eritrea and Arabia. *Int. J. Remote Sens.* **2006**, *27*, 2389–2409. [CrossRef]

60. Qari, M.; Madani, A.; Matsah, M.; Hamimi, Z. Utilization of ASTER and Landsat data in geologic mapping of basement rocks of Arafat Area, Saudi Arabia. *Arab. J. Sci. Eng.* **2008**, *33*, 99–116.

61. Pournamdari, M.; Hashim, M.; Pour, A. Spectral transformation of ASTER and Landsat TM bands for lithological mapping of Soghan ophiolite complex, South Iran. *Adv. Space Res.* **2015**, *54*, 694–709. [CrossRef]

62. Eslami, A.; Ghaderi, M.; Rajendran, S. Integration of ASTER and Landsat TM remote sensing data for chromite prospecting and lithological mapping in Neyriz ophiolite zone, south Iran. *Resour. Geol.* **2015**, *65*, 375–388. [CrossRef]

63. Adiri, Z.; El Harti, A.; Jellouli, A.; Maacha, L.; Bachaoui, E. Lithological mapping using Landsat 8 OLI and Terra ASTER multispectral data in the Bas Draa inlier, Moroccan Anti Atlas. *J. Appl. Remote Sens.* **2016**, *10*. [CrossRef]

64. Lamri, T.; Djemai, S.; Hamoudi, M.; Zoheir, B.; Benaouda, A.; Ouzegane, K.; Amara, M. Satellite imagery and airborne geophysics for geologic mapping of the Edembo area, Eastern Hoggar (Algerian Sahara). *J. Afr. Earth Sci.* **2016**, *115*, 143–158. [CrossRef]

65. Metelka, V.; Baratoux, L.; Jessell, M.; Barth, A.; Jezek, J.; Naba, S. Automated regolith landform mapping using airborne geophysics and remote sensing data, Burkina Faso, West Africa. *Remote Sens. Environ.* **2018**, *204*, 964–978. [CrossRef]

66. Yang, M.; Kang, L.; Chen, H.; Zhou, M.; Zhang, J. Lithological mapping of East Tianshan area using integrated data fused by Chinese GF-1 PAN and ASTER multi-spectral data. *Open Geosci.* **2018**, *10*, 532–543. [CrossRef]

67. Ge, W.; Cheng, Q.; Jing, L.; Armenakis, C.; Ding, H. Lithological discrimination using ASTER and Sentinel-2A in the Shibanjing ophiolite complex of Beishan orogenic in Inner Mongolia, China. *Adv. Space Res.* **2018**, *62*, 1702–1716. [CrossRef]

68. Hassan, S.; El Kazzaz, Y.; Taha, M. Late Neoproterozoic basement rocks of Meatiq area, Central Eastern Desert, Egypt: Petrography and remote sensing characterizations. *J. Afr. Earth Sci.* **2017**, *131*, 14–31. [CrossRef]

69. Ali-Bik, M.; Hassan, S.; Abou El Maaty, M.; Moustafa, A.; El Rahim, A.; Said, H. The late Neoproterozoic Pan-African low-grade metamorphic ophiolitic and island-arc assemblages at Gebel Zabara area, Central Eastern Desert, Egypt: Petrogenesis and remote sensing—Based geologic mapping. *J. Afr. Earth Sci.* **2018**, *144*. [CrossRef]

70. Hadigheh, S.; Ranjbar, H. Lithological Mapping in the Eastern Part of the Central Iranian Volcanic Belt Using Combined ASTER and IRS data. *J. Indian Soc. Remote Sens.* **2013**, *41*, 921–931. [CrossRef]

71. Lohrer, R.; Bertrams, M.; Eckmeier, E.; Protze, J. Mapping the distribution of weathered Pleistocene wadi deposits in Southern Jordan using ASTER, SPOT-5 data and laboratory spectroscopic analysis. *Catena* **2013**, *107*, 57–70. [CrossRef]

72. Boyce, J.; Keays, R.; Nicholls, I.; Hayman, P. Eruption centres of the Hamilton area of the Newer Volcanics Province, Victoria, Australia: Pinpointing volcanoes from a multifaceted approach to landform mapping. *Austral. J. Earth Sci.* **2014**, *61*, 735–754. [CrossRef]

73. Soltaninejad, A.; Ranjbar, H.; Honarmand, M.; Dargahi, S. Evaporite mineral mapping and determining their source rocks using remote sensing data in Sirjan playa, Kerman, Iran. *Carbonates Evaporites* **2018**, *33*, 255–274. [CrossRef]

74. Chen, Q.; Zhao, Z.; Jiang, Q.; Tan, S.; Tian, Y. Identification of metamorphic rocks in Wuliangshan Mountains (Southwest China) using ASTER data. *Arab. J. Geosci.* **2018**, *11*. [CrossRef]

75. Wang, R.; Lin, J.; Zhao, B.; Xiao, Z. Integrated Approach for Lithological Classification Using ASTER Imagery in a Shallowly Covered Region-The Eastern Yanshan Mountain of China. *IEEE J. Sel. Top. Appl. Earth Obs. Remote Sens.* **2018**, *11*, 4791–4807. [CrossRef]

76. Kurata, K.; Yamaguchi, Y. Integration and Visualization of Mineralogical and Topographical Information Derived from ASTER and DEM Data. *Remote Sens.* **2019**, *11*, 162. [CrossRef]

77. Crosta, A.; De Souza, C.; Azevedo, F. Targeting key alteration minerals in epithermal deposits in Patagonia, Argentina, using ASTER imagery and principal component analysis. *Int. J. Remote Sens.* **2003**, *24*, 4233–4240. [CrossRef]

78. Galvao, L.; Almeida, R.; Vitorello, I. Spectral discrimination of hydrothermally altered materials using ASTER short-wave infrared bands: Evaluation in a tropical savannah environment. *Int. J. Appl. Earth Obs. Geoinform.* **2005**, *7*, 107–114. [CrossRef]

79. Moore, F.; Rastmanesh, F.; Asadi, H.; Modabberi, S. Mapping mineralogical alteration using principal-component analysis and matched filter processing in the Takab area, north-west Iran, from ASTER data. *Int. J. Remote Sens.* **2008**, *29*, 2851–2867. [CrossRef]

80. Rockwell, B.; Hofstra, A. Identification of quartz and carbonate minerals across northern Nevada using ASTER thermal infrared emissivity data—Implications for geologic mapping and mineral resource investigations in well-studied and frontier areas. *Geosphere* **2008**, *4*, 218–246. [CrossRef]

81. Hosseinjani, M.; Tangestani, M. Mapping alteration minerals using sub-pixel unmixing of ASTER data in the Sarduiyeh area, SE Kerman, Iran. *Int. J. Dig. Earth* **2011**, *4*, 487–504. [CrossRef]

82. Honarmand, M.; Ranjbar, H.; Shahabpour, J. Application of Principal Component Analysis and Spectral Angle Mapper in the Mapping of Hydrothermal Alteration in the Jebal-Barez Area, Southeastern Iran. *Res. Geol.* **2012**, *62*, 119–139. [CrossRef]

83. Kruse, F.; Boardman, A.; Lefkoff, A.; Heidebrecht, K.; Shapiro, A.; Goetz, A. The spectral image processing system (SIPS)—Interactive visualization and analysis of imaging spectrometer data. *Remote Sens. Environ.* **1993**, *44*, 145–163. [CrossRef]

84. Abdi, M.; Karimpour, M. Application of Spectral Angle Mapper Classification to Discriminate Hydrothermal Alteration in Southwest Birjand, Iran, Using Advanced Spaceborne Thermal Emission and Reflection Radiometer Image Processing. *Acta Geol. Sin.* **2012**, *86*, 1289–1296. [CrossRef]

85. Abbaszadeh, M.; Hezarkhani, A. Enhancement of hydrothermal alteration zones using the spectral feature fitting method in Rabor area, Kerman. *Iran. Arab. J. Geosci.* **2013**, *6*, 1957–1964. [CrossRef]

86. Zadeh, M.; Tangestani, M.; Roldan, F.; Yusta, I. Mineral Exploration and Alteration Zone Mapping Using Mixture Tuned Matched Filtering Approach on ASTER Data at the Central Part of Dehaj-Sarduiyeh Copper Belt, SE Iran. *IEEE J. Sel. Top. Appl. Earth Obs. Remote Sens.* **2014**, *7*, 284–289. [CrossRef]

87. Available online: https://ww2w.harrisgeospatial.com.docs/MTMF.html (accessed on 20 March 2019).

88. Tayebi, M.; Tangestani, M.; Vincent, R. Alteration mineral mapping with ASTER data by integration of coded spectral ratio imaging and SOM neural network model. *Turk. J. Earth Sci.* **2014**, *23*, 627–644. [CrossRef]

89. Modabberi, S.; Ahmadi, A.; Tangestani, M. Sub-pixel mapping of alunite and jarosite using ASTER data; a case study from north of Semnan, north central Iran. *Ore Geol. Rev.* **2017**, *80*, 429–436. [CrossRef]

90. Rowan, L.; Hook, S.; Abrams, M.; Mars, J. Mapping hydrothermally altered rocks at Cuprite, Nevada, using the advanced Spaceborne Thermal Emission and Reflection Radiometer (ASTER), a new satellite-imaging system. *Econ. Geol.* **2003**, *98*, 1019–1027. [CrossRef]

91. Bhadra, B.; Pathak, S.; Karunakar, G.; Sharma, J. ASTER Data Analysis for Mineral Potential Mapping Around Sawar-Malpura Area, Central Rajasthan. *J. Indian Soc. Remote Sens.* **2013**, *41*, 391–404. [CrossRef]

92. Liu, L.; Zhou, J.; Yin, F.; Feng, M.; Zhang, B. The Reconnaissance of Mineral Resources through ASTER Data-Based Image Processing, Interpreting and Ground Inspection in the Jiafushaersu Area, West Junggar, China. *J. Earth Sci.* **2014**, *25*, 397–406. [CrossRef]

93. Popov, K.; Bakardjiev, D. Identification of hydrothermal alteration areas in the Panagyurishte ore region by satellite ASTER spectral data. *Comptes Rendus de L'Academie Bulgare des Sciences* **2014**, *67*, 1547–1554.

94. Son, Y.; Kang, M.; Yoon, W. Lithological and mineralogical survey of the Oyu Tolgoi region, Southeastern Gobi, Mongolia using ASTER reflectance and emissivity data. *Int. J. Appl. Earth Obs. Geoinform.* **2014**, *26*, 205–216. [CrossRef]

95. Canbaz, O.; Gursoy, O.; Gokce, A. Detecting Clay Minerals in Hydrothermal Alteration Areas with Integration of ASTER Image and Spectral Data in Kosedag-Zara (Sivas), Turkey. *J. Geol. Soc. India* **2018**, *91*, 483–488. [CrossRef]

96. Mahanta, P.; Maiti, S. Regional scale demarcation of alteration zone using ASTER imageries in South Purulia Shear Zone, East India: Implication for mineral exploration in vegetated regions. *Ore Geol. Rev.* **2018**, *102*, 846–861. [CrossRef]

97. Abdelkareem, M.; El-Baz, F. Characterizing hydrothermal alteration zones in Hamama area in the central Eastern Desert of Egypt by remotely sensed data. *Geocarto Int.* **2018**, *33*, 1307–1325. [CrossRef]

98. Pour, A.; Hashim, M.; Park, Y.; Hong, J. Mapping alteration mineral zones and lithological units in Antarctic regions using spectral bands of ASTER remote sensing data. *Geocarto Int.* **2018**, *33*, 1281–1306. [CrossRef]

99. Cudahy, T. Satellite ASTER Geoscience Product Notes for Australia. In *Australian ASTER Geoscience Product Notes, Version 1*; CSIR0 ePublish No. EP-30-07-12-44; CSIRO: Canberra, Australia, 7 August 2012.

100. Testa, F.; Villanueva, C.; Cooke, D.; Zhang, L. Lithological and hydrothermal alteration mapping of epithermal, porphyry and tourmaline breccia districts in the Argentine Andes using ASTER imagery. *Remote Sens.* **2018**, *10*, 203. [CrossRef]

101. Hubbard, B.; Crowley, J.; Zimbelman, D. Comparative alteration mineral mapping using visible to shortwave infrared (0.4-2.4 mu m) Hyperion, ALI, and ASTER imagery. *IEEE Trans. Geosci. Remote Sens.* **2003**, *41*, 1401–1410. [CrossRef]

102. Bedini, E. Mineral mapping in the Kap Simpson complex, central East Greenland, using HyMap and ASTER remote sensing data. *Adv. Space Res.* **2011**, *47*, 60–73. [CrossRef]

103. Honarmand, M.; Ranjbar, H.; Shahabpour, J. Combined use of ASTER and ALI data for hydrothermal alteration mapping in the northwestern part of the Kerman magmatic arc, Iran. *Int. J. Remote Sens.* **2013**, *34*, 2023–2046. [CrossRef]

104. Pour, A.; Hashim, M. ASTER, ALI and Hyperion sensors data for lithological mapping and ore minerals exploration. *Springerplus* **2014**, *3*. [CrossRef]

105. Ramos, Y.; Goita, K.; Peloquin, S. Mapping advanced argillic alteration zones with ASTER and Hyperion data in the Andes Mountains of Peru. *J. Appl. Remote Sens.* **2016**, *10*. [CrossRef]

106. Gemail, K.; Abd-El Rahman, N.; Ghiath, B. Integration of ASTER and airborne geophysical data for mineral exploration and environmental mapping: A case study, Gabal Dara, North Eastern Desert, Egypt. *Environ. Earth Sci.* **2016**, *75*. [CrossRef]

107. Lampinen, H.; Laukamp, C.; Occhipinti, S.; Metelka, V. Delineating Alteration Footprints from Field and ASTER SWIR Spectra, Geochemistry, and Gamma-Ray Spectrometry above Regolith-Covered Base Metal Deposits-An Example from Abra, Western Australia. *Econ. Geol.* **2017**, *8*, 1977–2003. [CrossRef]

108. Liu, L.; Feng, J.; Han, L.; Zhou, J.; Xu, X.; Liu, R. Mineral mapping using spaceborne Tiangong-1 hyperspectral imagery and ASTER data: A case study of alteration detection in support of regional geological survey at Jintanzi-Malianquan area, Beishan, Gansu Province, China. *Geol. J.* **2018**, *53*, 372–383. [CrossRef]

109. Hu, B.; Xu, Y.; Wan, B.; Wu, X.; Yi, G. Hydrothermally altered mineral mapping using synthetic application of Sentinel-2A MSI, ASTER and Hyperion data in the Duolong area, Tibetan Plateau, China. *Ore Geol. Rev.* **2018**, *101*, 384–397.

110. Volesky, J.; Stern, R.; Johnson, P. Geological control of massive sulfide mineralization in the Neoproterozoic Wadi Bidah shear zone, southwestern Saudi Arabia, inferences from orbital remote sensing and field studies. *Precambrian Res.* **2003**, *123*, 235–247. [CrossRef]

111. Karimpour, M.; Mazhari, N.; Shafaroudi, A. Discrimination of Different Erosion Levels of Porphyry Cu Deposits using ASTER Image Processing in Eastern Iran: A Case Study in the Maherabad, Shadan, and Chah Shaljami Areas. *Acta Geol. Sin.* **2014**, *88*, 1195–1213. [CrossRef]

112. Mohebi, A.; Mirnejad, H.; Lentz, D. Controls on porphyry Cu mineralization around Hanza Mountain, southeast of Iran: An analysis of structural evolution from remote sensing, geophysical, geochemical and geological data. *Ore Geol. Rev.* **2015**, *69*, 187–198. [CrossRef]

113. Sojdehee, M.; Rasa, I.; Nezafati, M.; Abedini, M. Application of spectral analysis to discriminate hydrothermal alteration zones at Daralu copper deposit, SE Iran. *Arab. J. Geosci.* **2016**, *9*. [CrossRef]

114. Farahbakhsh, E.; Shirmard, H.; Bahroudi, A. Fusing ASTER and QuickBird-2 Satellite Data for Detailed Investigation of Porphyry Copper Deposits Using PCA; Case Study of Naysian Deposit, Iran. *J. Indian Soc. Remote Sens.* **2016**, *44*, 525–537. [CrossRef]

115. Yousefi, S.J.; Ranjbar, H.; Alirezaei, S.; Dargahi, S. Discrimination of Sericite Phyllic and Quartz-Rich Phyllic Alterations by Using a Combination of ASTER TIR and SWIR Data to Explore Porphyry Cu Deposits Hosted by Granitoids, Kerman Copper Belt, Iran. *J. Indian Soc. Remote Sens.* **2018**, *46*, 717–727. [CrossRef]

116. Yousefi, S.; Ranjbar, H.; Alirezaei, S.; Dargahi, S. Comparison of hydrothermal alteration patterns associated with porphyry Cu deposits hosted by granitoids and intermediate-mafic volcanic rocks, Kerman Magmatic Arc, Iran: Application of geological, mineralogical and remote sensing data. *J. Afr. Earth Sci.* **2018**, *142*, 112–123. [CrossRef]

117. Safari, M.; Maghsoudi, A.; Pour, A. Application of Landsat-8 and ASTER satellite remote sensing data for porphyry copper exploration: A case study from Shahr-e-Babak, Kerman, south of Iran. *Geocarto Int.* **2018**, *33*, 1186–1201. [CrossRef]

118. Pour, A.; Hashim, M. Identifying areas of high economic-potential copper mineralization using ASTER data in the Urumieh-Dokhtar Volcanic Belt, Iran. *Adv. Space Res.* **2012**, *49*, 753–769. [CrossRef]

119. Pour, A.; Hashim, M. Identification of hydrothermal alteration minerals for exploring of porphyry copper deposit using ASTER data, SE Iran. *J. Asian Earth Sci.* **2011**, *42*, 1309–1323. [CrossRef]

120. Honarpazhouh, J.; Hassanipak, A.; Shabani, K. Integration of stream sediment geochemical and ASTER data for porphyry copper deposit exploration in Khatun Abad, NW Iran. *Arch. Min. Sci.* **2013**, *58*, 37–54.

121. Pazand, K.; Sarvestani, J.; Ravasan, M. Hydrothermal Alteration Mapping Using ASTER Data for Reconnaissance Porphyry Copper Mineralization in the Ahar Area, NW Iran. *J. Indian Soc. Remote Sens.* **2013**, *41*, 379–389. [CrossRef]

122. Alimohammadi, M.; Alirezaei, S.; Kontak, D. Application of ASTER data for exploration of porphyry copper deposits: A case study of Daraloo-Sarmeshk area, southern part of the Kerman copper belt, Iran. *Ore Geol. Rev.* **2015**, *70*, 290–304. [CrossRef]

123. Yazdi, Z.; Rad, A.; Ajayebi, K. Analysis and modeling of geospatial datasets for porphyry copper prospectivity mapping in Chahargonbad area, Central Iran. *Arab. J. Geosci.* **2015**, *8*, 8237–8248. [CrossRef]

124. Saadat, S. Comparison of various knowledge-driven and logistic-based mineral prospectivity methods to generate Cu and Au exploration targets Case study: Feyz-Abad area (North of Lut block, NE Iran). *J. Min. Environ.* **2017**, *8*, 611–619.

125. Zadeh, M.; Honarmand, M. A remote sensing-based discrimination of high- and low-potential mineralization for porphyry copper deposits; a case study from Dehaj-Sarduiyeh copper belt, SE Iran. *Eur. J. Remote Sens.* **2017**, *50*, 332–342. [CrossRef]

126. Mars, J.; Rowan, L. Regional mapping of phyllic- and argillic-altered rocks in the Zagros magmatic arc, Iran, using Advanced Spaceborne Thermal Emission and Reflection Radiometer (ASTER) data and logical operator algorithms. *Geosphere* **2006**, *2*, 161–186. [CrossRef]

127. Carrino, T.; Crosta, A.; Toledo, C.; Silva, A.; Silva, J. Geology and Hydrothermal Alteration of the Chapi Chiara Prospect and Nearby Targets, Southern Peru, Using ASTER Data and Reflectance Spectroscopy. *Econ. Geol.* **2015**, *110*, 73–90. [CrossRef]

128. Ibrahim, W.; Watanabe, K.; Yonezu, K. Structural and litho-tectonic controls on Neoproterozoic base metal sulfide and gold mineralization in North Hamisana shear zone, South Eastern Desert, Egypt: The integrated field, structural, Landsat 7 ETM + and ASTER data approach. *Ore Geol. Rev.* **2016**, *79*, 62–77. [CrossRef]

129. Zhang, T.; Yi, G.; Li, H. Integrating Data of ASTER and Landsat-8 OLI (AO) for Hydrothermal Alteration Mineral Mapping in Duolong Porphyry Cu-Au Deposit, Tibetan Plateau, China. *Remote Sens.* **2016**, *8*, 890. [CrossRef]

130. Rajendran, S.; Nasir, S. Characterization of ASTER spectral bands for mapping of alteration zones of volcanogenic massive sulphide deposits. *Ore Geol. Rev.* **2017**, *88*, 317–335. [CrossRef]

131. Zhang, N.; Zhou, K. Identification of hydrothermal alteration zones of the Baogutu porphyry copper deposits in northwest China using ASTER data. *J. Appl. Remote Sci.* **2017**, *11*. [CrossRef]

132. Dai, J.; Qu, X.; Song, Y. Porphyry Copper Deposit Prognosis in the Middle Region of the Bangonghu-Nujiang Metallogenic Belt, Tibet, Using ASTER Remote Sensing Data. *Resour. Geol.* **2018**, *68*, 65–82. [CrossRef]

133. Amer, R.; El Mezayen, A.; Hasanein, M. ASTER spectral analysis for alteration minerals associated with gold mineralization. *Ore Geol. Rev.* **2016**, *75*, 239–251.

134. Salem, S.; Soliman, N.; Ramadan, T.; Greiling, R. Exploration of new gold occurrences in the alteration zones at the Barramiya District, Central Eastern Desert of Egypt using ASTER data and geological studies. *Arab. J. Geosci.* **2014**, *7*, 1717–1731. [CrossRef]

135. Salem, S.; Soliman, N. Exploration of gold at the east end of Wadi Allaqi, South Eastern Desert, Egypt, using remote sensing techniques. *Arab. J. Geosci.* **2015**, *8*, 9271–9282. [CrossRef]

136. Hasan, E.; Fagin, T.; El Alfy, Z. Spectral Angle Mapper and aeromagnetic data integration for gold-associated alteration zone mapping: A case study for the Central Eastern Desert Egypt. *Int. J. Remote Sens.* **2016**, *37*, 1762–1776. [CrossRef]

137. Salem, S.; El Sharkawi, M.; El-Alfy, Z.; Soliman, N.; Ahmed, S. Exploration of gold occurrences in alteration zones at Dungash district, Southeastern Desert of Egypt using ASTER data and geochemical analyses. *J. Afr. Earth Sci.* **2016**, *117*, 389–400. [CrossRef]

138. Abdelnasser, A.; Kumral, M.; Zoheir, B.; Karaman, M. REE geochemical characteristics and satellite-based mapping of hydrothermal alteration in Atud gold deposit, Egypt. *J. Afr. Earth Sci.* **2018**, *145*, 317–330. [CrossRef]

139. Salem, S.; El Sharkawi, M.; El Alfy, Z.; Ahmed, S. The use of ASTER data and geochemical analyses for the exploration of gold at Samut area, South Eastern Desert of Egypt. *Arab. J. Geosci.* **2018**, *11*, 11–18. [CrossRef]

140. Hot Copper. Available online: https://hotcopper.com.au/threads/ann-aster-interpretation-identifies-new-target-a.2066859/ (accessed on 3 January 2019).

141. Western Australia Satellite Technology and Applications Consortium (WASTAC). 2014 Annual Report; 2014; 56p. Available online: https://www.wastac.wa.gov.au (accessed on 23 August 2018).

142. Zhang, X.; Pamer, M.; Duke, N. Lithologic and mineral information extraction for gold exploration using ASTER data in the south Chocolate Mountains (California). *ISPRS J. Photogramm. Remote Sens.* **2007**, *62*, 271–282. [CrossRef]

143. De Palomera, P.; van Ruitenbeek, F.; Carranza, E. Prospectivity for epithermal gold-silver deposits in the Deseado Massif, Argentina. *Ore Geol. Rev.* **2015**, *71*, 484–501. [CrossRef]

144. Wells, M.; Laukamp, C.; Hancock, E. Reflectance spectroscopic characterisation of mineral alteration footprints associated with sediment-hosted gold mineralisation at Mt Olympus (Ashburton Basin, Western Australia). *Aust. J. Earth Sci.* **2016**, *63*, 987–1002.

145. Yao, K.; Pradhan, B.; Idrees, M. Identification of Rocks and Their Quartz Content in Gua Musang Goldfield Using Advanced Spaceborne Thermal Emission and Reflection Radiometer Imagery. *J. Sens.* **2017**. [CrossRef]

146. Yousefi, T.; Aliyari, F.; Abedini, A.; Calagari, A. Integrating geologic and Landsat-8 and ASTER remote sensing data for gold exploration: A case study from Zarshuran Carlin-type gold deposit, NW Iran. *Arab. J. Geosci.* **2018**, *11*, 482–499. [CrossRef]

147. Rani, K.; Guha, A.; Mondal, S.; Pal, S.; Kumar, K. ASTER multispectral bands, ground magnetic data, ground spectroscopy and space-based EIGEN6C4 gravity data model for identifying potential zones for gold sulphide mineralization in Bhukia, Rajasthan, Indian. *J. Appl. Geophys.* **2019**, *160*, 28–46. [CrossRef]

148. Rajendran, S.; Thirunavukkarasu, A.; Balamurugan, G.; Shankar, K. Discrimination of iron ore deposits of granulite terrain of Southern Peninsular India using ASTER data. *J. Asian Earth Sci.* **2011**, *41*, 99–106. [CrossRef]

149. Huang, S.; Chen, S.; Zhang, Y. Comparison of altered mineral information extracted from ETM plus, ASTER and Hyperion data in aguas Claras iron ore, Brazil. *IET Image Proc.* **2019**, *13*, 355–364. [CrossRef]

150. Duuring, P.; Hagemann, S.; Novikova, Y.; Cudahy, T.; Laukamp, C. Targeting Iron Ore in Banded Iron Formations Using ASTER Data: Weld Range Greenstone Belt, Yilgarn Craton, Western Australia. *Econ. Geol.* **2012**, *107*, 585–597. [CrossRef]

151. Mansouri, E.; Feizi, F.; Rad, A.; Arian, M. Remote-sensing data processing with the multivariate regression analysis method for iron mineral resource potential mapping: A case study in the Sarvian area, Central Iran. *Solid Earth* **2018**, *9*, 373–384. [CrossRef]

152. Mazhari, N.; Shafaroudi, A.; Ghaderi, M. Detecting and mapping different types of iron mineralization in Sangan mining region, NE Iran, using satellite image and airborne geophysical data. *Geosci. J.* **2017**, *21*, 137–148. [CrossRef]

153. Moghtaderi, A.; Moore, F.; Ranjbar, H. Application of ASTER and Landsat 8 imagery data and mathematical evaluation method in detecting iron minerals contamination in the Chadormalu iron mine area, central Iran. *J. Appl. Remote Sci.* **2017**, *11*, 016027. [CrossRef]

154. Cardoso-Fernandes, J.; Teodoro, A.; Lima, A. Remote sensing data in lithium (Li) exploration: A new approach for the detection of Li-bearing pegmatites. *Int. J. Appl. Earth Obs. Geoinform.* **2019**, *76*, 10–25. [CrossRef]

155. Shawky, M.; El-Arafy, R.; El Zalaky, M. Validating (MNF) transform to determine the least inherent dimensionality of ASTER image data of some uranium localities at Central Eastern Desert. Egypt. *J. Afr. Earth Sci.* **2019**, *149*, 441–450. [CrossRef]

156. Hewson, R.; Cudahy, T.; Drake-Brockman, J.; Meyers, J.; Hashemi, A. Mapping geology associated with manganese mineralisation using spectral sensing techniques at Woodie Woodie, East Pilbara. *Explor. Geophys.* **2006**, *37*, 389–400. [CrossRef]

157. Othman, A.; Gloaguen, R. Improving Lithological Mapping by SVM Classification of Spectral and Morphological Features: The Discovery of a New Chromite Body in the Mawat Ophiolite Complex (Kurdistan, NE Iraq). *Remote Sens.* **2014**, *6*, 6867–6896. [CrossRef]

158. Fu, B.; Zheng, G.; Ninomiya, Y.; Wang, C.; Sun, G. Mapping hydrocarbon-induced mineralogical alteration in the northern Tian Shan using ASTER multispectral data. *Terra Nova* **2007**, *19*, 225–231. [CrossRef]

159. Shi, B.; Fu, B.; Ninomiya, Y.; Wang, C. Multispectral remote sensing mapping for hydrocarbon seepage-induced lithologic anomalies in the Kuqa foreland basin, south Tian Shan. *J. Asian Earth Sci.* **2012**, *46*, 70–77. [CrossRef]

160. Soydan, H.; Koz, A.; Duzgun, H. Identification of hydrocarbon microseepage induced alterations with spectral target detection and unmixing algorithms. *Int. J. Appl. Earth Obs. Geoinform.* **2019**, *74*, 209–221. [CrossRef]

161. Pena, S.; Abdelsalam, M. Orbital remote sensing for geological mapping in southern Tunisia: Implication for oil and gas exploration. *J. Afr. Earth Sci.* **2006**, *44*, 203–219. [CrossRef]

162. Petrovic, A.; Khan, S.; Chafetz, H. Remote detection and geochemical studies for finding hydrocarbon-induced alterations in Lisbon Valley, Utah. *Mar. Petrol. Geol.* **2008**, *25*, 696–705. [CrossRef]

163. Mars, J. Mineral and lithologic mapping capability of WorldView 3 data at Mountain Pass, California, using true- and false-color composite images, band rations, and logical operator algorithms. *Econ. Geol.* **2018**, *113*, 1587–1601. [CrossRef]

164. Abrams, M. ASTER: Data products for the high spatial resolution imager on NASA's EOS-AM1 platform. *Int. J. Remote Sens.* **2000**, *21*, 847–861. [CrossRef]

165. Data Processing Levels. Available online: https://science.nasa.gov/earth-science/earth-science-data/data-processing-levels-for-eosdis-data-products (accessed on 15 August 2018).

166. ASTER Data Products. Available online: https://lpdaac.usgs.gov/dataset_discovery/aster/aster_products_table (accessed on 9 August 2018).

Review

The Spatial and Spectral Resolution of ASTER Infrared Image Data: A Paradigm Shift in Volcanological Remote Sensing

Michael S. Ramsey * and Ian T.W. Flynn

Department of Geology and Environmental Science, University of Pittsburgh, 4107 O'Hara Street, Pittsburgh, PA 15260-3332, USA; ITF2@pitt.edu
* Correspondence: mramsey@pitt.edu

Received: 26 December 2019; Accepted: 12 February 2020; Published: 23 February 2020

Abstract: During the past two decades, the Advanced Spaceborne Thermal Emission and Reflection Radiometer (ASTER) instrument on the Terra satellite has acquired nearly 320,000 scenes of the world's volcanoes. This is ~10% of the data in the global ASTER archive. Many of these scenes captured volcanic activity at never before seen spatial and spectral scales, particularly in the thermal infrared (TIR) region. Despite this large archive of data, the temporal resolution of ASTER is simply not adequate to understand ongoing eruptions and assess the hazards to local populations in near real time. However, programs designed to integrate ASTER into a volcanic data sensor web have greatly improved the cadence of the data (in some cases, to as many as 3 scenes in 48 h). This frequency can inform our understanding of what is possible with future systems collecting similar data on the daily or hourly time scales. Here, we present the history of ASTER's contributions to volcanology, highlighting unique aspects of the instrument and its data. The ASTER archive was mined to provide statistics including the number of observations with volcanic activity, its type, and the average cloud cover. These were noted for more than 2000 scenes over periods of 1, 5 and 20 years.

Keywords: ASTER; thermal infrared data; volcanic processes; image archive; future concepts

---

1. Introduction

1.1. ASTER Instrument and History

The Advanced Spaceborne Thermal Emission and Reflection Radiometer (ASTER) instrument was launched on the NASA Terra satellite on 18 December 1999. Prior to the instrument beginning its operational phase on 4 March 2000, ASTER acquired several images including the first thermal infrared (TIR) image on 6 February (Figure 1). That image was of Erta Ale volcano (Ethiopia), and initiated a long association of ASTER data with volcanological observations. ASTER was developed and built in Japan under the Japanese Ministry of Economy, Trade and Industry (METI), and is one of five Earth observing instruments on Terra. The combined science team of Japanese and United States investigators has changed over the years, however always maintaining a strong volcanological component [1,2]. During the past two decades, the data from ASTER have been applied to numerous questions and scales of surface processes, most notably for volcanic activity, e.g., [3–6].

ASTER was designed to observe the surface at multiple spatial and spectral resolutions as well as from different viewing geometries. It is actually a suite of three instruments with independent bore-sighted telescopes, originally having 14 spectral channels in the visible/near-infrared (VNIR), the shortwave infrared (SWIR), and the thermal infrared (TIR) regions [7]. The VNIR instrument (0.52–0.86 μm) has three spectral channels at a spatial resolution of 15 m/pixel paired with one channel oriented in a backward look direction for the creation of digital elevation models (DEMs). The SWIR instrument (1.6–2.43 μm) unfortunately failed in 2008, but originally had six channels at a spatial resolution of 30 m/pixel. Finally, and perhaps most important for many aspects of volcanological remote sensing, the TIR instrument (8.13–11.65 μm) has five channels at a spatial resolution of 90 m/pixel. During the lifetime of the mission, ASTER has acquired over 3.5 million individual scenes—approximately 22% of which were collected at night. Here, we describe the two-decade history of ASTER, specific programs to improve the observational frequency of volcanoes, and present examples of those data.

Figure 1. The first Advanced Spaceborne Thermal Emission and Reflection Radiometer (ASTER) multispectral thermal infrared (TIR) image acquired on 6 February 2000, before the start of the instrument's science operational phase. Erta Ale volcano, Ethiopia (summit location: 13.60°N, 40.67°E), is shown using a decorrelation stretch (DCS) of the TIR bands 14, 12, 10 in red, green, and blue, respectively. Color variations are mainly caused by rock and soil compositional differences, and only possible with the ASTER multispectral TIR data. The blues and purple colors are indicative of dominantly basaltic lava flows. The small cluster of white pixels in the lower central part of the image is the summit lava lake thermal anomaly. The ASTER DCS data are overlain on a visible Google Earth image for context. Image credit: NASA/METI/AIST/Japan Space Systems, and U.S./Japan ASTER Science Team.

1.2. Twenty Years of Volcanic Studies Using ASTER Data

The ASTER Science Team created useful derived science products that have benefited numerous scientific fields such as volcanology. These include a robust temperature emissivity separation algorithm for the first orbital high spatial resolution multispectral TIR data [8], programs such as the Science Team Acquisition Request (STAR) to acquire data focused on important science questions of individual users/teams, as well as the ability to generate DEMs [9]. These individual-scene DEMs of 60 by 60 km were later composited globally into the ASTER Global DEM or GDEM—version 3 of which was released in August 2019 [10,11].

Arguably, volcanology is one of the scientific disciplines on which ASTER data have had the greatest impact. In order to quantify this impact and gather all volcanology related studies together into one reference document, we have performed an extensive literature review (Appendix A). This found 271 peer-reviewed publications from 1 January 1995 to 1 December 2019, an average of nearly 11 per year (Figure 2). Papers published prior to the Terra launch were considered precursory studies, commonly describing how the future ASTER data would be used for certain volcanic studies. This list of 271 publications was subdivided into 12 categories based on the volcanic focus of the papers. The category names and number of papers in those categories (shown in parentheses) are: Analogs (3), Calibration (4), Lava Flows (4), Gas/Plumes (30), Geothermal (9), Mapping (54), Modeling (16), Monitoring (96), Operational (4), Other (4), Precursory (16), and Topography (31). Based on this categorization, ASTER data have been primarily used to monitor volcanic activity; somewhat surprising considering the lower temporal frequency of the data. This speaks to the need for future high spatial, high spectral resolution data at temporal resolutions far better than the nominal 16 day equatorial repeat time of ASTER and Landsat, or even the 5 day resolution of the Sentinel-2 constellation. This same finding was also brought forward in the most recent Decadal Survey for Earth Science, which noted data such as these are critical for addressing two of the most important science questions related to natural hazards [12].

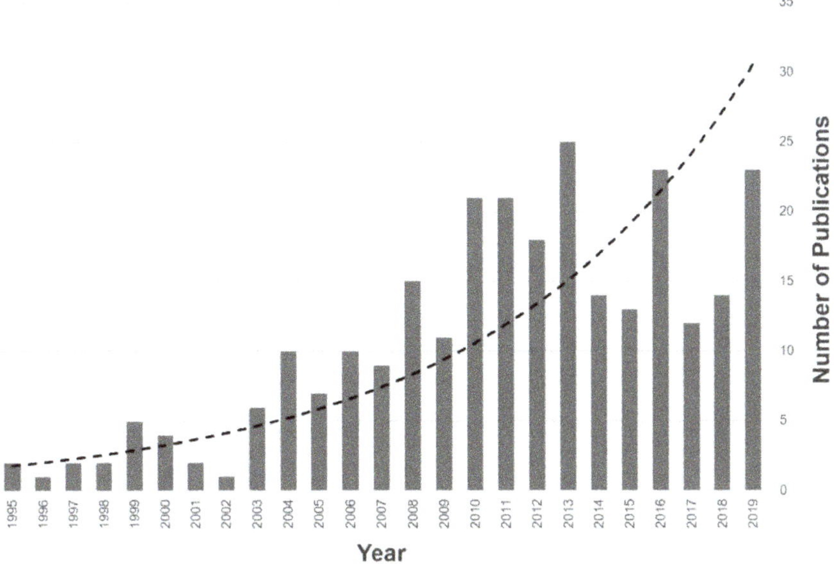

Figure 2. The number of volcano-related publications per year that have incorporated ASTER data in some aspect of the work. Through 2016, the growth has been roughly exponential (dashed line). That growth has declined somewhat in the last several years, although still remaining respectable. The total number of publications shown is 271, spanning the last 25 years.

2. Background

2.1. Volcanological Remote Sensing

The data from spaceborne sensors used to detect, monitor, and even forecast eruptions have been analyzed since the earliest days of the satellite era, e.g., [13–15]. Those early studies using the data available at the time focused mostly on hot spot detection and temperature measurements with TIR data. The studies continued to expand despite the fact that no sensor launched by any country has ever been specifically devoted to volcano science. Scientific studies grew ever more complex with the launch of new sensors providing better spatial, temporal and spectral data, arguably creating the field of spaceborne volcanology. The ability to extract critical information from subtle phases of precursory activity in order to perform the detailed spectral mapping of the erupted products grew exponentially [16]. Many of these studies describe the detection of a new thermal anomaly at a quiescent volcano, which gave rise to models of the sub-pixel temperature distribution. High temporal temperature data allowed more accurate modeling of lava and gas flux rates as well as chronological descriptions of each eruptive phase. Ramsey and Harris [17] summarized the history of satellite-based TIR research of active volcanoes into four broad themes: (1) thermal detection, (2) analysis of sub-pixel components, (3) heat/mass flux studies, and (4) eruption chronologies. Ramsey [2] added a fifth theme, the creation of sensor webs consisting of integrated data from multiple sensors to improve the spatial and/or temporal resolution.

Volcanology, as is the case for many other disciplines relying on orbital image data, adapted to the fundamental technological divide of the availability of high temporal/low spatial resolution versus that of low temporal/high spatial resolution data. Volcanological processes operating at the minute to hourly time scale (e.g., lava effusion, plume emplacement, drifting ash clouds) require data very different than those acquired on the time scale of days to weeks. The former falls under a class of TIR sensors designed primarily for weather and atmospheric studies and includes sensors such as the Advanced Very High Resolution Radiometer (AVHRR), the Along Track Scanning Radiometer (ATSR), the Moderate Resolution Imaging Spectroradiometer (MODIS), the Visible Infrared Imaging Radiometer Suite (VIIRS), as well as instruments on the Geostationary Operational Environmental Satellites (GOES). These sensors are commonly designed with wide swath widths, a limited number of spectral bands, and spatial resolutions of 1.0 km/pixel or larger, which result in temporal frequencies of minutes to hours. Modeling the data to extract information below the scale of the pixel have made these datasets invaluable for both the rapid detection of new activity as well as the analysis of time-scale dependent eruptive processes [18–20].

TIR data of the Earth's surface evolved from very course spatial resolution to the sub-100 m scale, and from one spectral channel for temperature measurements to five for the ASTER and ECOsystem Spaceborne Thermal Radiometer Experiment on Space Station (ECOSTRESS) instruments. This class of sensors includes instruments on the current Landsat platforms as well as older systems like ASTER and newer ones such as ECOSTRESS. These commonly have a larger number of spectral bands than the weather class of sensors, spatial resolutions of 100 m/pixel or better, but a temporal frequency of days to weeks. The improved spatial and spectral resolution does provide for studies of smaller-scale volcanic processes with detection of much smaller temperature variations, e.g., [3,5,21]. These sensors are also excellent for detecting early precursory activity despite the infrequent coverage, e.g., [22]. However, the data cannot be used to describe the high-frequency changes ongoing during an eruption despite providing a detailed "snapshot in time" of that activity.

2.2. The 2018 Decadal Survey Recommendations

Land surface image data at ever-improving spatial, spectral and temporal scales, which also span a wide wavelength range from the VNIR to the TIR, have greatly improved our understanding of geological and biological processes operating at those observational scales. This has been recognized for decades with Landsat data, which have been improved spatially and spectrally as changes were made to the instrument design. Following its launch, ASTER data greatly increased both the spatial and spectral ranges, providing new capabilities from that of Landsat, significant especially for volcanology.

The need to continue (and improve upon) this class of measurement was recognized in the first Decadal Survey (DS) for Earth Science in 2007 [23]. In that report, a notional list of mission concepts was proposed, which included the Hyperspectral Infrared Imager (HyspIRI) mission that paired a hyperspectral visible/shortwave infrared (VSWIR) instrument with a multispectral TIR instrument. Many years of planning, design and science concept studies were later performed. Ultimately, however, the mission was never selected by NASA to move forward and the concept study was formally ended with the publication of the second Decadal Survey in 2018 [12].

The second report focused on science questions and key observables, around which new mission concepts could be structured. Notably, the lack of future global TIR and VSWIR image data was once again brought forward in the new DS report. For example, the requirement for infrared measurements spanned all of the working group panels, being mentioned over 190 times in the report and 12 times in the requests for information from the general science community. Focusing specifically on the reports of the Ecosystems and the Earth Surface and Interior panels, which contributed most directly to the need for TIR data, those data were key inputs for 14 different science objectives—half of which were designated as "most" or "very" important, the top categories. The science and applications summary includes the need for measurements of surface geology, active geologic processes such as natural disasters, surface/water temperature, as well as functional traits of vegetation and ecosystems. The Earth Surface and Interior panel, for example, focused two of its top-level science questions on natural disasters: data prior to the event and the outcomes following. TIR measurements are noted as vital for several of these disasters, including volcanoes, landslides and wildfires. Although temperature is an important measurement, the need for vast improvements in TIR spatial, spectral and temporal resolution data was made clear. TIR data acquired in 1–2 channels at 100 m resolution every 2 weeks no longer satisfies the requirements deemed important by the science community for the future. Hyperspectral TIR coupled with vastly improved temporal resolution at spatial scales exceeding current ASTER capabilities were recommended. The DS report lists several Designated Observables (DOs), including the Surface Biology and Geology (SBG) DO. A mission concept designed to address this DO will likely include some combination of these recommended scales of TIR data, which will allow far more detailed volcanic measurements than currently possible with either ASTER or Landsat data.

3. The Methodology of ASTER Volcano Observations

3.1. The Need for More Routine ASTER Volcano Observations

3.1.1. The ASTER Volcano Science Team Acquisition Request (STAR)

The recognition that ASTER data would provide a fundamentally new tool for volcanic observations was documented even before the Terra launch [7]. During that development period, the ASTER Science Team (AST) foresaw that the instrument would eventually provide data at spatial and spectral scales never before observed, routine data at night, the ability to point off-nadir to improve temporal revisit time, and therefore, required daily observation schedules. Being a scheduled instrument, unlike many other nadir-viewing systems, brought both a higher level of mission complexity as well as unique opportunities for Earth observations. Scheduling allowed specific ground targeting, a focus on larger-scale global processes, as well as an important goal of creating a global map of ASTER data.

One observational and scheduling strategy developed was the creation of the science team acquisition request or STAR. The STARs are a series of globally distributed regions of interest (ROIs) over targets with high scientific value. The ROIs had associated attributes such as seasonality, instrument gain settings, number of observation attempts per year, etc. These targets were integrated into the daily scheduling so that both STAR-focused scenes were acquired together with the many other required observations during any given orbital period. The STAR's assured a priority set of observations for high-interest science such as monitoring the global land ice inventory, change detection of large urban environments, and volcanic activity [24–26].

The ASTER Volcano STAR's scheduling plan was designed to allow routine observations over the world's volcanoes, which were then made available to the scientific community as quickly as possible [26]. The original plan divided the global list of the approximately 1000 active and/or potentially active volcanoes into high, medium and low priority classes. These divisions, initially dubbed class A, B, and C, respectively, varied according to the historical frequency of their eruption activity. Class A consisted of volcanoes that had several recorded eruptions during the prior decade; class B volcanoes had several recorded eruptions during the past several decades; and class C consisted of the remainder of volcanoes that had not seen activity in the prior century. The STAR designated that class A targets would be observed every 48 days during the day and every 32 days at night. Class B targets were to be observed every 3 months, both in the day and night. Finally, class C targets were to be observed once every 6 months [27,28]. Although this plan continues to provide regular data of all the volcanoes on Earth, many eruptions and the precursory activity prior, were missed due to this schedule rigidity. For the Volcano STAR, ASTER acquires ~16,000 scenes per year on average, which is a combination of day and nighttime data, for a total of ~320,000 individual ASTER volcano scenes over the mission lifetime. Therefore, for the 964 individual volcano ROIs in the ASTER Volcano STAR, each volcano is observed ~16 times per year on average. Although an improvement over the nominal observational schedule, this frequency is still not enough to allow rapid response observations nor discrimination of short-timescale activity, especially considering that some percentage of these scenes are dominated by clouds.

3.1.2. The ASTER Urgent Request Protocol (URP)

Because of the lack of an adequate temporal sampling of the very restless and actively erupting volcanoes with the volcano STAR, the ASTER Urgent Request Protocol (URP) Program was proposed [2,29]. Simply, the URP is a means to improve the number of observations at the most active volcanic centers around the world. The URP integrates ASTER into a sensor web construct where all scales of activity at an erupting volcano can be captured [29,30]. The initial and most straight-forward implementation of this approach for the URP uses detection of thermally elevated pixels in high temporal resolution data to subsequently trigger more rapid scheduling and acquisition of the higher spatial/spectral resolution data from ASTER. With such a system in place, the high-frequency activity can be continually imaged throughout the eruption, with the high spatial resolution data ideal for capturing small scale changes. These ASTER data also serve as validation for the low spatial, high temporal resolution data.

The URP program has been in place as part of the ASTER sensor's operational scheduling since 2005 [2,29], responsible for over 5000 additional scenes of active volcanoes during that time (one new scene on average every day). Perhaps more importantly, the URP can be triggered manually if precursory activity is noted based on ground-based observations or reports. This allows pre-eruption data to be acquired that significantly add to the monitoring process [22].

Later expansion of the URP Program increased the original monitored area from the northern Pacific region to the entire globe [2]. The URP currently operates with two global monitoring systems using MODIS data: MODVOLC [31] and Middle InfraRed Observation of Volcanic Activity (MIROVA) [32] as well as AVHRR data focused in the north Pacific region using the Okmok algorithm [33]. New triggering systems are now being tested that will integrate ground-based thermal camera data as the source for new URP data. A trial system at Mt. Etna volcano in Italy has been ongoing since mid-2019 and will expand to Piton de la Fournaise volcano on Réunion Island in 2020 using seismic alerts as the triggering source. The addition of the URP to the ASTER observation schedule is a vast improvement from the original volcano STAR [3,27]. Importantly, however, the URP operates in tandem with the volcano STAR. The volcano STAR data represent additional scenes for the very active volcanoes, thus supplementing the URP archive. Conversely, the volcano STAR data are commonly the only information for the less active, non-thermally elevated targets, and therefore continue to represent an important source for global volcano data.

3.2. Volcano Data Archives

3.2.1. The ASTER Image Database for Volcanoes

The entire ASTER archive is available at the Land Processes Distributed Active Archive Center (LP DAAC), which can be searched using tools such as EarthDATA and GloVis (Table 1). Other web-based data repositories, however, have been created specifically for the volcano data products. The first of these is the Image Database for Volcanoes (IDV), which was created by M. Urai, and is located in and served from Japan [27,34]. A new volcano image is commonly added within a week of acquisition. The database contains all ASTER images of the 964 target volcanoes that comprise the ASTER Volcano STAR. A 20 km^2 area centered on the geographic location of volcano is shown and stored in the database. Links to each volcano are also displayed and upon selection, the best VNIR image, geographic information, and a table ordered by date of all the acquisitions are shown. The chronological list has thumbnails of the each of the three ASTER imaging systems (where available) along with metadata links for access and download.

Table 1. Volcano-specific data archives and search tools available for all ASTER data.

Site Name	Site Web Address and Relevant Information
Land Processes Distributed Active Archive Center (LP DAAC)	https://lpdaac.usgs.gov/data/get-started-data/collection-overview/missions/aster-overview/
	Contains the entire ASTER archive from 2000 to present, including the on-demand higher-level data products
Image Database for Volcanoes (IDV)	https://gbank.gsj.jp/vsidb/image/index-E.html
	Contains a 20 km^2 area around each of the 964 volcanoes in the ASTER Volcano STAR database, for the entire mission from 2000 to present
ASTER Volcano Archive (AVA)	http://ava.jpl.nasa.gov
	Contains the full ASTER scene in multiple formats for ~1500 volcanoes, in addition to other sensor data and derived products. Archive currently spans from 2000 to 2017
EarthDATA Search Tool	https://earthdata.nasa.gov/
	Searchable visual archive for the entire ASTER mission from 2000 to present, including the on-demand higher-level data products
GloVis Search Tool	https://glovis.usgs.gov/
	Searchable visual archive for the entire ASTER mission from 2000 to present

Urai [34] conducted an initial evaluation of the ASTER Volcano STAR performance using data in the IDV. As expected, volcanoes in Iceland and Kamchatka were observed the most frequently. At the time of the study, 77% of all the volcanoes in the database had at least one daytime image with <10% cloud cover. The IDV remains active as of the time of this writing, and as of the latest IDV update on 1 December 2019, the average number of scenes acquired had grown to 146.3 (daytime) and 185.4 (nighttime) per volcano.

3.2.2. The ASTER Volcano Archive (AVA)

The second online volcano database to appear was the AVA, which is housed and served in the United States at the Jet Propulsion Laboratory [27,28]. The AVA is the largest dedicated archive of web-accessible volcano images now containing image data other than ASTER as well as secondary derived products for each volcano (Table 1). The archive provides capabilities to inventory and monitor properties and processes over time. These include the spectral signatures for volcanic emissions (e.g., eruption columns and plumes) and surficial deposits (e.g., lava flows, pyroclastic flows), as well as eruption precursor data. AVA has helped to improve the global monitoring and access to archival ASTER image data of more targets (~1500) than are in the ASTER Volcano STAR. Much like the IDV, volcanoes can be searched by name or location, and thumbnail images. Unlike the IDV, AVA data can be downloaded as full ASTER scenes in geoTIFF, KML, or HDF formats. These data can be also combined with the DEM files and ancillary data in an easily accessible format, which is useful in organizing the multispectral geological analysis of a particular volcano [35]. As of this writing, however, no new data have been added to the archive since late 2017. This is scheduled to be reactivated sometime in 2020.

3.3. Operational Structure of the Ongoing ASTER Volcano Observations

All ASTER data acquired as part of ongoing routine Earth observations (i.e., the Volcano STAR, global map imaging, etc.) are archived at the LP DAAC. These data are stored in several formats including raw radiance at sensor, from which numerous level 2 data products (e.g., digital elevation model, surface TIR temperature), can be ordered as on-demand products. Image data that fulfill the requirements of the Volcano STAR are also assessed weekly and stored in the IDV and the AVA archives, the latter until 2017. These volcano archives allow quick data searches on particular volcanoes and visualizations of the latest data, whereas the LP DAAC archive contains all the data, all possible level 2 options and the most up to date processing levels for each on-demand product.

Data collected as part of the URP Program are also eventually stored in the LP DAAC ASTER archive and are pulled over to the individual volcano archives as well. However, because of their expedited classification, these data are processed quickly into the Level 1BE (expedited) format and staged on the expedited data system webpage at the LP DAAC for immediate download and assessment. Within 2 h of the data being acquired, all scientists involved with the URP Program are automatically notified by email and have immediate web-based access to the new scene. Any significant changes detected in the data of a particular eruption are disseminated to the responsible monitoring agencies, local scientists working on the eruption, as well as the global community through e-mail and mailing lists. More detailed science analysis is then commonly performed over time and with the arrival of new data. All new, newly scheduled, and ongoing volcano observations of each URP volcano are tracked both through a database and email system as well as web-based map tool showing all the current targets color coded by ASTER observations status (Figure 3). Each color coded pin is clickable allowing a query of the latest observations and links to the metadata for each scene. Also shown is the current position of the Terra satellite.

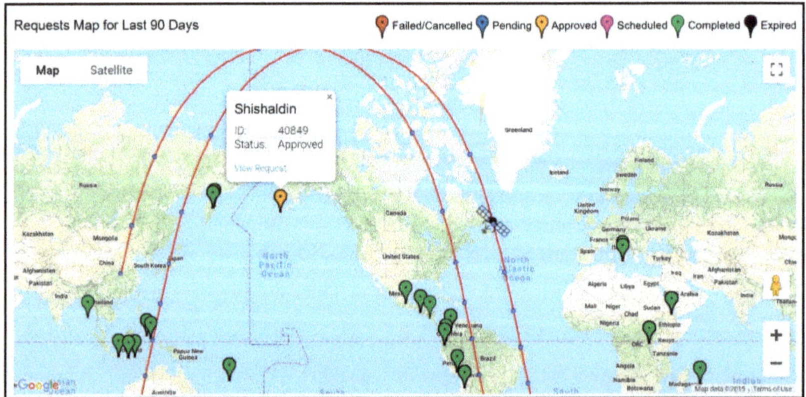

Figure 3. The ASTER Urgent Request Protocol (URP) Program scheduling interface using Google Maps, and maintained at the LP DAAC. All currently monitored URP target volcanoes are shown with a color-coded pin. These are updated automatically when new targets are triggered or existing ones have a change of status. Each pin is clickable to display more information and links to other metadata. The position of the Terra spacecraft and its orbit tracks are also shown.

3.4. Statistical Analysis

All ASTER scenes used in this study were acquired from the LP DAAC through the NASA EarthData web-based search tool using an initial search refinement by date. The refined datasets were then downloaded as ASTER Level 2 surface kinetic temperature (AST_08) products. This data product allows the most efficient way to identify volcanic activity by temperature within a given scene. Scenes were automatically scanned and later visually inspected for any thermal variations. If no activity was initially identified, a more detailed investigation for thermally elevated pixels was performed for each scene. If this process yielded a negative result, the scene was classified as having no detectable volcanic activity. If volcanic (thermal) activity was present, then the type of activity was classed as a plume, lava flow, other flow, or summit hot spot. The cloud cover percentage was also visually determined for all the scenes investigated. This was then compared to the ASTER Cloud Cover Assessment Algorithm (ACCAA) value for future assessment analysis.

4. Results

4.1. ASTER Capabilities and Observation Strategies

The unique instrument characteristics that make ASTER particularly well suited for volcanic observations include multispectral TIR data, routine TIR data at night, high spatial resolution data, variable gain settings to limit data saturation, off-axis pointing, and generation of along-track digital elevation models [3,36]. For example, the multispectral TIR data at a relatively high spatial resolution allowed a variety of surface materials to be distinguished and a better understanding of thermal and compositional mixing at the sub-100 m pixel scale [6,37,38]. The following are examples particularly relevant to each of the ASTER instrument characteristics. For a complete list of all ASTER-specific volcanological papers, see Appendix A.

4.1.1. Routine TIR Data at Night: Elevated Temperatures at Fuego Volcano, Guatemala

Unlike other orbiting instruments of similar spatial and temporal resolutions, ASTER routinely acquires TIR data at night. In certain special observational modes (e.g., large and highly radiant lava flows), VNIR nighttime data can also be acquired. Nighttime TIR have the advantage of reduced residual solar heating, reduced thermal topographic effects, and typically have lower cloud percentages compared to daytime scenes [3]. Subtle thermal anomalies are therefore more easily identified in nighttime TIR data. A new study by Flynn and Ramsey [39] of Fuego volcano from (1 January 2000 until 30 April 2018) found that ASTER acquired 308 scenes—193 of which were collected at night. Of those nighttime scenes, 109 had visible volcanic activity present, with 47 being summit hot spots and 62 showing either lava flows or pyroclastic density currents (PDCs). These data were compared to the Guatemalan monitoring agency's weekly reports to determine the specific volcanic event that could have caused the thermal detections in the ASTER data. This data synthesis is then used to create volcanic hazard maps and cross check datasets for missed events. Without the nighttime TIR data, many of these volcanic events may not have been observed.

4.1.2. Multispectral TIR Data: SO_2 Plumes from Lascar Volcano, Chile

The multispectral resolution of the ASTER TIR data enabled Henney et al., [40] to detect and measure very low SO_2 concentrations (<1 g/m^2) released from Lascar volcano in December 2004 (Figure 4). Of note by the authors was the high spatial resolution and radiometric sensitivity, which along with the needed multispectral resolution allowed these retrievals of very small SO_2 burdens, which are not possible with any other orbital instrument. The TIR results were also compared to ground-based measurements made in the ultraviolet (UV) spectral region during a coordinated overpass of ASTER and found to be well within the error of those instruments.

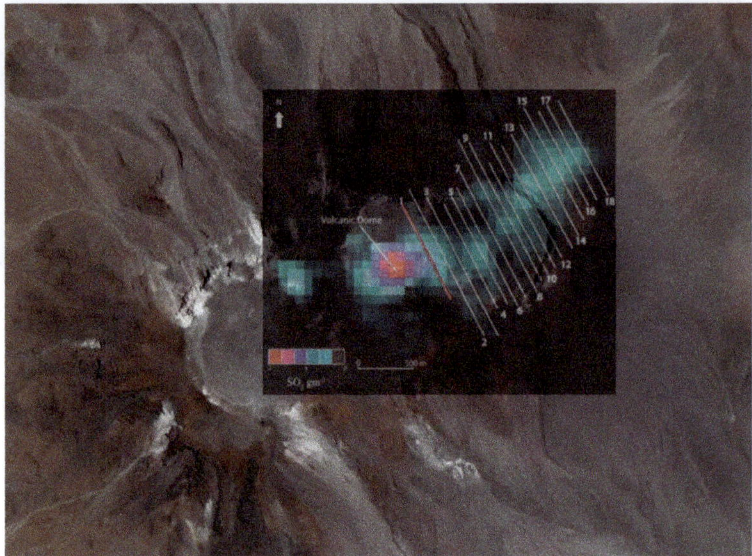

Figure 4. The map of the very low SO$_2$ burden retrieved using ASTER TIR data acquired at Lascar volcano, Chile (summit location: 23.37°S, 67.73°W) on 7 December 2004. Image is draped on a visible Google Earth image for context. Numbered lines indicate transects taken through the plume and reported in the original study. SO$_2$ results were created using the Map_SO$_2$ software of Realmuto [41] and each color bin corresponds to ~0.3 g/m^2 of SO$_2$. Figure modified from Henney et al. [40].

4.1.3. High Spatial Resolution Data: TIR Analysis of Hawaiian Volcanoes

The high spatial resolution of the ASTER data, in particular in the TIR, was documented by Patrick and Witzke [42] for long-term mapping and monitoring of the active or potentially active volcanoes on the islands of Hawai'i and Maui. The goal of the study was to determine the baseline thermal behavior over ten years (2000–2010) in order to assess thermal changes that may precede a future eruption. They used cloud-free kinetic temperature ASTER data acquired for the five major subaerial volcanoes in Hawai'i (Kīlauea, Mauna Loa, Hualālai, Mauna Kea, and Haleakalā). The data were geolocated and stacked to create time-averaged thermal maps and to extract temperature trends over the study period. Conspicuous thermal areas were found on the summits and rift zones of Kīlauea, Mauna Loa, and the small pit craters on Hualālai. No thermal areas were detected on Haleakalā or Mauna Kea. One limiting factor noted for the lack of detections was the pixel size of the ASTER TIR, which despite being one of the highest from orbit, was seen as still too large to detect possible subtle thermal changes as well as to identify small-scale, low-temperature thermal activity.

4.1.4. Variable Gain Settings (VNIR/SWIR): High Temperature Monitoring of Klyuchevskoy Volcano, Russia

The ASTER VNIR and SWIR (no longer functioning) subsystems both have/had the ability to acquire data at different data gain settings. The VNIR has three settings (low, normal, and high), whereas the SWIR had four (low1, low2, normal, and high). These gain settings must be set prior to scheduling an observation, so some degree of advanced knowledge of the target is required. They were created to limit data saturation in regions of excessively high or low radiance (e.g., bright glacier and clouds surfaces, dark water body surfaces, etc.). Typically, these gain settings are uniformly set across all bands in the VNIR and the SWIR (when it was operating). However, a unique scenario was created for the volcano STAR whereby every other SWIR band was alternated between the normal and the low2 gain settings in order to maximize the possibility of capturing some unsaturated data of a future volcanic target that may have lava on the surface [25]. The saturation pixel-integrated brightness temperature for the SWIR ranged from 86 °C in band 9 (*wavelength = 2.336 μm and gain = high*) to 467 °C in band 4 (*wavelength = 0.804 μm and gain = low2*). For the VNIR, the range is 669 °C in band 3 (*wavelength = 0.807 μm and gain = high*) to 1393 °C in band 1 (*wavelength = 0.556 μm and gain = low*). By comparison, the TIR has only one gain setting and a saturation pixel-integrated brightness temperature of 97 °C.

These variable gain settings have proven quite useful for deriving accurate pixel-integrated brightness temperatures of large, highly radiant lava flows over the ASTER mission lifetime. For example, Rose and Ramsey [43] describe the use of all three ASTER subsystems to monitor the emplacement of multiple long lava flows at Klyuchevskoy volcano, Kamchatka during the 2005 and 2007 eruptions. In 2007, temperatures were extracted from 4 January to 7 June 2007 and fit into the volcanic warning color codes over that period (Figure 5). For the first 4.5 months of the eruption, TIR data remained unsaturated and the SWIR-derived temperatures, acquired in high-gain mode, were only detectable in the highest SWIR wavelengths. This corresponded to the time period that Klyuchevskoy was designated with yellow and orange color codes, signifying increasing levels of restlessness, but no eruption. From 26 April to 10 May, all TIR and SWIR-derived temperatures became saturated due to the emplacement of new open-channel lava flow that was radiant enough to be detected at the 15 m VNIR spatial scale. VNIR-derived temperatures (852–895 °C) were detected for the first time in the eruption. These detections took place one week before the alert level was raised to red (signifying an active eruption), attesting to the importance of high-repeat, non-saturated data for the monitoring of these more remote volcanoes.

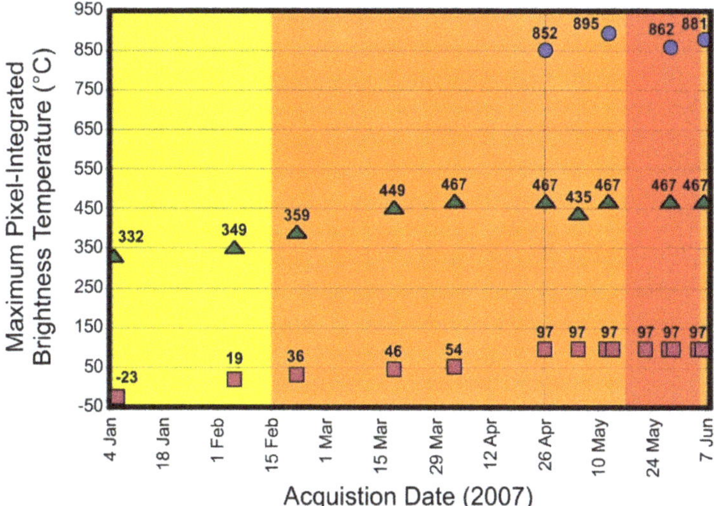

Figure 5. Maximum ASTER-derived pixel-integrated brightness temperatures detected during the 2007 eruption of Klyuchevskoy volcano, Russia. Thermal infrared (TIR) temperatures are denoted with purple squares; shortwave infrared (SWIR) temperatures with green triangles; and visible/near-infrared (VNIR) temperatures with blue circles. The background colors represent the volcanic color code issued by the monitoring agencies (yellow = elevated unrest above known background levels; orange = heightened unrest with increased likelihood of eruption; red = eruption is forecast to be imminent). TIR and SWIR temperatures become saturated (and VNIR temperatures become measureable) near the time shown by the thin vertical line. This saturation indicates that a significant amount of highly radiant lava is on the surface and occurred despite adjustments to the SWIR gain settings. Importantly, this was more than two weeks before the color code is changed to red. Modified from Rose and Ramsey [43].

4.1.5. Off-Axis Pointing Capability: Improved Observational Frequency at Piton de la Fournaise Volcano, France

During the April–May 2018 eruption of Piton de la Fournaise volcano on Réunion Island, satellite-based surveillance of the thermal activity and emitted plumes was paired with near-real-time flow modeling to create an ensemble-based approach in response to the crisis [44]. This combined effort of four institutions in several countries using data from numerous sensors to model and forecast lava flow advance was done as a proof of concept in order to assist the small staff of the volcano observatory on the island. Rapid data acquisition was critical and provided by MODIS via the MIROVA system to determine lava discharge rates. The less frequent but higher spatial resolution ASTER data were important both for determining the length, shape and direction of the flow using the TIR data (Figure 6), as well as the precise location of the vent using the VNIR data. The TIR images served as a validation for the MODIS data and the predicted down slope flow modeling, whereas the VNIR-derived vent location was used as the initiation point for the flow modeling, critical for model accuracy. During the 35 day eruption, a total of 11 ASTER images were acquired. This average of ~1 image every 3 days was a significant improvement over the nominal 16 day repeat time for targets close to the equator. The improved temporal resolution was only made possible with the ASTER URP Program and the off-nadir pointing capability of 8° nominally and up to 24° for the VNIR.

Figure 6. Piton de la Fournaise volcano, Réunion Island (summit location: 21.24°S, 55.71°E) during the April–May 2018 eruption crisis. Image/data modified from Harris et al. [44]. A color-coded Moderate Resolution Imaging Spectroradiometer (MODIS) radiance image produced by the Middle InfraRed Observation of Volcanic Activity (MIROVA) monitoring system is draped over a 3D Google Earth visible image for context. These KML image products are routinely produced by MIROVA. Draped on both data sets is the nighttime ASTER TIR data acquired as part of the URP Program. The ASTER image, acquired 6 days after the triggering MODIS detection, shows the spatial details of the propagating lava flow as the brighter white pixels within the colorized MODIS pixels.

4.1.6. Generation of Along-Track Digital Elevation Models (DEMs): Volcanoes in Guatemala, New Zealand, and Mexico

The ability to produce single 60 by 60 km scene DEMs from any daytime VNIR data is another significant capability of the instrument. Importantly, these data provided digital topography for most of the Earth's land surface at 10 m accuracy early in the mission when such information was not available from other data sources such as radar. Later in the mission, these individual-scene DEMs were compiled into a seamless global data set known as the ASTER Global DEM or GDEM. The GDEM has better signal to noise and removes areas of clouds or other artificial errors commonly found in the single-scene DEMs (Figure 7). The final version (v.3) was released in August 2019 having corrections made to minor areas of known errors.

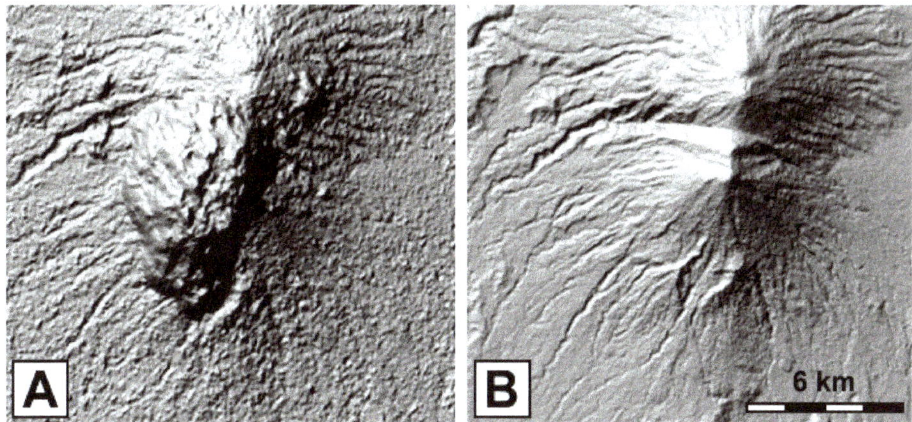

Figure 7. Examples of ASTER-derived DEMs for Fuego Volcano, Guatemala (summit location: 14.48°N, 90.88°W). (**A**) Single-scene DEM from VNIR data acquired on 10 December 2018. Clouds are causing the erroneous topographic high that obscures the summit and central crater. The pixel to pixel noise is also clearly visible. (**B**) ASTER GDEM v3 using only cloud-free scenes and greatly improving the signal to noise of the DEM.

Whereas the GDEM provides an excellent baseline for volcanological studies reliant upon topography, it does not account for changes to the surface, either during the period used for the GDEM creation (2000–2011) as they would have been averaged out or not included in subsequent data. Dynamic topography is a characteristic of most active volcanoes and accurate knowledge of that topography becomes important for any type of lava or pyroclastic flow forecast modeling. Therefore, current single-scene DEMs continue to be an important dataset for volcanological studies using ASTER. Stevens et al., [45] noted this point in their study examining the accuracy of the ASTER single-scene DEM for Ruapehu and Taranaki volcanoes in New Zealand. They found an average root-mean-squared (RMS) error of ~10 m for the ASTER single-scene DEM compared to digitized 1:50,000 scale topographic maps, and later noted that these data will continue to be relevant for future surface change even as global DEM products from radar systems came online. A later study by Huggel et al., [46] evaluated the ASTER DEM against such a global dataset, the Shuttle Radar Topography Mission (SRTM) DEM. The data were used for lahar modeling on Popocatépetl Volcano, Mexico. They found that although the higher number of errors in the ASTER DEM affected certain models more, both the ASTER and SRTM DEMs were feasible for lahar modeling, but that verification and sensitivity analysis of the chosen DEM is fundamental to deriving accurate hazard maps from the modeled inundation areas.

4.2. URP-Specific Results

ASTER data for a subset of volcanoes over several different time intervals were examined for this study and the significant statistics were compiled. The volcanoes were all continuously (or nearly so) active throughout these periods. The goal of this statistical analysis was to determine the improvement of the URP over the nominal Volcano STAR observations. Further details such as the number of cloudy scenes, the number of scenes with confirmed activity and the days between successive observations were also calculated. The interval periods were: one year (1 June 2018 to 31 May 2019) for 10 volcanoes;

the approximate five-year period of the global URP observations (1 January 2014–31 May 2019) for four of those volcanoes; and the nearly entire period of the ASTER mission (13 April 2000–31 May 2019) for one of those volcanoes.

The average percentage of scenes for all the periods that contained detectable volcanic activity was 47.6%, with a range of 19%–69%. For this same period and number of scenes, the average cloud percentage was 50.5%, with a range from 16% to 84%. Finally, during the last five years of the implementation of the global URP Program, the average days between observations at the four volcanoes studied was 6.5 days, compared to 20.2 days with the Volcano STAR Program, a nearly 200% improvement.

4.2.1. One-Year Analysis

Ten volcanoes were chosen for a statistical analysis over a recent one-year period (Table 2). The volcanoes chosen represent a range of eruptive styles, compositions, and latitudinal distributions (Figure 8). All volcanoes were active during the study period and that activity triggered the ASTER URP observations. The number of triggers (derived from MODIS data by the MODVOLC and MIROVA systems) varied from 36 for Popocatépetl to 837 for the Nyamuragira-Nyiragongo volcanoes. The number of triggers is determined by several factors—the most important of which is the style of activity (i.e., persistently active lava lake versus an intermittently active lava dome) and size of the thermal feature on the ground. These combine to determine the overall emitted radiant power. Other factors contributing to the number of ASTER URP triggers also include the latitude of the target together with the average daily cloud cover. Higher latitude volcanoes will have more overpasses due to the converging orbit tracks of the MODIS sensors and hence, more triggers. Persistently cloudy targets will commonly mask thermal activity, lowering the number of triggers. Average cloud percentage was determined by visual inspection of each URP scene done at the same time as the inspection for volcanic activity.

Table 2. Statistical analysis of ten URP-monitored volcanoes active during one year (1 June 2018 to 31 May 2019). Each ASTER scene was inspected for confirmed/detected volcanic activity (e.g., thermally elevated pixels, presence of a plume) and average cloud cover. The percentage of those URP scenes with confirmed activity is also given and generally correlates to the average cloud percentage.

	Number of URP Triggers	Number of URP Scenes	Detected Volcanic Activity and (%)	Avg. Cloud
Ambrym	138	21	4 (19.0%)	62.7%
Erebus	824	36	16 (44.4%)	84.2%
Erta Ale	760	29	13 (44.8%)	16.0%
Fuego	532	30	16 (53.3%)	32.9%
Nyamuragira-Nyiragongo	837	42	23 (54.8%)	61.9%
Piton de la Fournaise	456	22	5 (22.7%)	44.5%
Popocatépetl	36	22	9 (40.9%)	31.5%
Sangeang Api	251	27	11 (40.7%)	46.8%
Shiveluch	623	29	20 (69.0%)	55.0%
Yasur	209	18	1 (5.6%)	77.6%

These MODVOLC and MIROVA triggers resulted in 276 new ASTER scenes during the one-year period with confirmed volcanic activity in ~45% of those scenes, with the remainder being obscured by clouds. The frequency of ASTER URP scenes increases at the higher latitude volcanoes on the list (e.g., Erebus and Shiveluch) as expected. However, Erta Ale at a much lower latitude had a similar number of observations to those two targets, a function of nearly continuous thermally elevated activity present there during the one-year period.

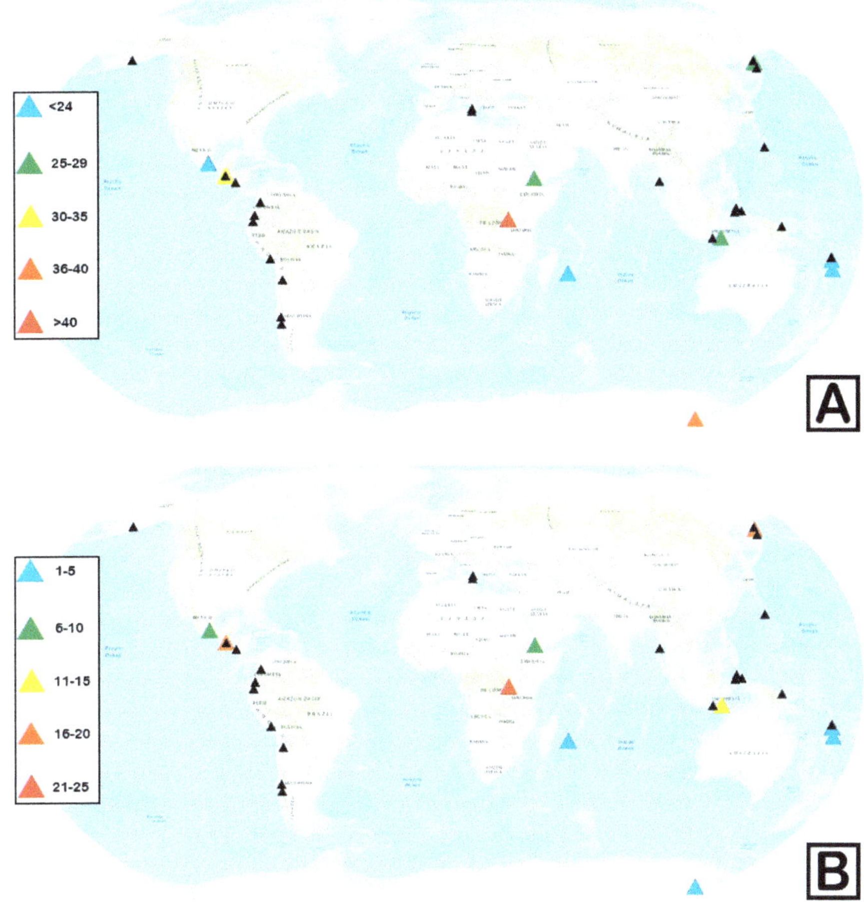

Figure 8. Map of the statistics for the 10 volcanoes detailed in Table 2, which were compiled over 1 year (1 June 2018 to 31 May 2019). Small black triangles indicate active URP-monitored volcanoes at the time, but not analyzed for this study. (**A**) Number of ASTER scenes acquired. (**B**) Number of those ASTER scenes with visible volcanic activity.

4.2.2. Five-Year Analysis

Four of the volcanoes in the prior list were then selected for a longer study period of slightly over five years (1 January 2014–31 May 2019). This period begins when the ASTER URP Program became fully global with the integration of the MODVOLC triggering system. The time frame, therefore, provides the most complete global higher temporal resolution ASTER data for any thermally elevated volcano. As such, the statistics for these observations are the most comprehensive analysis available for the planning of future orbital TIR systems, for example (e.g., expected long-term cloud cover, optimal temporal frequency, etc.).

The data for the four volcanoes chosen (Ambrym, Erta Ale, Popocatépetl, Shiveluch) are shown in Table 3. With the exception of Shiveluch, the others are more equatorial. The four do, however, represent a range of compositions (e.g., dacite, andesite, and basalt) and styles (e.g., dome-forming, lava lakes, and flows). A total of 1001 ASTER scenes were downloaded and analyzed for these four targets. Volcanic activity was confirmed in 535 of these (53.5%) with the remainder being too cloud-covered or having no obvious activity visible in the image. Erta Ale had the highest percent of observed activity

(80.2%) whereas Ambrym had the lowest (29.7%), a clear function of the semi-arid, nearly cloud-free location of Erta Ale compared to the tropical, nearly cloud-covered location of Ambrym. Also analyzed was the average days between an observation for the routine Volcano STAR (20.2 days/scene) as compared to those data from the URP Program (6.5 days/scene). This was an average improvement of 215%, with the best improvement (261%) for Ambrym volcano. The URP Program excels at acquiring more data of targets that are commonly cloudy or close to the equator, and therefore typically missed by the routine STAR observations.

Table 3. Statistical analysis of four volcanoes from Table 2 active during the longer global URP period (1 January 2014–31 May 2019). Each ASTER scene was inspected for confirmed/detected volcanic activity (e.g., thermally elevated pixels, presence of a plume) and average cloud cover. The percentage of those URP scenes with confirmed activity is also given and generally correlates to the average cloud percentage. The time between scenes decreases dramatically for the URP operations as compared to the ASTER Volcano STAR. The last column shows this improvement percentage in the number of days between scenes using the URP versus the STAR scheduling.

	ASTER Scenes	Detected Activity/%	Avg. Cloud	Days/Scene (URP)	Days/Scene (STAR)	% Change (URP/STAR)
Ambrym	239	71 (29.7%)	62.0%	6.4	23.0	261.0%
Erta Ale	243	195 (80.2%)	38.6%	4.4	13.2	198.6%
Popocatépetl	313	153 (48.9%)	19.0%	7.0	24.4	247.1%
Shiveluch	206	117 (56.8%)	62.4%	8.0	20.3	153.8%

4.2.3. Twenty-Year Analysis

Finally, one volcano (Shiveluch) was selected for analysis for nearly the entire period (13 April 2000–31 May 2019) that Terra has been in orbit (Table 4). The first ASTER image acquired of Shiveluch was on 13 April 2000, only 40 days after the start of the operational phase. These early scenes of the volcanoes of Kamchatka also captured the products of a large eruption at Bezymianny volcano described by Ramsey and Dehn [47]. Shiveluch has been persistently active nearly the past twenty years (e.g., [6]) and therefore provides a good target for long-term analysis. It was also imaged more frequently early in the mission because of its location at higher latitudes as well as being one of the northern Pacific volcanoes originally monitored in the first phase of the URP Program, which began in 2005 and relying upon AVHRR data for detection triggering [29].

Table 4. Statistical analysis of a single volcano (Shiveluch) from Table 2 active during most of the entire ASTER orbital period (13 April 2000–31 May 2019). Each ASTER scene was inspected for confirmed/detected volcanic activity (e.g., thermally elevated pixels, presence of a plume, etc.) and average cloud cover. The percentage of those URP scenes with confirmed activity is also given and generally correlates to the average cloud percentage.

	ASTER Scenes	Detected Activity/%	Activity (plumes)	Activity (flows)	Activity (hot spot)	Avg. Cloud	Days/Scene (URP)
Shiveluch	815	430 (52.8%)	53	109	268	62.8%	5.1

The analysis results are shown in Table 4. A total of 815 confirmed scenes were downloaded and interrogated. Similar to the prior results, volcanic activity was confirmed in 430 (52.8%) of the scenes. We further analyzed the type of activity seen in those clear scenes. Plumes were noted in 53 (12.3%) scenes, flows in 109 (25.3%), and summit hot spots in 268 (62.3%) scenes. We also examined the impact of the URP Program to acquire more rapid data of eruption activity. The average time between ASTER Volcano STAR observations at Shiveluch is 8.8 days, much lower than the more equatorial volcanoes (~20 days). Implementation of the URP Program, however, reduced this temporal resolution slightly to 6.1 days. If one were to only consider the data following a new URP triggering event, this value decreases further to 5.1 days. If volcanic activity wanes and then renews, the first ASTER image acquired after a new triggering event may come many days to weeks after the prior image was acquired. Therefore, this lower interval is likely a better metric of the URP operational behavior. It is also close to the theoretical minimum frequency of ASTER for high latitude targets.

5. Discussion

5.1. ASTER as a Volcanological Instrument

ASTER, unlike almost any instrument designed or chosen by a space agency in the past several decades, was first and foremost a geological sensor. The wavelength regions chosen, the placement of the multispectral bands, the multispatial resolution, and the DEM capability are all critically important for geologic surface mapping. Although it was not designed specifically for volcanology, the geologic heritage of ASTER makes the instrument's data well suited for active volcanic monitoring and data collection. Perhaps the only capability missing was a mid-infrared channel for high-temperature detections. This was offset for the first 8 years of the mission with the 2 μm region SWIR data. As originally noted by Pieri and Abrams [3] and later described by Ramsey [2] and herein, six important instrument/data characteristics made ASTER particularly well suited for spaceborne volcanology. These are the multispectral TIR data, routine TIR data at night, high spatial resolution data, variable gain settings in the VNIR and SWIR, off-axis pointing, and generation of DEMs. Individually, these characteristics all have been used for specific volcanic studies. Where combined, these capabilities become much more powerful for addressing pressing questions in volcanology [45,48].

5.2. Toward an Improving Temporal Resolution

ASTER is able to provide data with improved radiometric resolutions (absolute reflectance $\leq 1.3\%$ and absolute temperature ≤ 0.3 K) [7]. Moreover, it provides data at improved spatial (15–90 m) and spectral resolutions (14 wavelength band total, with 5 in the TIR and 6 in the SWIR). This multiscale spatial resolution across a wide wavelength range provides a unique tool for volcanological applications. For example, the wavelength range coupled with the variable gain settings allows the acquisition of unsaturated data over a wide temperature range (−73–1393 °C). Furthermore, saturation in one wavelength region (e.g., the TIR) can be compensated using the nested pixels from a shorter wavelength region (e.g., the 9 SWIR or 36 VNIR pixels) to correct for that saturation (e.g., [38]). These data enable thermal anomalies of only a few degrees above the background temperature, as well as sub-pixel, highly radiant anomalies of only several m^2, to be determined.

For the past 15 years, an ASTER-focused program called the Urgent Request Protocol (URP) has combined the rapid detection capability of higher temporal resolution instruments like MODIS with the high spatial resolution scheduled observations of ASTER. These observations have improved our knowledge of multi-year volcanic monitoring, ongoing eruption behavior, and post-eruption change. They have also become important for capturing points-in-time during any ongoing lava flow or plume forming eruption. More commonly, the ASTER URP data (as well as the entire ASTER volcanic archive) are being used for operational response to new eruptions; determining thermal trends months prior to an eruption; inferring the emplacement of new lava lobes; and mapping the constituents of volcanic plumes, to name a few (see Appendix A). These all require the higher spatial resolution data of ASTER, and in most cases, its multispectral capability.

A framework like the URP sensor web only works in its current form if there is a large enough radiant target to be detected by the operational systems using MODIS data like MODVOLC and MIROVA. The MIROVA system applies an enhanced thermal index (ETI) and therefore detects more subtle thermal signals as compared to MODVOLC [32]. Even with this enhanced detection, however, there will be a significant number of thermal events and numerous volcanoes missed every year because they are not large enough (spatially or radiantly) to be detected in a 1 km MODIS pixel. It is these smaller signals that are frequently important as they indicate the onset of renewed activity, sometimes months to years prior to an eruption [22].

There are some volcanoes whose eruptions are long-lived enough and/or whose character changes throughout their ongoing eruptions, which allow sporadic larger signals to be detected with MODIS. These then trigger subsequent ASTER data, which captures more subtle, transient activity. The best examples of such activity are typically dome-forming eruptions. Larger thermal signals are caused by dome collapse exposing the hotter material or periodic larger dome-destroying eruptions. With renewed dome growth, the cooler carapace insulates the hotter interior and limits detection by MODIS. One of the best examples of this eruptive style is Sheveluch volcano, Russia. In a recent set of three observations by ASTER over a one week period, small changes in the dome's thermal output and moderately sized block-and-ash flows were captured (Figure 9). Although interesting from a volcanological perspective for Sheveluch, it is these relatively high-frequency ASTER URP data, captured over the one to multi-year timeframe, which greatly improve our understanding of how eruptions proceed and how volcanoes reawaken.

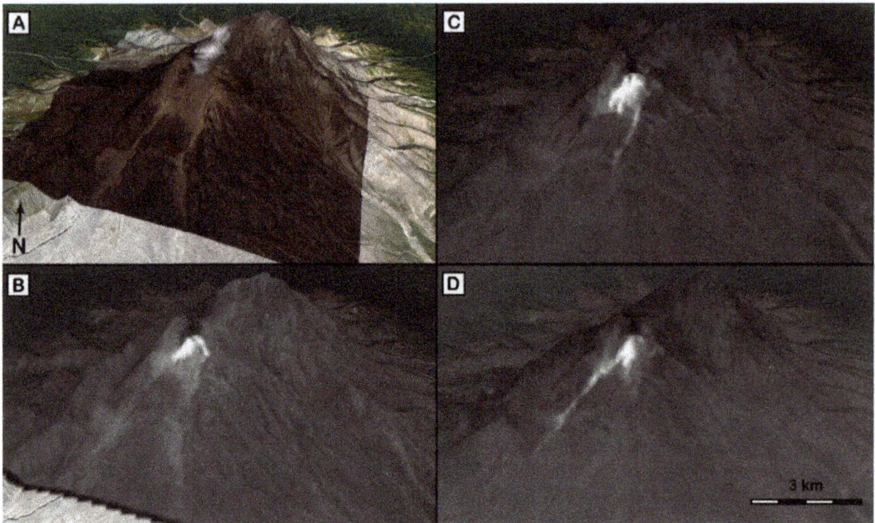

Figure 9. An example of the improved time series made possible by the ASTER URP Program, here showing the ASTER data draped over a 3D Google Earth visible image viewed looking NW toward Sheveluch volcano, Russia (summit location: 56.65°N, 161.36°E). (**A**) ASTER VNIR image acquired on 20 April 2019. The small white plume is visible drifting to the north. (**B**) ASTER TIR data from the same date. Two thermally elevated regions are seen on the summit lava dome. (**C**) ASTER nighttime TIR image acquired ~36 h later (21 April 2019) showing a significant increase in the area of the thermally elevated region on the dome and a ~2 km flow down the eastern side of the valley. (**D**) ASTER nighttime TIR data acquired on 28 April 2019 showing the continued, but somewhat lower, thermal activity on the dome and a new debris flow extending ~3 km down the western side of the valley.

6. Conclusions

Despite the two-decade archive of multispectral, multispatial resolution ASTER data (or perhaps because of its success), there have been no approved follow-on instruments by NASA or other space agencies that have similar spatial and spectral scales. Some of this land imaging is filled by sensors on the Landsat, Sentinel-2, and SPOT satellites, but the data gap discrepancy looms largest in the TIR. Currently, only the limited mission lifetime ECOSTRESS instrument on the ISS is the most similar. This gap was also noted in both the 2007 and 2018 Decadal Surveys for NASA Earth Science. In 2007, a notional mission called HyspIRI was recommended, from which ECOSTRESS derives its heritage.

In 2018, the focus was on specific science questions and important designated observables. TIR data was again highlighted numerous times.

What will arise from the 2018 recommendations is yet to be formulated. TIR instruments will likely get smaller and more numerous using uncooled detectors, becoming CubeSat compatible and operating in a dense sensor web network for improved response times. Although perhaps not designed specifically for volcanology, the data from these instruments will become critical for volcanic crisis response and enable never before possible measurements such as the global inventory of volcanic degassing, thermal precursory trends at every volcano, and accurate temperatures of small activity, which can be used as input to predictive flow and hazard assessment models.

This next generation of high spatial, high spectral TIR data captured at ever-improving temporal resolutions will only become possible because of the ASTER instrument design and mission success, most notably for volcanic remote sensing. For example, with the Terra spacecraft's converging orbits at the poles, higher latitude volcanoes like those in Kamchatka, Iceland, and Antarctica are imaged more routinely. The addition of off-axis pointing in the ASTER design and the later establishment of the STAR and URP Programs further improved this temporal resolution to near the theoretical maximum. Multispectral TIR data are now being routinely acquired that allow us to track subtle thermal anomalies, precursory activity, explosive events, plumes, and the percentage of obscuring clouds. These data can help inform future instrument and mission design. For example, looking again at the data in Table 2, one can calculate a "miss rate percentage" between the number of MODIS-based triggers and the number of ASTER scenes acquired from those triggers. This rate varies from a low of 39.8% (Popocatépetl volcano) to a maximum of 96.2% (Erta Ale volcano), with an average of 87.6%. These high values demonstrate the amount of high spatial resolution data that are theoretically being missed because there is not a TIR sensor or sensor system with the spatial/spectral resolution of ASTER and the temporal resolution of MODIS. Perhaps even more critically, with an average cloud percentage near 50% based on our analysis of the ASTER URP archive, half of the current observation attempts fail to detect surface activity. This further highlights the need for improved temporal resolution. Ultimately, these results provide a baseline for future TIR orbital concepts that could respond to the 2018 Decadal Survey recommendations for the TIR data critically needed to address key science questions.

Author Contributions: The following contributions were provided by each author: conceptualization, M.S.R.; data curation, I.T.W.F. and M.S.R.; formal analysis, M.S.R. and I.T.W.F.; funding acquisition, M.S.R.; investigation, M.S.R. and I.T.W.F.; methodology, M.S.R.; project administration, M.S.R.; resources, M.S.R.; supervision, M.S.R.; validation, M.S.R. and I.T.W.F.; visualization, M.S.R.; writing—original draft, M.S.R.; writing—review and editing, M.S.R. and I.T.W.F. All authors have read and agreed to the published version of the manuscript.

Funding: This work was supported by NASA and several successful proposals to the Science of Terra and Aqua Program over the years, with the current being 80NSSC18K1001.

Acknowledgments: The authors wish to thank the members of the ASTER Science Team for helpful discussions, comments, and support during the past 20 years of the mission—in particular, David Pieri at the Jet Propulsion Laboratory, who, along with Minoru Urai at the Geologic Survey of Japan, headed up the early days of volcano science with ASTER. We would like to thank Ben McKeeby for his assistance with the ASTER archival search and creation of Figure 8, as well as Tyler Leggett, Marco Michelini, James Thompson, and Daniel Williams for their assistance with the painstaking reference checking for the Appendix A. We apologize in advance for any papers inadvertently left off this list; it was not intentional. We intend to keep this list updated on the lead author's laboratory website (http://ivis.eps.pitt.edu/archives/ASTER/). Please send any future or past/missing references to the lead author and they will be added to the online table. Finally, the authors would like to thank the three anonymous reviewers, whose comments and suggestions helped to improve the quality of this manuscript.

Conflicts of Interest: The authors declare no conflict of interest.

Appendix A

Table A1. Reference list of 271 volcanological publications that use, contain, or make mention of (in the case of the Precursory category) ASTER data. The list spans 25 years, from 1995 to 2019, and is subdivided by category with references therein listed in chronological order. The full citations appear below.

Publication Category	Reference List
Analogs	Davies et al., 2008; Price et al., 2016; Ramsey et al., 2016.
Calibration	Barreto et al., 2010; Vaughan et al., 2010; Blackett & Wooster, 2011; Thompson et al., 2019.
Gas/Plumes	Corradini et al., 2003; Urai, 2003; Urai, 2004; Ino et al., 2005; Iwashita et al., 2006; Pugnaghi et al., 2006; Kearney et al., 2008; Campion et al., 2010; Diaz et al., 2010; Kobayashi et al., 2010b; Camiz et al., 2010; Spinetti et al., 2011; Campion et al., 2012; Henney et al., 2012; Abrams et al., 2013; Pieri et al., 2013; Spinetti et al 2013; Campion, 2014; Diaz et al., 2015; Stebel et al., 2015; Carn et al., 2016; Realmuto & Berk, 2016; Robertson et al., 2016; Xi et al., 2016; Kern et al., 2017; Moussallam et al., 2017; Troncoso et al., 2017; Williams & Ramsey, 2019; Williams et al., 2019; Laiolo et al., 2019.
Geothermal	Hellman & Ramsey, 2004; Viramonte et al., 2005; Vaughan et al., 2012a; Vaughan et al., 2012b; Silvestri et al., 2016; Braddock et al., 2017; Caudron et al., 2018; Mia et al., 2018a; Mia et al., 2018b.
Lava Flows	Wright et al., 2010; Favalli et al., 2012; Wadge et al., 2012; Head et al., 2013.
Mapping	Hubbard et al., 2003; Rowan et al., 2003; Watanabe & Matsue, 2003; Byrnes et al., 2004: Torres et al., 2004; Dmochowski, 2005; Tralli et al., 2005; Capra, 2006; Mars & Rowan, 2005; Rowan et al., 2006; Coolbaugh et al., 2007; Davila et al., 2007; Hubbard et al., 2007; Kervyn et al., 2007; Carter et al., 2008; Kervyn et al., 2008c; Saepuloh et al., 2008; Schneider et al., 2008; Carter & Ramsey, 2009; Baliatan & Obille, 2009; Bogie et al., 2010; Brandmeier, 2010; Kobayashi et al., 2010a; Piscini et al., 2010; Chadwick et al., 2011; Davila-Hernandez et al., 2011; Mars & Rowan, 2011; Wadge & Burt, 2011; Wantim et al., 2011; Diaz-Castellon et al., 2012; Amici et al., 2013; Graettinger et al., 2013; Lara et al., 2013; Watt et al., 2013; Boyce et al., 2014; Mars, 2014; Tayebi et al., 2014; Castruccia & Clavero, 2015; Ramsey, 2015; Selles et al., 2015; Folguera et al., 2016; Oikonomidis et al., 2016; Prambada et al., 2016; Suminar et al., 2016; Ulusoy, 2016; Yulianto & Sofan, 2016; Ali-Bik et al., 2017; Bustos et al., 2017; Takarada, 2017; Auer et al., 2018; Godoy et al., 2018; Krippner et al., 2018; Aufaristama et al., 2019b; Fu et al., 2019; Pallister et al., 2019.
Modeling	Favalli et al., 2006; Huggel et al., 2007; Huggel et al., 2008; Carter et al., 2009; Favalli et al., 2009; Joyce et al., 2009b; Munoz-Salinas et al., 2009; Capra et al., 2011; Sosio et al., 2012; Worni et al., 2012; Wantim et al., 2013; Rose et al., 2014; Rose & Ramsey, 2015; Carr et al., 2019; Ramsey et al., 2019; Rogic et al., 2019.
Monitoring	Tsu et al., 2001; Urai et al., 2001; Ellrod et al., 2002; Mattiolo et al., 2004; Pieri & Abrams, 2004; Ramsey & Flynn, 2004; Ramsey & Dehn, 2004; Patrick et al., 2005; Pieri & Abrams, 2005; Wright et al., 2005; Gogu et al., 2006; Vaughan & Hook, 2006; Carter et al., 2007; Permenter & Oppenheimer, 2007; Vaughan et al., 2007; Hirn et al., 2008; Joyce et al., 2008; Kervyn et al., 2008b; Moran et al., 2008; Sincioco, 2008; Tunk & Bernard, 2008; Vaughan et al., 2008; Ji et al., 2009; Joyce et al., 2009a; Rose & Ramsey, 2009; Zlotnicki et al., 2009; Bailey et al., 2010; Carter & Ramsey, 2010; Coombs et al., 2010; Ferguson et al., 2010; Ganas et al., 2010; Ji et al., 2010; Murphy et al., 2010; Thomas & Watson, 2010; Wessels et al., 2010; Urai & Pieri, 2010; Grishin, 2011; Mathieu et al., 2011; Murphy et al., 2011; Rybin et al., 2011; Saepuloh et al., 2011; Urai, 2011; Urai & Ishizuka, 2011; Gutierrez et al., 2012; Hooper et al., 2012; Jousset at al., 2012; Patrick & Orr, 2012; Ramsey et al., 2012; Solikhin et al., 2012; Bleick et al., 2013; Buongiorno et al., 2013; Colvin et al., 2013; Dvigalo et al., 2013; Girina, 2013; Jay et al., 2013; Murphy et al., 2013; Ramsey & Harris, 2013; Roverato et al., 2013; Saepuloh et al., 2013; Wessels et al., 2013; West, 2013; Delgado et al., 2014; McGimsey et al., 2014; Moyano et al., 2014; Pritchard et al., 2014; Smets et al., 2014; Worden et al., 2014; Jay et al., 2015; Mars et al., 2015; Volynets et al., 2015; Whelley et al., 2015; Brothelande et al., 2016; Carr et al., 2016; Naranjo et al., 2016; Patrick et al., 2016; Rathnam & Ramashri, 2016a; Rathnam & Ramashri, 2016b; Reath et al., 2016; Blackett, 2017; Furtney et al., 2018; Girina et al., 2018; Girona et al., 2018; Harris et al., 2018; Plank et al., 2018; Wadge et al., 2018; Aufaristama et al., 2019a; Caputo et al., 2019; Gray et al., 2019; Harris et al., 2019; Henderson et al., 2019; Kaneko et al., 2019; Mannini et al, 2019; Mia et al., 2019; Reath et al., 2019; Sekertekin & Arslan, 2019; Silvestri et al., 2019.
Operational	Duda et al., 2009; Patrick & Witzke, 2011; Abrams et al., 2015; Ramsey, 2016.
Other	Scholte et al., 2003; Patrick et al., 2004; Mantas et al., 2011; Rivera et al., 2014.

Table A1. *Cont.*

Publication Category	Reference List
Precursory	Fujisada, 1995; Pieri et al., 1995; Oppenheimer, 1996; Oppenheimer, 1997; Realmuto et al., 1997; Oppenheimer et al., 1998; Yamaguchi et al., 1998; Glaze et al., 1999; Harris et al., 1999; Ramsey & Fink, 1999; Urai et al., 1999; Wright et al., 1999; Flynn et al., 2000; Realmuto, 2000; Realmuto & Worden, 2000; Wright et al., 2000.
Topography	Stevens et al., 2004; Kass, 2005; Kervyn et al., 2006; Pavez et al., 2006; Urai et al., 2007; Kervyn et al., 2008a; Arellano-Baeza et al., 2009; Gilichinsky et al., 2010; Inbar et al., 2011; Volker et al., 2011; Zouzias et al., 2011; Ebmeier et al., 2012; Fornaciai et al., 2012; Grosse et al., 2012; Camiz et al., 2013; Ebmeier et al., 2013; Le Corvec et al., 2013; Pritchard et al., 2013; Hamlyn et al., 2013; Kim & Lees, 2014; Albino et al., 2015; Walter et al., 2015; Kereszturi & Procter, 2016; Bannari et al., 2017; Camiz et al., 2017; Girod et al., 2017; Holohan et al., 2017; Aisyah et al., 2018; Raharimahefa & Rasoazanamparany, 2018; Deng et al., 2019; Morgado et al., 2019.

Complete Reference List for Table A1.

Abrams, M.; Pieri, D.; Realmuto, V.; Wright, R. Using EO-1 Hyperion Data as HyspIRI Preparatory Data Sets for Volcanology Applied to Mt Etna, Italy. *IEEE J. Sel. Top. Appl. Earth Obs. Remote Sens.* **2013**, *6* (2), 375–385. https://doi.org/10.1109/JSTARS.2012.2224095.

Abrams, M.; Tsu, H.; Hulley, G.; Iwao, K.; Pieri, D.; Cudahy, T.; Kargel, J. The Advanced Spaceborne Thermal Emission and Reflection Radiometer (ASTER) after Fifteen Years: Review of Global Products. *Int. J. Appl. Earth Obs. Geoinf.* **2015**, *38*, 292–301. https://doi.org/10.1016/j.jag.2015.01.013.

Aisyah, N.; Iguchi, M.; Subandriyo; Budisantoso, A.; Hotta, K.; Sumarti, S. Combination of a Pressure Source and Block Movement for Ground Deformation Analysis at Merapi Volcano Prior to the Eruptions in 2006 and 2010. *J. Volcanol. Geotherm. Res.* **2018**, *357*, 239–253. https://doi.org/10.1016/j.jvolgeores.2018.05.001.

Albino, F.; Smets, B.; D'Oreye, N.; Kervyn, F. High-Resolution TanDEM-X DEM: An Accurate Method to Estimate Lava Flow Volumes at Nyamulagira Volcano (D. R. Congo). *J. Geophys. Res. Solid Earth.* **2015**, *120* (6), 4189–4207. https://doi.org/10.1002/2015JB011988.

Ali-Bik, M. W.; Abd El Rahim, S. H.; Wahab, W. A.; Abayazeed, S. D.; Hassan, S. M. Geochemical Constraints on the Oldest Arc Rocks of the Arabian-Nubian Shield: The Late Mesoproterozoic to Late Neoproterozoic (?) Sa'al Volcano-Sedimentary Complex, Sinai, Egypt. *Lithos.* **2017**, *284–285*, 310–326. https://doi.org/10.1016/j.lithos.2017.03.031.

Amici, S.; Piscini, A.; Buongiorno, M. F.; Pieri, D. Geological Classification of Volcano Teide by Hyperspectral and Multispectral Satellite Data. *Int. J. Remote Sens.* **2013**, *34* (9–10), 3356–3375. https://doi.org/10.1080/01431161.2012.716913.

Arellano-Baeza, A. A.; García, R. V.; Trejo-Soto, M.; Martinez Bringas, A. Use of the ASTER Satellite Images for Evaluation of Structural Changes in the Popocatépetl Volcano Related to Microseismicity. *Adv. Sp. Res.* **2009**, *43* (2), 224–230. https://doi.org/10.1016/j.asr.2008.03.021.

Auer, A.; Belousov, A.; Belousova, M. Deposits, Petrology and Mechanism of the 2010–2013 Eruption of Kizimen Volcano in Kamchatka, Russia. *Bull. Volcanol.* **2018**, *80* (4), 4. https://doi.org/10.1007/s00445-018-1199-z.

Aufaristama, M.; Hoskuldsson, A.; Jonsdottir, I.; Ulfarsson, M. O.; Erlangga, I. G. D.; Thordarson, T. Thermal Model of Lava in Mt. Agung during December 2017 Episodes Derived from Integrated SENTINEL 2A and ASTER Remote Sensing Datasets. In *IOP Conference Series: Earth and Environmental Science*; 2019a; Vol. 311, p 1. https://doi.org/10.1088/1755-1315/311/1/012016.

Aufaristama, M.; Hoskuldsson, A.; Ulfarsson, M. O.; Jonsdottir, I.; Thordarson, T. The 2014-2015 Lava Flow Field at Holuhraun, Iceland: Using Airborne Hyperspectral Remote Sensing for Discriminating the Lava Surface. *Remote Sens.* **2019b**, *11* (5), 5. https://doi.org/10.3390/rs11050476.

Bailey, J. E.; Dean, K. G.; Dehn, J.; Webley, P. W. Integrated Satellite Observations of the 2006 Eruption of Augustine Volcano. *US Geol. Surv. Prof. Pap.* **2010**, No. 1769, 481–503. https://doi.org/10.3133/pp176920.

Bannari, A.; Kadhem, G.; Hameid, N.; El-Battay, A. Small Islands DEMs and Topographic Attributes Analysis: A Comparative Study among SRTM-V4.1, ASTER-V2.1, High Topographic Contours Map and DGPS. *J. Earth Sci. Eng.* **2017**, *7* (2), 90–119. https://doi.org/10.17265/2159-581x/2017.02.003.

Barreto, Á.; Arbelo, M.; Hernández-Leal, P. A.; Núez-Casillas, L.; Mira, M.; Coll, C. Evaluation of Surface Temperature and Emissivity Derived from ASTER Data: A Case Study Using Ground-Based Measurements at a Volcanic Site. *J. Atmos. Ocean. Technol.* **2010**, *27* (10), 1677–1688. https://doi.org/10.1175/2010JTECHA1447.1.

Blackett, M. An Overview of Infrared Remote Sensing of Volcanic Activity. *J. Imaging* **2017**, *3* (2). https://doi.org/10.3390/jimaging3020013.

Blackett, M.; Wooster, M. J. Evaluation of SWIR-Based Methods for Quantifying Active Volcano Radiant Emissions Using NASA EOS-ASTER Data. *Geomatics, Nat. Hazards Risk.* **2011**, *2* (1), 51–78. https://doi.org/10.1080/19475705.2010.541501.

Bleick, H. A.; Coombs, M. L.; Cervelli, P. F.; Bull, K. F.; Wessels, R. L. Volcano-Ice Interactions Precursory to the 2009 Eruption of Redoubt Volcano, Alaska. *J. Volcanol. Geotherm. Res.* **2013**, *259*, 373–388. https://doi.org/10.1016/j.jvolgeores.2012.10.008.

Bogie, I.; Sugiono, S. R.; Malike, D. Volcanic Landforms That Mark the Successfully Developed Geothermal Systems of Java, Indonesia Identified from ASTER Satellite Imagery. In *Proceedings World Geothermal Congress*, 2010; pp1-9.

Boyce, J. A.; Keays, R. R.; Nicholls, I. A.; Hayman, P. Eruption Centres of the Hamilton Area of the Newer Volcanics Province, Victoria, Australia: Pinpointing Volcanoes from a Multifaceted Approach to Landform Mapping. *Aust. J. Earth Sci.* **2014**, *61* (5), 735–754. https://doi.org/10.1080/08120099.2014.923508.

Braddock, M.; Biggs, J.; Watson, I. M.; Hutchison, W.; Pyle, D. M.; Mather, T. A. Satellite Observations of Fumarole Activity at Aluto Volcano, Ethiopia: Implications for Geothermal Monitoring and Volcanic Hazard. *J. Volcanol. Geotherm. Res.* **2017**, *341*, 70–83. https://doi.org/10.1016/j.jvolgeores.2017.05.006.

Brandmeier, M. Remote Sensing of Carhuarazo Volcanic Complex Using ASTER Imagery in Southern Peru to Detect Alteration Zones and Volcanic Structures - a Combined Approach of Image Processing in ENVI and ArcGIS/ArcScene. *Geocarto Int.* **2010**, *25* (8), 629–648. https://doi.org/10.1080/10106049.2010.519787.

Brothelande, E.; Lénat, J. F.; Chaput, M.; Gailler, L.; Finizola, A.; Dumont, S.; Peltier, A.; Bachèlery, P.; Barde-Cabusson, S.; Byrdina, S.; et al. Structure and Evolution of an Active Resurgent Dome Evidenced by Geophysical Investigations: The Yenkahe Dome-Yasur Volcano System (Siwi Caldera, Vanuatu). *J. Volcanol. Geotherm. Res.* **2016**, *322*, 241–262. https://doi.org/10.1016/j.jvolgeores.2015.08.021.

Buongiorno, M. F.; Pieri, D.; Silvestri, M. Thermal Analysis of Volcanoes Based on 10 Years of ASTER Data on Mt. Etna. In *Thermal Infrared Remote Sensing: Sensors, Methods, Applications*; Kuenzer, C., Dech, S., Eds.; Springer Netherlands: Dordrecht, **2013**; pp 409–428. https://doi.org/10.1007/978-94-007-6639-6_20.C.

Bustos, E.; Báez, W. A.; Chiodi, A. L.; Arnosio, J. M.; Norini, G.; Groppell, G. Using Optical Imagery Data for Lithological Mapping of Composite Volcanoes in High Arid Puna Plateau. Tuzgle Volcano Case Study. *Rev. la Asoc. Geol. Argentina.* **2017**, *74* (3), 357–371.

Byrnes, J. M.; Ramsey, M. S.; Crown, D. A. Surface Unit Characterization of the Mauna Ulu Flow Field, Kilauea Volcano, Hawai'i, Using Integrated Field and Remote Sensing Analyses. *J. Volcanol. Geotherm. Res.* **2004**, *135* (1–2), 169–193. https://doi.org/10.1016/j.jvolgeores.2003.12.016.

Camiz, S.; Poscolieri, M.; Roverato, M. Comparison of Three Andean Volcanic Complexes through Multidimensional Analyses of Geomorphometric Data. G. *Bianchini al., EICES.* **2013**, 52–74.

Camiz, S.; Poscolieri, M.; Roverato, M. Geomorphometric Comparative Analysis of Latin-American Volcanoes. *J. South Am. Earth Sci.* **2017**, *76*, 47–62. https://doi.org/10.1016/j.jsames.2017.02.011.

Campion, R. New Lava Lake at Nyamuragira Volcano Revealed by Combined ASTER and OMI SO2 Measurements. *Geophys. Res. Lett.* **2014**, *41* (21), 7485–7492. https://doi.org/10.1002/2014GL061808.

Campion, R.; Salerno, G. G.; Coheur, P. F.; Hurtmans, D.; Clarisse, L.; Kazahaya, K.; Burton, M.; Caltabiano, T.; Clerbaux, C.; Bernard, A. Measuring Volcanic Degassing of SO2 in the Lower Troposphere with ASTER Band Ratios. *J. Volcanol. Geotherm. Res.* **2010**, *194* (1–3), 42–54. https://doi.org/10.1016/j.jvolgeores.2010.04.010.

Campion, R.; Martinez-Cruz, M.; Lecocq, T.; Caudron, C.; Pacheco, J.; Pinardi, G.; Hermans, C.; Carn, S.; Bernard, A. Space- and Ground-Based Measurements of Sulphur Dioxide Emissions from Turrialba Volcano (Costa Rica). *Bull. Volcanol.* **2012**, *74* (7), 1757–1770. https://doi.org/10.1007/s00445-012-0631-z.

Capra, L. Volcanic Natural Dams Associated with Sector Collapses: Textural and Sedimentological Constraints on Their Stability. *Ital. J. Eng. Geol. Environ.* **2006**, *2006*, 279–294. https://doi.org/10.1007/978-3-642-04764-0_9.

Capra, L.; Manea, V. C.; Manea, M.; Norini, G. The Importance of Digital Elevation Model Resolution on Granular Flow Simulations: A Test Case for Colima Volcano Using TITAN2D Computational Routine. *Nat. Hazards* **2011**, *59* (2), 665–680. https://doi.org/10.1007/s11069-011-9788-6.

Caputo, T.; Sessa, E. B.; Silvestri, M.; Buongiorno, M. F.; Musacchio, M.; Sansivero, F.; Vilardo, G. Surface Temperature Multiscale Monitoring by Thermal Infrared Satellite and Ground Images at Campi Flegrei Volcanic Area (Italy). *Remote Sens.* **2019**, *11* (9), 1007. https://doi.org/10.3390/rs11091007.

Carn, S. A.; Clarisse, L.; Prata, A. J. Multi-Decadal Satellite Measurements of Global Volcanic Degassing. *J. Volcanol. Geotherm. Res.* **2016**, *311*, 99–134. https://doi.org/10.1016/j.jvolgeores.2016.01.002.

Carr, B. B.; Clarke, A. B.; Vanderkluysen, L. The 2006 Lava Dome Eruption of Merapi Volcano (Indonesia): Detailed Analysis Using MODIS TIR. *J. Volcanol. Geotherm. Res.* **2016**, *311*, 60–71. https://doi.org/10.1016/j.jvolgeores.2015.12.004.

Carr, B. B.; Clarke, A. B.; Vanderkluysen, L.; Arrowsmith, J. R. Mechanisms of Lava Flow Emplacement during an Effusive Eruption of Sinabung Volcano (Sumatra, Indonesia). *J. Volcanol. Geotherm. Res.* **2019**, *382*, 137–148. https://doi.org/10.1016/j.jvolgeores.2018.03.002.

Carter, A. J.; Ramsey, M. S. ASTER- and Field-Based Observations at Bezymianny Volcano: Focus on the 11 May 2007 Pyroclastic Flow Deposit. *Remote Sens. Environ.* **2009**, *113* (10), 2142–2151. https://doi.org/10.1016/j.rse.2009.05.020.

Carter, A.; Ramsey, M. S. Long-Term Volcanic Activity at Shiveluch Volcano: Nine Years of ASTER Spaceborne Thermal Infrared Observations. *Remote Sens.* **2010**, *2* (11), 2571–2583. https://doi.org/10.3390/rs2112571.

Carter, A. J.; Ramsey, M. S.; Belousov, A. B. Detection of a New Summit Crater on Bezymianny Volcano Lava Dome: Satellite and Field-Based Thermal Data. *Bull. Volcanol.* **2007**, *69* (7), 811–815. https://doi.org/10.1007/s00445-007-0113-x.

Carter, A. J.; Girina, O.; Ramsey, M. S.; Demyanchuk, Y. V. ASTER and Field Observations of the 24 December 2006 Eruption of Bezymianny Volcano, Russia. *Remote Sens. Environ.* **2008**, *112* (5), 2569–2577. https://doi.org/10.1016/j.rse.2007.12.001.

Carter, A. J.; Ramsey, M. S.; Durant, A. J.; Skilling, I. P.; Wolfe, A. Micron-Scale Roughness of Volcanic Surfaces from Thermal Infrared Spectroscopy and Scanning Electron Microscopy. *J. Geophys. Res. Solid Earth.* **2009**, *114* (2), 13. https://doi.org/10.1029/2008JB005632.

Castruccio, A.; Clavero, J. Lahar Simulation at Active Volcanoes of the Southern Andes: Implications for Hazard Assessment. *Nat. Hazards.* **2015**, *77* (2), 693–716. https://doi.org/10.1007/s11069-015-1617-x.

Caudron, C.; Bernard, A.; Murphy, S.; Inguaggiato, S.; Gunawan, H. Volcano-Hydrothermal System and Activity of Sirung Volcano (Pantar Island, Indonesia). *J. Volcanol. Geotherm. Res.* **2018**, *357*, 186–199. https://doi.org/10.1016/j.jvolgeores.2018.04.011.

Chadwick, W. W.; Jónsson, S.; Geist, D. J.; Poland, M.; Johnson, D. J.; Batt, S.; Harpp, K. S.; Ruiz, A. The May 2005 Eruption of Fernandina Volcano, Galápagos: The First Circumferential Dike Intrusion Observed by GPS and InSAR. *Bull. Volcanol.* **2011**, *73* (6), 679–697. https://doi.org/10.1007/s00445-010-0433-0.

Colvin, A.; Rose, W. I.; Varekamp, J. C.; Palma, J. L.; Escobar, D.; Gutierrez, E.; Montalvo, F.; Maclean, A. Crater Lake Evolution at Santa Ana Volcano (El Salvador) Following the 2005 Eruption. *Spec. Pap. Geol. Soc. Am.* **2013**, *498*, 23–43. https://doi.org/10.1130/2013.2498(02).

Coolbaugh, M. F.; Kratt, C.; Fallacaro, A.; Calvin, W. M.; Taranik, J. V. Detection of Geothermal Anomalies Using Advanced Spaceborne Thermal Emission and Reflection Radiometer (ASTER) Thermal Infrared Images at Bradys Hot Springs, Nevada, USA. *Remote Sens. Environ.* **2007**, *106* (3), 350–359. https://doi.org/10.1016/j.rse.2006.09.001.

Coombs, M. L.; Bull, K. F.; Vallance, J. W.; Schneider, D. J.; Thoms, E. E.; Wessels, R. L.; McGimsey, R. G. Timing, Distribution, and Volume of Proximal Products of the 2006 Eruption of Augustine Volcano. *US Geol. Surv. Prof. Pap.* **2010**, No. 1769, 145–185.

Coppola, D.; Campion, R.; Laiolo, M.; Cuoco, E.; Balagizi, C.; Ripepe, M.; Cigolini, C.; Tedesco, D. Birth of a Lava Lake: Nyamulagira Volcano 2011–2015. *Bull. Volcanol.* **2016**, *78* (3), 1–13. https://doi.org/10.1007/s00445-016-1014-7.

Corradini, S.; Pugnaghi, S.; Teggi, S.; Buongiorno, M. F.; Bogliolo, M. P. Will ASTER See the Etna SO2 Plume? *Int. J. Remote Sens.* **2003**, *24* (6), 1207–1218. https://doi.org/10.1080/01431160210153084.

Davies, A. G.; Calkins, J.; Scharenbroich, L.; Vaughan, R. G.; Wright, R.; Kyle, P.; Castaño, R.; Chien, S.; Tran, D. Multi-Instrument Remote and in Situ Observations of the Erebus Volcano (Antarctica) Lava Lake in 2005: A Comparison with the Pele Lava Lake on the Jovian Moon Io. *J. Volcanol. Geotherm. Res.* **2008**, *177* (3), 705–724. https://doi.org/10.1016/j.jvolgeores.2008.02.010.

Davila, N.; Capra, L.; Gavilanes-Ruiz, J. C.; Varley, N.; Norini, G.; Vazquez, A. G. Recent Lahars at Volcán de Colima (Mexico): Drainage Variation and Spectral Classification. *J. Volcanol. Geotherm. Res.* **2007**, *165* (3–4), 127–141. https://doi.org/10.1016/j.jvolgeores.2007.05.016.

Dávila-Hernández, N.; Lira, J.; Capra-Pedol, L.; Zucca, F. A Normalized Difference Lahar Index Based on Terra/Aster and Spot 5 Images: An Application at Colima Volcano, Mexico. *Rev. Mex. Ciencias Geol.* **2011**, *28* (3), 630–644.

Delgado, F.; Pritchard, M.; Lohman, R.; Naranjo, J. A. The 2011 Hudson Volcano Eruption (Southern Andes, Chile): Pre-Eruptive Inflation and Hotspots Observed with InSAR and Thermal Imagery. *Bull. Volcanol.* **2014**, *76* (5), 1–19. https://doi.org/10.1007/s00445-014-0815-9.

Deng, F.; Rodgers, M.; Xie, S.; Dixon, T. H.; Charbonnier, S.; Gallant, E. A.; López Vélez, C. M.; Ordoñez, M.; Malservisi, R.; Voss, N. K.; et al. High-Resolution DEM Generation from Spaceborne and Terrestrial Remote Sensing Data for Improved Volcano Hazard Assessment — A Case Study at Nevado Del Ruiz, Colombia. *Remote Sens. Environ.* **2019**, *233*, 11134. https://doi.org/10.1016/j.rse.2019.111348.

Diaz, J. A.; Pieri, D.; Arkin, C. R.; Gore, E.; Griffin, T. P.; Fladeland, M.; Bland, G.; Soto, C.; Madrigal, Y.; Castillo, D.; et al. Utilization of in Situ Airborne MS-Based Instrumentation for the Study of Gaseous Emissions at Active Volcanoes. *Int. J. Mass Spectrom.* **2010**, *295* (3), 105–112. https://doi.org/10.1016/j.ijms.2010.04.013.

Diaz, J. A.; Pieri, D.; Wright, K.; Sorensen, P.; Kline-Shoder, R.; Arkin, C. R.; Fladeland, M.; Bland, G.; Buongiorno, M. F.; Ramirez, C.; et al. Unmanned Aerial Mass Spectrometer Systems for In-Situ Volcanic Plume Analysis. *J. Am. Soc. Mass Spectrom.* **2015**, *26* (2), 292–304. https://doi.org/10.1007/s13361-014-1058-x.

Díaz-Castellón, R.; Hubbard, B. E.; Carrasco-Núñez, G.; Rodríguez-Vargas, J. L. The Origins of Late Quaternary Debris Avalanche and Debris Flow Deposits from Cofre de Perote Volcano, México. *Geosphere.* **2012**, *8* (4), 950–971. https://doi.org/10.1130/GES00709.1.

Duba, K.A.; Ramsey, M.; Wessels, R.; Dehn, J. Optical Satellite Volcano Monitoring: A Multi-Sensor Rapid Response System. In *Geoscience and Remote Sensing*; 2009; pp 473–496. https://doi.org/10.5772/8303.

Dvigalo, V. N.; Melekestsev, I. V.; Shevchenko, A. V.; Svirid, I. Y. The 2010-2012 Eruption of Kizimen Volcano: The Greatest Output (from the Data of Remote-Sensing Observations) for Eruptions in Kamchatka in the Early 21st Century Part I. The November 11, 2010 to December 11, 2011 Phase. *J. Volcanol. Seismol.* **2013**, *7* (6), 345–361. https://doi.org/10.1134/S074204631306002X.

Ebmeier, S. K.; Biggs, J.; Mather, T. A.; Elliott, J. R.; Wadge, G.; Amelung, F. Measuring Large Topographic Change with InSAR: Lava Thicknesses, Extrusion Rate and Subsidence Rate at Santiaguito Volcano, Guatemala. *Earth Planet. Sci. Lett.* **2012**, *335–336*, 216–225. https://doi.org/10.1016/j.epsl.2012.04.027.

Ebmeier, S. K.; Biggs, J.; Mather, T. A.; Amelung, F. Applicability of InSAR to Tropical Volcanoes: Insights from Central America. *Geol. Soc. Spec. Publ.* **2013**, *380* (1), 15–37. https://doi.org/10.1144/SP380.2.

Ellrod, G. P.; Helz, R. L.; Wadge, G. *Volcanic Hazards Assessment*; CEOS. NOAA. gov, 2002.

Favalli, M.; Chirico, G. D.; Papale, P.; Pareschi, M. T.; Coltelli, M.; Lucaya, N.; Boschi, E. Computer Simulations of Lava Flow Paths in the Town of Goma, Nyiragongo Volcano, Democratic Republic of Congo. *J. Geophys. Res. Solid Earth.* **2006**, *111* (6). https://doi.org/10.1029/2004JB003527.

Favalli, M.; Chirico, G. D.; Papale, P.; Pareschi, M. T.; Boschi, E. Lava Flow Hazard at Nyiragongo Volcano, D.R.C. 1. Model Calibration and Hazard Mapping. *Bull. Volcanol.* **2009**, *71* (4), 363–374. https://doi.org/10.1007/s00445-008-0233-y.

Favalli, M.; Tarquini, S.; Papale, P.; Fornaciai, A.; Boschi, E. Lava Flow Hazard and Risk at Mt. Cameroon Volcano. *Bull. Volcanol.* **2012**, *74* (2), 423–439. https://doi.org/10.1007/s00445-011-0540-6.

Ferguson, D. J.; Barnie, T. D.; Pyle, D. M.; Oppenheimer, C.; Yirgu, G.; Lewi, E.; Kidane, T.; Carn, S.; Hamling, I. Recent Rift-Related Volcanism in Afar, Ethiopia. *Earth Planet. Sci. Lett.* **2010**, *292* (3–4), 409–418.

Flynn, L. P.; Harris, A. J. L.; Rothery, D. A.; Oppenheimer, C. High-Spatial-Resolution Thermal Remote Sensing of Active Volcanic Features Using Landsat and Hyperspectral Data. *Geophys. Monogr. Ser.* **2000**, *116*, 161–177. https://doi.org/10.1029/GM116p0161.

Folguera, A.; Rojas Vera, E.; Vélez, L.; Tobal, J.; Orts, D.; Agusto, M.; Caselli, A.; Ramos, V. A. A Review of the Geology, Structural Controls, and Tectonic Setting of Copahue Volcano, Southern Volcanic Zone, Andes, Argentina. In *Copahue Volcano*; Springer: Berlin, 2016; pp 3–22. https://doi.org/10.1007/978-3-662-48005-2_1.

Fornaciai, A.; Favalli, M.; Karátson, D.; Tarquini, S.; Boschi, E. Morphometry of Scoria Cones, and Their Relation to Geodynamic Setting: A DEM-Based Analysis. *J. Volcanol. Geotherm. Res.* **2012**, *217–218*, 56–72. https://doi.org/10.1016/j.jvolgeores.2011.12.012.

Fu, H.; Fu, B.; Ninomiya, Y.; Shi, P. New Insights of Geomorphologic and Lithologic Features Onwudalianchi Volcanoes in the Northeastern China from the ASTER Multispectral Data. *Remote Sens.* **2019**, *11* (22), 2663. https://doi.org/10.3390/rs11222663.

Fujisada, H. Design and Performance of ASTER Instrument. Proceedings of SPIE. *Int. Soc. Opt. Eng.* **1995**, *2583* (2583), Int. Soc. Opt. Eng.

Furtney, M. A.; Pritchard, M. E.; Biggs, J.; Carn, S. A.; Ebmeier, S. K.; Jay, J. A.; McCormick Kilbride, B. T.; Reath, K. A. Synthesizing Multi-Sensor, Multi-Satellite, Multi-Decadal Datasets for Global Volcano Monitoring. *J. Volcanol. Geotherm. Res.* **2018**, *365*, 38–56. https://doi.org/10.1016/j.jvolgeores.2018.10.002.

Ganas, A.; Lagios, E.; Petropoulos, G.; Psiloglou, B. Thermal Imaging of Nisyros Volcano (Aegean Sea) Using ASTER Data: Estimation of Radiative Heat Flux. *Int. J. Remote Sens.* **2010**, *31* (15), 4033–4047. https://doi.org/10.1080/01431160903140837.

Gilichinsky, M.; Melnikov, D.; Melekestsev, I.; Zaretskaya, N.; Inbar, M. Morphometric Measurements of Cinder Cones from Digital Elevation Models of Tolbachik Volcanic Field, Central Kamchatka. *Can. J. Remote Sens.* **2010**, *36* (4), 287–300. https://doi.org/10.5589/m10-049.

Girina, O. A. Chronology of Bezymianny Volcano Activity, 1956-2010. *J. Volcanol. Geotherm. Res.* **2013**, *263*, 22–41. https://doi.org/10.1016/j.jvolgeores.2013.05.002.

Girina, O. A.; Loupian, E. A.; Sorokin, A. A.; Mel'Nikov, D. V.; Manevich, A. G.; Manevich, T. M. Satellite and Ground-Based Observations of Explosive Eruptions on Zhupanovsky Volcano, Kamchatka, Russia in 2013 and in 2014–2016. *J. Volcanol. Seismol.* **2018**, *12* (1), 1–15. https://doi.org/10.1134/S0742046318010049.

Girod, L.; Nuth, C.; Kääb, A.; McNabb, R.; Galland, O. MMASTER: Improved ASTER DEMs for Elevation Change Monitoring. *Remote Sens.* **2017**, *9* (7), 704. https://doi.org/10.3390/rs9070704.

Girona, T.; Huber, C.; Caudron, C. Sensitivity to Lunar Cycles Prior to the 2007 Eruption of Ruapehu Volcano /704/4111 /704/2151/598 /141 Article. *Sci. Rep.* **2018**, *8* (1), 1476. https://doi.org/10.1038/s41598-018-19307-z.

Glaze, L. S.; Wilson, L.; Mouginis-Mark, P. J. Volcanic Eruption Plume Top Topography and Heights as Determined from Photoclinometric Analysis of Satellite Data. *J. Geophys. Res. Solid Earth.* **1999**, *104* (B2), 2989–3001. https://doi.org/10.1029/1998jb900047.

Godoy, B.; Lazcano, J.; Rodríguez, I.; Martínez, P.; Parada, M. A.; Le Roux, P.; Wilke, H. G.; Polanco, E. Geological Evolution of Paniri Volcano, Central Andes, Northern Chile. *J. South Am. Earth Sci.* **2018**, *84*, 184–200. https://doi.org/10.1016/j.jsames.2018.03.013.

Gogu, R. C.; Dietrich, V. J.; Jenny, B.; Schwandner, F. M.; Hurni, L. A Geo-Spatial Data Management System for Potentially Active Volcanoes - GEOWARN Project. *Comput. Geosci.* **2006**, *32* (1), 29–41. https://doi.org/10.1016/j.cageo.2005.04.004.

Graettinger, A. H.; Ellis, M. K.; Skilling, I. P.; Reath, K.; Ramsey, M. S.; Lee, R. J.; Hughes, C. G.; McGarvie, D. W. Remote Sensing and Geologic Mapping of Glaciovolcanic Deposits in the Region Surrounding Askja (Dyngjufjöll) Volcano, Iceland. *Int. J. Remote Sens.* **2013**, *34* (20), 7178–7198. https://doi.org/10.1080/01431161.2013.817716.

Gray, D. M.; Burton-Johnson, A.; Fretwell, P. T. Evidence for a Lava Lake on Mt. Michael Volcano, Saunders Island (South Sandwich Islands) from Landsat, Sentinel-2 and ASTER Satellite Imagery. *J. Volcanol. Geotherm. Res.* **2019**, *379*, 60–71. https://doi.org/10.1016/j.jvolgeores.2019.05.002.

Grishin, S. Y. Environmental Impact of the Powerful Eruption of Sarychev Peak Volcano (Kuril Islands, 2009) According to Satellite Imagery. *Izv. - Atmos. Ocean Phys.* **2011**, *47* (9), 1028–1031. https://doi.org/10.1134/S0001433811090064.

Grosse, P.; van Wyk de Vries, B.; Euillades, P. A.; Kervyn, M.; Petrinovic, I. A. Systematic Morphometric Characterization of Volcanic Edifices Using Digital Elevation Models. *Geomorphology.* **2012**, *136* (1), 114–131. https://doi.org/10.1016/j.geomorph.2011.06.001.

Gutiérrez, F. J.; Lemus, M.; Parada, M. A.; Benavente, O. M.; Aguilera, F. A. Contribution of Ground Surface Altitude Difference to Thermal Anomaly Detection Using Satellite Images: Application to Volcanic/Geothermal Complexes in the Andes of Central Chile. *J. Volcanol. Geotherm. Res.* **2012**, *237–238*, 69–80. https://doi.org/10.1016/j.jvolgeores.2012.05.016.

Hamlyn, J. E.; Keir, D.; Wright, T. J.; Neuberg, J. W.; Goitom, B.; Hammond, J. O. S.; Pagli, C.; Oppenheimer, C.; Kendall, J. M.; Grandin, R. Seismicity and Subsidence Following the 2011 Nabro Eruption, Eritrea: Insights into the Plumbing System of an off-Rift Volcano. *J. Geophys. Res. Solid Earth.* **2014**, *119* (11), 8267–8282. https://doi.org/10.1002/2014JB011395.

Harris, A. J. L.; Wright, R.; Flynn, L. P. Remote Monitoring of Mount Erebus Volcano, Antarctica, Using Polar Orbiters: Progress and Prospects. *Int. J. Remote Sens.* **1999**, *20* (15–16), 3051–3071. https://doi.org/10.1080/014311699211615.

Harris, A.; Chevrel, M.; Coppola, D.; Ramsey, M.; Hrysiewicz, A.; Thivet, S.; Villeneuve, N.; Favalli, M.; Peltier, A.; Kowalski, P.; et al. Validation of an Integrated Satellite-Data-Driven Response to an Effusive Crisis: The April–May 2018 Eruption of Piton de La Fournaise. *Ann. Geophys.* **2019**, *61* (Vol 61 (2018)). https://doi.org/10.4401/ag-7972.

Head, E. M.; Maclean, A. L.; Carn, S. A. Mapping Lava Flows from Nyamuragira Volcano (1967-2011) with Satellite Data and Automated Classification Methods. *Geomatics, Nat. Hazards Risk.* **2013**, *4* (2), 119–144. https://doi.org/10.1080/19475705.2012.680503.

Hellman, M. J.; Ramsey, M. S. Analysis of Hot Springs and Associated Deposits in Yellowstone National Park Using ASTER and AVIRIS Remote Sensing. *J. Volcanol. Geotherm. Res.* **2004**, *135* (1–2), 195–219. https://doi.org/10.1016/j.jvolgeores.2003.12.012.

Henderson, S. T.; Pritchard, M. E.; Cooper, J. R.; Aoki, Y. Remotely Sensed Deformation and Thermal Anomalies at Mount Pagan, Mariana Islands. *Front. Earth Sci.* **2019**, *7*, 238. https://doi.org/10.3389/feart.2019.00238.

Henney, L. A.; Rodríguez, L. A.; Watson, I. M. A Comparison of SO 2 Retrieval Techniques Using Mini-UV Spectrometers and ASTER Imagery at Lascar Volcano, Chile. *Bull. Volcanol.* **2012**, *74* (2), 589–594. https://doi.org/10.1007/s00445-011-0552-2.

Hirn, B.; Di Bartola, C.; Ferrucci, F. Spaceborne Monitoring 2000-2005 of the Pu'u 'O'o-Kupaianaha (Hawaii) Eruption by Synergetic Merge of Multispectral Payloads ASTER and MODIS. *IEEE Trans. Geosci. Remote Sens.* **2008**, *46* (10), 2848–2856. https://doi.org/10.1109/TGRS.2008.2001033.

Holohan, E. P.; Sudhaus, H.; Walter, T. R.; Schöpfer, M. P. J.; Walsh, J. J. Effects of Host-Rock Fracturing on Elastic-Deformation Source Models of Volcano Deflation. *Sci. Rep.* **2017**, *7* (1), 10970. https://doi.org/10.1038/s41598-017-10009-6.

Hooper, A.; Prata, F.; Sigmundsson, F. Remote Sensing of Volcanic Hazards and Their Precursors. In *Proceedings of the IEEE*; 2012; Vol. 100, pp 2908–2930. https://doi.org/10.1109/JPROC.2012.2199269.

Hubbard, B. E.; Crowley, J. K.; Zimbelman, D. R. Comparative Alteration Mineral Mapping Using Visible to Shortwave Infrared (0.4–2.4 Mm) Hyperion, ALI, and ASTER Imagery. *Comp. Gen. Pharmacol.* **2003**, *41* (6), 1401–1410.

Hubbard, B. E.; Sheridan, M. F.; Carrasco-Núñez, G.; Díaz-Castellón, R.; Rodríguez, S. R. Comparative Lahar Hazard Mapping at Volcan Citlaltépetl, Mexico Using SRTM, ASTER and DTED-1 Digital Topographic Data. *J. Volcanol. Geotherm. Res.* **2007**, *160* (1–2), 99–124. https://doi.org/10.1016/j.jvolgeores.2006.09.005.

Huggel, C.; Caplan-Auerbach, J.; Waythomas, C. F.; Wessels, R. L. Monitoring and Modeling Ice-Rock Avalanches from Ice-Capped Volcanoes: A Case Study of Frequent Large Avalanches on Iliamna Volcano, Alaska. *J. Volcanol. Geotherm. Res.* **2007**, *168* (1–4), 114–136. https://doi.org/10.1016/j.jvolgeores.2007.08.009.

Huggel, C.; Schneider, D.; Miranda, P. J.; Delgado Granados, H.; Kääb, A. Evaluation of ASTER and SRTM DEM Data for Lahar Modeling: A Case Study on Lahars from Popocatépetl Volcano, Mexico. *J. Volcanol. Geotherm. Res.* **2008**, *170* (1–2), 99–110. https://doi.org/10.1016/j.jvolgeores.2007.09.005.

Inbar, M.; Gilichinsky, M.; Melekestsev, I.; Melnikov, D.; Zaretskaya, N. Morphometric and Morphological Development of Holocene Cinder Cones: A Field and Remote Sensing Study in the Tolbachik Volcanic Field, Kamchatka. *J. Volcanol. Geotherm. Res.* **2011**, *201* (1–4), 301–311. https://doi.org/10.1016/j.jvolgeores.2010.07.013.

Ino, N.; Toshiaki, Y.; Kinoshita, K. Regional Characteristics of High Concentration Events of Volcanic Gas at Miyakejima. *J. Japan Soc. Nat. Disaster Sci.* **2005**, *23*, 505–520.

Iwashita, K.; Asaka, T.; Nishikawa, H.; Kondoh, T.; Tahara, T. Vegetation Biomass Change of the Bosoh Peninsula. Impacted by the Volcano Fumes from the Miyakejima. *Adv. Sp. Res.* **2006**, *37* (4), 734–740. https://doi.org/10.1016/j.asr.2004.12.071.

Jay, J. A.; Welch, M.; Pritchard, M. E.; Mares, P. J.; Mnich, M. E.; Melkonian, A. K.; Aguilera, F.; Naranjo, J. A.; Sunagua, M.; Clavero, J. Volcanic Hotspots of the Central and Southern Andes as Seen from Space by ASTER and MODVOLC between the Years 2000 and 2010. *Geol. Soc. Spec. Publ.* **2013**, *380* (1), 161–185. https://doi.org/10.1144/SP380.1.

Jay, J. A.; Delgado, F. J.; Torres, J. L.; Pritchard, M. E.; Macedo, O.; Aguilar, V. Deformation and Seismicity near Sabancaya Volcano, Southern Peru, from 2002 to 2015. *Geophys. Res. Lett.* **2015**, *42* (8), 2780–2788. https://doi.org/10.1002/2015GL063589.

Ji, L. Y.; Xu, J. D.; Lin, X. D.; Luan, P. Application of Satellite Thermal Infrared Remote Sensing in Monitoring Magmatic Activity of Changbaishan Tianchi Volcano. *Chinese Sci. Bull.* **2010**, *55* (24), 2731–2737. https://doi.org/10.1007/s11434-010-3232-2.

Jousset, P.; Pallister, J.; Boichu, M.; Buongiorno, M. F.; Budisantoso, A.; Costa, F.; Andreastuti, S.; Prata, F.; Schneider, D.; et al. The 2010 Explosive Eruption of Java's Merapi Volcano-A "100-Year" Event. *J. Volcanol. Geotherm. Res.* **2012**, *241–242*, 121–135. https://doi.org/10.1016/j.jvolgeores.2012.06.018.

Joyce, K.; Samsonov, S.; Jolly, G. Satellite Remote Sensing of Volcanic Activity in New Zealand. In Proceedings of the 2008 2nd Workshop on USE of Remote Sensing Techniques for Monitoring Volcanoes and Seismogenic Areas, USEReST; 2008; Vol. 2008, pp 1–4. https://doi.org/10.1109/USEREST. 2008.4740346.

Joyce, K. E.; Samsonov, S.; Jolly, G. Temperature, Color and Deformation Monitoring of Volcanic Regions in New Zealand. In *International Geoscience and Remote Sensing Symposium (IGARSS)*; 2009a; Vol. 1; pp 1-17. https://doi.org/10.1109/IGARSS.2009.5416924.

Joyce, K. E.; Samsonov, S.; Manville, V.; Jongens, R.; Graettinger, A.; Cronin, S. J. Remote Sensing Data Types and Techniques for Lahar Path Detection: A Case Study at Mt Ruapehu, New Zealand. *Remote Sens. Environ.* **2009b**, *113* (8), 1778–1786. https://doi.org/10.1016/j.rse.2009.04.001.

Kaneko, T.; Maeno, F.; Yasuda, A. Observation of the Eruption Sequence and Formation Process of a Temporary Lava Lake during the June–August 2015 Mt. Raung Eruption, Indonesia, Using High-Resolution and High-Frequency Satellite Image Datasets. *J. Volcanol. Geotherm. Res.* **2019**, *377*, 17–32. https://doi.org/10.1016/j.jvolgeores.2019.03.016.

Kass, M. A. Slope Stability Analysis of the Iliamna Volcano, Alaska, Using Aster TIR, SRTM Dem, and Aeromagnetic Data. In *Proceedings of the Symposium on the Application of Geophyics to Engineering and Environmental Problems, SAGEEP*; Alaska, 2005; Vol. 2, pp 883–891. https://doi.org/10.4133/1.2923545.

Kearney, C. S.; Dean, K.; Realmuto, V. J.; Watson, I. M.; Dehn, J.; Prata, F. Observations of SO2 Production and Transport from Bezymianny Volcano, Kamchatka Using the MODerate Resolution Infrared Spectroradiometer (MODIS). *Int. J. Remote Sens.* **2008**, *29* (22), 6647–6665. https://doi.org/10. 1080/01431160802168392.

Kereszturi, G.; Procter, J. Error in Topographic Attributes for Volcanic Hazard Assessment of the Auckland Volcanic Field (New Zealand). *New Zeal. J. Geol. Geophys.* **2016**, *59* (2), 286–301. https: //doi.org/10.1080/00288306.2015.1130155.

Kern, C.; Masias, P.; Apaza, F.; Reath, K. A.; Platt, U. Remote Measurement of High Preeruptive Water Vapor Emissions at Sabancaya Volcano by Passive Differential Optical Absorption Spectroscopy. *J. Geophys. Res. Solid Earth.* **2017**, *122* (5), 3540–3564. https://doi.org/10.1002/2017JB014020.

Kervyn, M.; Goossens, R.; Jacobs, P.; Ernst, G. G. J. ASTER DEMs for Volcano Topographic Mapping: Accuracy and Limitations. In *IAMG 2006 - 11th International Congress for Mathematical Geology: Quantitative Geology from Multiple Sources*, 2006; pp 1-8.

Kervyn, M.; Kervyn, F.; Goossens, R.; Rowland, S. K.; Ernst, G. G. J. Mapping Volcanic Terrain Using High-Resolution and 3D Satellite Remote Sensing. *Geol. Soc. Spec. Publ.* **2007**, *283*, 5–30. https://doi.org/10.1144/SP283.2.

Kervyn, M.; Ernst, G. G. J.; Goossens, R.; Jacobs, P. Mapping Volcano Topography with Remote Sensing: ASTER vs. SRTM. *Int. J. Remote Sens.* **2008a**, *29* (22), 6515–6538. https://doi.org/10.1080/ 01431160802167949.

Kervyn, M.; Ernst, G. G. J.; Harris, A. J. L.; Belton, F.; Mbede, E.; Jacobs, P. Thermal Remote Sensing of the Low-Intensity Carbonatite Volcanism of Oldoinyo Lengai, Tanzania. *Int. J. Remote Sens.* **2008b**, *29* (22), 6467–6499. https://doi.org/10.1080/01431160802167105.

Kervyn, M.; Ernst, G. G. J.; Klaudius, J.; Keller, J.; Mbede, E.; Jacobs, P. Remote Sensing Study of Sector Collapses and Debris Avalanche Deposits at Oldoinyo Lengai and Kerimasi Volcanoes, Tanzania. *Int. J. Remote Sens.* **2008c**, *29* (22), 6565–6595. https://doi.org/10.1080/01431160802168137.

Kim, K.; Lees, J. M. Local Volcano Infrasound and Source Localization Investigated by 3D Simulation. *Seismol. Res. Lett.* **2014**, *85* (6), 1177–1186. https://doi.org/10.1785/0220140029.

Kobayashi, C.; Orihashi, Y.; Hiarata, D.; Naranjo, J. A.; Kobayashi, M.; Anma, R. Compositional Variations Revealed by ASTER Image Analysis of the Viedma Volcano, Southern Andes Volcanic Zone. *Andean Geol.* **2010a**, *37* (2), 433–441.

Kobayashi, C.; Orihashi, Y.; Hiarata, D.; Naranjo, J. A.; Kobayashi, M.; Anma, R. Spaceborne ASTER Image Analyses Revealed Compositional Variation of the Viedma Volcano in Andean Austral Volcanic Zone. *Andean Geol.* **2010b**, *37* (2), 433–441. https://doi.org/10.5027/andgeov37n2-a09.

Krippner, J. B.; Belousov, A. B.; Belousova, M. G.; Ramsey, M. S. Parametric Analysis of Lava Dome-Collapse Events and Pyroclastic Deposits at Shiveluch Volcano, Kamchatka, Using Visible and Infrared Satellite Data. *J. Volcanol. Geotherm. Res.* **2018**, *354*, 115–129. https://doi.org/10.1016/j.jvolgeores.2018.01.027.

Laiolo, M.; Massimetti, F.; Cigolini, C.; Ripepe, M.; Coppola, D. Long-Term Eruptive Trends from Space-Based Thermal and SO2 Emissions: A Comparative Analysis of Stromboli, Batu Tara and Tinakula Volcanoes. *Bull. Volcanol.* **2018**, *80* (9), 9. https://doi.org/10.1007/s00445-018-1242-0.

Lara, L. E.; Moreno, R.; Amigo, Á.; Hoblitt, R. P.; Pierson, T. C. Late Holocene History of Chaiten Volcano: New Evidence for a 17th Century Eruption. *Andean Geol.* **2013**, *40* (2), 249–261. https://doi.org/10.5027/andgeoV40n2-a04.

Le Corvec, N.; Spörli, K. B.; Rowland, J.; Lindsay, J. Spatial Distribution and Alignments of Volcanic Centers: Clues to the Formation of Monogenetic Volcanic Fields. *Earth-Science Rev.* **2013**, *124*, 96–114. https://doi.org/10.1016/j.earscirev.2013.05.005.

Mannini, S.; Harris, A. J. L.; Jessop, D. E.; Chevrel, M. O.; Ramsey, M. S. Combining Ground- and ASTER-Based Thermal Measurements to Constrain Fumarole Field Heat Budgets: The Case of Vulcano Fossa 2000–2019. *Geophys. Res. Lett.* **2019**, *46*, 2019. https://doi.org/10.1029/2019GL084013.

Mantas, V. M.; Pereira, A. J. S. C.; Morais, P. V. Plumes of Discolored Water of Volcanic Origin and Possible Implications for Algal Communities. The Case of the Home Reef Eruption of 2006 (Tonga, Southwest Pacific Ocean). *Remote Sens. Environ.* **2011**, *115* (6), 1341–1352. https://doi.org/10.1016/j.rse.2011.01.014.

Mars, J. C. Regional Mapping of Hydrothermally Altered Igneous Rocks along the Urumieh-Dokhtar, Chagai, and Alborz Belts of Western Asia Using Advanced Spaceborne. *US Geol. Surv.* **2014**, *2010-5090-*, 1–46. https://dx.doi.org/10.3133/sir20105090O.

Mars, J. C.; Rowan, L. C. Regional Mapping of Phyllic- and Argillic-Altered Rocks in the Zagros Magmatic Arc, Iran, Using Advanced Spaceborne Thermal Emission and Reflection Radiometer (ASTER) Data and Logical Operator Algorithms. *Geosphere* **2006**, *2* (3), 161–186. https://doi.org/10.1130/GES00044.1.

Mars, J. C.; Rowan, L. C. ASTER Spectral Analysis and Lithologic Mapping of the Khanneshin Carbonatite Volcano, Afghanistan. *Geosphere.* **2011**, *7* (1), 276–289. https://doi.org/10.1130/GES00630.1.

Mars, J. C.; Hubbard, B.; Pieri, D.; Linick, J. Alteration, Slope-Classified Alteration, and Potential Lahar Inundation Maps of Volcanoes for the Advanced Spaceborne Thermal Emission and Reflection Radiometer (ASTER) Volcanoes Archive. *US Geol. Surv.* **2015**, *2015-5035*.

Mathieu, L.; Kervyn, M.; Ernst, G. G. J. Field Evidence for Flank Instability, Basal Spreading and Volcano-Tectonic Interactions at Mt Cameroon, West Africa. *Bull. Volcanol.* **2011**, *73* (7), 851–867. https://doi.org/10.1007/s00445-011-0458-z.

Mattiou, G. S.; Young, S. R.; Voight, B.; Steven, R.; Sparks, J.; Shalev, E.; Sacks, S.; Malin, P.; Linde, A.; Johnston, W.; et al. Prototype Pbo Instrumentation of Calipso Project Captures World-Record Lava Dome Collapse on Montserrat Volcano. *Eos (Washington. DC).* **2004**, *85* (34), 317–325. https://doi.org/10.1029/2004eo340001.

McGimsey, R. G.; Neal, C.; Girina, O.; Chibisova, M.; Rybin, A. 2009 Volcanic Activity in Alaska , Kamchatka , and the Kurile Islands — Summary of Events and Response of the Alaska Volcano Observatory Scientific Investigations Report 2013 – 5213. *US Geol. Surv.* **2014**, 1–140.

Mia, M. B.; Fujimitsu, Y.; Nishijima, J. Exploration and Monitoring of Hatchobaru-Otake Geothermal Field Using ASTER Satellite Images. In *Grand Renewable Energy proceedings Japan council for Renewable Energy*; 2018a; Vol. 276, pp 1–4.

Mia, M. B.; Fujimitsu, Y.; Nishijima, J. Monitoring Thermal Activity of the Beppu Geothermal Area in Japan Using Multisource Satellite Thermal Infrared Data. *Geosci.* **2018b**, *8* (8), 306. https://doi.org/10.3390/geosciences8080306.

Mia, M. B.; Fujimitsu, Y.; Nishijima, J. Exploration of Hydrothermal Alteration and Monitoring of Thermal Activity Using Multi-Source Satellite Images: A Case Study of the Recently Active Kirishima Volcano Complex on Kyushu Island, Japan. *Geothermics* **2019**, *79*, 26–45. https://doi.org/10.1016/j.geothermics.2019.01.006.

Moran, S. C.; Freymueller, J. T.; LaHusen, R. G.; McGee, K. A.; Poland, M. P.; Power, J. A.; Schmidt, D. A.; Schneider, D. J.; Stephens, G.; Werner, C. A.; White, R.A. *Instrumentation Recommendations for Volcano Monitoring at US Volcanoes under the National Volcano Early Warning System*; 2008; Vol. 5114; pp 1-56.

Morgado, E.; Morgan, D. J.; Harvey, J.; Parada, M. Á.; Castruccio, A.; Brahm, R.; Gutiérrez, F.; Georgiev, B.; Hammond, S. J. Localised Heating and Intensive Magmatic Conditions Prior to the 22–23 April 2015 Calbuco Volcano Eruption (Southern Chile). *Bull. Volcanol.* **2019**, *81* (4). https://doi.org/10.1007/s00445-019-1280-2.

Moussallam, Y.; Peters, N.; Masias, P.; Apaza, F.; Barnie, T.; Ian Schipper, C.; Curtis, A.; Tamburello, G.; Aiuppa, A.; Bani, P.; et al. Magmatic Gas Percolation through the Old Lava Dome of El Misti Volcano. *Bull. Volcanol.* **2017**, *79* (6), 6. https://doi.org/10.1007/s00445-017-1129-5.

Moyano, J. E. R.; Ulrich, W. K.; Marfull, V. Short Chronological Analysis of the 2007-2009 Eruptive Cycle and Its Nested Cones Formation at Llaima Volcano. *J. Technol. Possibilism.* **2014**, *2* (3), 1–9.

Muñoz-Salinas, E.; Castillo-Rodríguez, M.; Manea, V.; Manea, M.; Palacios, D. Lahar Flow Simulations Using LAHARZ Program: Application for the Popocatépetl Volcano, Mexico. *J. Volcanol. Geotherm. Res.* **2009**, *182* (1–2), 13–22. https://doi.org/10.1016/j.jvolgeores.2009.01.030.

Murphy, S. W.; De Souza Filho, C. R.; Oppenheimer, C. The Spatial Extent of Thermal Anomalies at Lascar Volcano. *Int. Geosci. Remote Sens. Symp.* **2010**, 1557–1560. https://doi.org/10.1109/IGARSS.2010.5652858.

Murphy, S. W.; Filho, C. R. de S.; Oppenheimer, C. Monitoring Volcanic Thermal Anomalies from Space: Size Matters. *J. Volcanol. Geotherm. Res.* **2011**, *203* (1–2), 48–61. https://doi.org/10.1016/j.jvolgeores.2011.04.008.

Murphy, S. W.; Wright, R.; Oppenheimer, C.; Filho, C. R. S. MODIS and ASTER Synergy for Characterizing Thermal Volcanic Activity. *Remote Sens. Environ.* **2013**, *131*, 195–205. https://doi.org/10.1016/j.rse.2012.12.005.

Nakada, S.; Zaennudin, A.; Yoshimoto, M.; Maeno, F.; Suzuki, Y.; Hokanishi, N.; Sasaki, H.; Iguchi, M.; Ohkura, T.; Gunawan, H.; et al. Growth Process of the Lava Dome/Flow Complex at Sinabung Volcano during 2013–2016. *J. Volcanol. Geotherm. Res.* **2019**, *382*, 120–136. https://doi.org/10.1016/j.jvolgeores.2017.06.012.

Naranjo, M. F.; Ebmeier, S. K.; Vallejo, S.; Ramón, P.; Mothes, P.; Biggs, J.; Herrera, F. Mapping and Measuring Lava Volumes from 2002 to 2009 at El Reventador Volcano, Ecuador, from Field Measurements and Satellite Remote Sensing. *J. Appl. Volcanol.* **2016**, *5* (1), 1. https://doi.org/10.1186/s13617-016-0048-z.

Oikonomidis, D.; Albanakis, K.; Pavlides, S.; Fytikas, M. Reconstruction of the Paleo-Coastline of Santorini Island (Greece), after the 1613 BC Volcanic Eruption: A GIS-Based Quantitative Methodology. *J. Earth Syst. Sci.* **2016**, *125* (1), 1–11. https://doi.org/10.1007/s12040-015-0643-0.

Oppenheimer, C. Crater Lake Heat Losses Estimated by Remote Sensing. *Geophys. Res. Lett.* **1996**, *23* (14), 1793–1796. https://doi.org/10.1029/96GL01591.

Oppenheimer, C. Remote Sensing of the Colour and Temperature of Volcanic Lakes. *Int. J. Remote Sens.* **1997**, *18* (1), 5–37. https://doi.org/10.1080/014311697219259.

Oppenheimer, C.; Francis, P.; Burton, M.; Maciejewski, A. J. H.; Boardman, L. Remote Measurement of Volcanic Gases by Fourier Transform Infrared Spectroscopy. *Appl. Phys. B Lasers Opt.* **1998**, *67* (4), 505–515. https://doi.org/10.1007/s003400050536.

Paguican, E. M. R.; Lagmay, A. M. F.; Rodolfo, K. S.; Rodolfo, R. S.; Tengonciang, A. M. P.; Lapus, M. R.; Baliatan, E. G.; Obille, E. C. Extreme Rainfall-Induced Lahars and Dike Breaching, 30 November 2006, Mayon Volcano, Philippines. *Bull. Volcanol.* **2009**, *71* (8), 845–857. https://doi.org/10.1007/s00445-009-0268-8.

Pallister, J.; Wessels, R.; Griswold, J.; McCausland, W.; Kartadinata, N.; Gunawan, H.; Budianto, A.; Primulyana, S. Monitoring, Forecasting Collapse Events, and Mapping Pyroclastic Deposits at Sinabung Volcano with Satellite Imagery. *J. Volcanol. Geotherm. Res.* **2019**, *382*, 149–163. https://doi.org/10.1016/j.jvolgeores.2018.05.012.

Patrick, M. R.; Witzke, C.-N. Thermal Mapping of Hawaiian Volcanoes with ASTER Satellite Data. *U.S. Geol. Surv. Sci. Investig. Rep. 2011–5110* **2011**, 22 p.

Patrick, M. R.; Orr, T. R. Rootless Shield and Perched Lava Pond Collapses at Kīlauea Volcano, Hawai'i. *Bull. Volcanol.* **2012**, *74* (1), 67–78. https://doi.org/10.1007/s00445-011-0505-9.

Patrick, M.; Dean, K.; Dehn, J. Active Mud Volcanism Observed with Landsat 7 ETM+. *J. Volcanol. Geotherm. Res.* **2004**, *131* (3–4), 307–320. https://doi.org/10.1016/S0377-0273(03)00383-4.

Patrick, M. R.; Smellie, J. L.; Harris, A. J. L.; Wright, R.; Dean, K.; Izbekov, P.; Garbeil, H.; Pilger, E. First Recorded Eruption of Mount Belinda Volcano (Montagu Island), South Sandwich Islands. *Bull. Volcanol.* **2005**, *67* (5), 415–422. https://doi.org/10.1007/s00445-004-0382-6.

Patrick, M. R.; Kauahikaua, J.; Orr, T.; Davies, A.; Ramsey, M. Operational Thermal Remote Sensing and Lava Flow Monitoring at the Hawaiian Volcano Observatory. *Geol. Soc. Spec. Publ.* **2016**, *426* (1), 489–503. https://doi.org/10.1144/SP426.17.

Pavez, A.; Remy, D.; Bonvalot, S.; Diament, M.; Gabalda, G.; Froger, J. L.; Julien, P.; Legrand, D.; Moisset, D. Insight into Ground Deformations at Lascar Volcano (Chile) from SAR Interferometry, Photogrammetry and GPS Data: Implications on Volcano Dynamics and Future Space Monitoring. *Remote Sens. Environ.* **2006**, *100* (3), 307–320. https://doi.org/10.1016/j.rse.2005.10.013.

Permenter, J. L.; Oppenheimer, C. Volcanoes of the Tibesti Massif (Chad, Northern Africa). *Bull. Volcanol.* **2007**, *69* (6), 609–626. https://doi.org/10.1007/s00445-006-0098-x.

Pieri, D.; Abrams, M. ASTER Watches the World's Volcanoes: A New Paradigm for Volcanological Observations from Orbit. *J. Volcanol. Geotherm. Res.* **2004**, *135* (1–2), 13–28. https://doi.org/10.1016/j.jvolgeores.2003.12.018.

Pieri, D.; Abrams, M. ASTER Observations of Thermal Anomalies Preceding the April 2003 Eruption of Chikurachki Volcano, Kurile Islands, Russia. *Remote Sens. Environ.* **2005**, *99* (1–2), 84–94. https://doi.org/10.1016/j.rse.2005.06.012.

Pieri, D. C.; Crisp, J.; Kahle, A. B. Observing Volcanism and Other Transient Phenomena with ASTER. *J. Remote Sens. Soc. Japan.* **1995**, *15* (2), 148–153. https://doi.org/10.11440/rssj1981.15.148.

Pieri, D.; Diaz, J. A.; Bland, G.; Fladeland, M.; Madrigal, Y.; Corrales, E.; Alegria, O.; Alan, A.; Realmuto, V.; Miles, T.; et al. In Situ Observations and Sampling of Volcanic Emissions with NASA and UCR Unmanned Aircraft, Including a Case Study at Turrialba Volcano, Costa Rica. *Geol. Soc. Spec. Publ.* **2013**, *380* (1), 321–352. https://doi.org/10.1144/SP380.13.

Pinardi, G.; Campion, R.; Roozendael, M. Van; Fayt, C.; Geffen, J. Van; Galle, B. Comparison of Volcanic So2 Flux Measurements from Satellite and from the NOVAC Network. In *EUMETSAT conference*; 2010; Vol. 2, pp1-8.

Piscini, A.; Amici, S.; Fieri, D. Spectral Analysis of ASTER and Hyperion Data for Geological Classification of Volcano Teide. *International Geoscience and Remote Sensing Symposium (IGARSS)*. 2010, pp 2267–2270. https://doi.org/10.1109/IGARSS.2010.5652063.

Plank, S.; Nolde, M.; Richter, R.; Fischer, C.; Martinis, S.; Riedlinger, T.; Schoepfer, E.; Klein, D. Monitoring of the 2015 Villarrica Volcano Eruption by Means of DLR's Experimental TET-1 Satellite. *Remote Sens.* **2018**, *10* (9), 9. https://doi.org/10.3390/rs10091379.

Prambada, O.; Arakawa, Y.; Ikehata, K.; Furukawa, R.; Takada, A.; Wibowo, H. E.; Nakagawa, M.; Kartadinata, M. N. Eruptive History of Sundoro Volcano, Central Java, Indonesia since 34 Ka. *Bull. Volcanol.* **2016**, *78* (11), 11. https://doi.org/10.1007/s00445-016-1079-3.

Price, M. A.; Ramsey, M. S.; Crown, D. A. Satellite-Based Thermophysical Analysis of Volcaniclastic Deposits: A Terrestrial Analog for Mantled Lava Flows on Mars. *Remote Sens.* **2016**, *8* (2), 152. https://doi.org/10.3390/rs8020152.

Pritchard, M. E.; Jay, J. A.; Aron, F.; Henderson, S. T.; Lara, L. E. Subsidence at Southern Andes Volcanoes Induced by the 2010 Maule, Chile Earthquake. *Nat. Geosci.* **2013**, *6* (8), 632–636. https://doi.org/10.1038/ngeo1855.

Pritchard, M. E.; Henderson, S. T.; Jay, J. A.; Soler, V.; Krzesni, D. A.; Button, N. E.; Welch, M. D.; Semple, A. G.; Glass, B.; Sunagua, M.; et al. Reconnaissance Earthquake Studies at Nine Volcanic Areas of the Central Andes with Coincident Satellite Thermal and InSAR Observations. *J. Volcanol. Geotherm. Res.* **2014**, *280*, 90–103. https://doi.org/10.1016/j.jvolgeores.2014.05.004.

Pugnaghi, S.; Gangale, G.; Corradini, S.; Buongiorno, M. F. Mt. Etna Sulfur Dioxide Flux Monitoring Using ASTER-TIR Data and Atmospheric Observations. *J. Volcanol. Geotherm. Res.* **2006**, *152* (1–2), 74–90.

Raharimahefa, T.; Rasoazanamparany, C. Geomorphological Classification of Volcanic Cones in the Itasy Volcanic Field, Central Madagascar. *Int. J. Geol. Earth Sci.* **2018**, *4* (4), 14. https://doi.org/10.32937/ijges.4.4.2018.14-34.

Ramsey, M. S. Temperature and Textures of Ash Flow Surfaces: Sheveluch, Kamchatka, Russia (2004). In *Monitoring Volcanoes in the North Pacific*; Dean, K. G., Dehn, J., Eds.; **2015**; pp 79–100. https://doi.org/10.1007/978-3-540-68750-4.

Ramsey, M. S. Synergistic Use of Satellite Thermal Detection and Science: A Decadal Perspective Using ASTER. *Geol. Soc. Spec. Publ.* **2016**, *426* (1), 115–136. https://doi.org/10.1144/SP426.23.

Ramsey, M. S.; Fink, J. H. Estimating Silicic Lava Vesicularity with Thermal Remote Sensing: A New Technique for Volcanic Mapping and Monitoring. *Bull. Volcanol.* **1999**, *61* (1–2), 32–39. https://doi.org/10.1007/s004450050260.

Ramsey, M. S.; Flynn, L. P. Strategies, Insights, and the Recent Advances in Volcanic Monitoring and Mapping with Data from NASA's Earth Observing System. *J. Volcanol. Geotherm. Res.* **2004**, *135* (1–2), 1–11. https://doi.org/10.1016/j.jvolgeores.2003.12.015.

Ramsey, M.; Dehn, J. Spaceborne Observations of the 2000 Bezymianny, Kamchatka Eruption: The Integration of High-Resolution ASTER Data into near Real-Time Monitoring Using AVHRR. *J. Volcanol. Geotherm. Res.* **2004**, *135* (1–2), 127–146. https://doi.org/10.1016/j.jvolgeores.2003.12.014.

Ramsey, M. S.; Harris, A. J. L. Volcanology 2020: How Will Thermal Remote Sensing of Volcanic Surface Activity Evolve over the next Decade? *J. Volcanol. Geotherm. Res.* **2013**, *249*, 217–233. https://doi.org/10.1016/j.jvolgeores.2012.05.011.

Ramsey, M. S.; Wessels, R. L.; Anderson, S. W. Surface Textures and Dynamics of the 2005 Lava Dome at Shiveluch Volcano, Kamchatka. *Bull. Geol. Soc. Am.* **2012**, *124* (5–6), 678–689. https://doi.org/10.1130/B30580.1.

Ramsey, M. S.; Harris, A. J. L.; Crown, D. A. What Can Thermal Infrared Remote Sensing of Terrestrial Volcanoes Tell Us about Processes Past and Present on Mars? *J. Volcanol. Geotherm. Res.* **2016**, *311*, 198–216. https://doi.org/10.1016/j.jvolgeores.2016.01.012.

Ramsey, M. S.; Chevrel, M. O.; Coppola, D.; Harris, A. J. L. The Influence of Emissivity on the Thermo-Rheological Modeling of the Channelized Lava Flows at Tolbachik Volcano. *Ann. Geophys.* **2019**, *62* (Special Issue), 2. https://doi.org/10.4401/ag-8077.

Rathnam, S. M.; Ramashri, T. Identification of Volcano Hotspots in Multi Spectral ASTER Satellite Images Using DTCWT Image Fusion and ANFIS Classifier. *Am. J. Eng. Res. (AJER),* **2016a**, *12* (12), 21–31.

Rathnam, S. M.; Ramashri, T. Identification of Volcano Hotspots Using Multispectral ASTER Satellite Images. *Int. J. Comput. Sci. Inf. Secur.* **2016b**, *14* (10), 138.

Realmuto, V. J.; Sutton, A. J.; Elias, T. Multispectral Thermal Infrared Mapping of Sulfur Dioxide Plumes: A Case Study from the East Rift Zone of Kilauea Volcano, Hawaii. *J. Geophys. Res. Solid Earth.* **1997**, *102* (B7), 15057–15072. https://doi.org/10.1029/96jb03916.

Realmuto, V. J. The potential use of the earth observing system data to monitor the passive emissions of sulfur dioxide from volcanoes. Geophys. Monogr. Ser. 2000, 116, 101-115. https://doi.org/10.1029/GM116pp0101.

Realmuto, V. J.; Worden, H. M. Impact of Atmospheric Water Vapor on the Thermal Infrared Remote Sensing of Volcanic Sulfur Dioxide Emissions: A Case Study from the Pu'u 'O' Vent of Kilauea Volcano, Hawaii. *J. Geophys. Res. Solid Earth.* **2000**, *105* (B9), 21497–21507. https://doi.org/10.1029/2000jb900172.

Realmuto, V. J.; Berk, A. Plume Tracker: Interactive Mapping of Volcanic Sulfur Dioxide Emissions with High-Performance Radiative Transfer Modeling. *J. Volcanol. Geotherm. Res.* **2016**, *327*, 55–69. https://doi.org/10.1016/j.jvolgeores.2016.07.001.

Reath, K.; Ramsey, M. S.; Dehn, J.; Webley, P. W. Predicting Eruptions from Precursory Activity Using Remote Sensing Data Hybridization. *J. Volcanol. Geotherm. Res.* **2016**, *321*, 18–30. https://doi.org/10.1016/j.jvolgeores.2016.04.027.

Reath, K.; Pritchard, M. E.; Moruzzi, S.; Alcott, A.; Coppola, D.; Pieri, D. The AVTOD (ASTER Volcanic Thermal Output Database) Latin America Archive. *J. Volcanol. Geotherm. Res.* **2019**, *376*, 62–74. https://doi.org/10.1016/j.jvolgeores.2019.03.019.

Rivera, M.; Thouret, J. C.; Samaniego, P.; Le Pennec, J. L. The 2006-2009 Activity of the Ubinas Volcano (Peru): Petrology of the 2006 Eruptive Products and Insights into Genesis of Andesite Magmas, Magma Recharge and Plumbing System. *J. Volcanol. Geotherm. Res.* **2014**, *270*, 122–141. https://doi.org/10.1016/j.jvolgeores.2013.11.010.

Robertson, E.; Biggs, J.; Edmonds, M.; Clor, L.; Fischer, T. P.; Vye-Brown, C.; Kianji, G.; Koros, W.; Kandie, R. Diffuse Degassing at Longonot Volcano, Kenya: Implications for CO2 Flux in Continental Rifts. *J. Volcanol. Geotherm. Res.* **2016**, *327*, 208–222. https://doi.org/10.1016/j.jvolgeores.2016.06.016.

Rogic, N.; Cappello, A.; Ferrucci, F. Role of Emissivity in Lava Flow 'Distance-to-Run' Estimates from Satellite-Based Volcano Monitoring. *Remote Sens.* **2019**, *11* (6), 662. https://doi.org/10.3390/rs11060662.

Rose, S.; Ramsey, M. S. The 2005 Eruption of Kliuchevskoi Volcano: Chronology and Processes Derived from ASTER Spaceborne and Field-Based Data. *J. Volcanol. Geotherm. Res.* **2009**, *184* (3–4), 367–380. https://doi.org/10.1016/j.jvolgeores.2009.05.001.

Rose, S. R.; Ramsey, M. S. The 2005 and 2007 Eruptions of Klyuchevskoy Volcano, Russia: Behavior and Effusion Mechanisms. In *Monitoring Volcanoes in the North Pacific: Observations from Space*; Dean, K. G., Dehn J. Dehn, J., Eds.; - 540-24125-6, 389 pp: DVD), Springer-Praxis Books, ISBN, **2015**; pp 973–978.

Rose, S. R.; Watson, I. M.; Ramsey, M. S.; Hughes, C. G. Thermal Deconvolution: Accurate Retrieval of Multispectral Infrared Emissivity from Thermally-Mixed Volcanic Surfaces. *Remote Sens. Environ.* **2014**, *140*, 690–703. https://doi.org/10.1016/j.rse.2013.10.009.

Roverato, M.; Capra, L.; Sulpizio, R. First Evidence of Hydromagmatism at Colima Volcano (Mexico). *J. Volcanol. Geotherm. Res.* **2013**, *249*, 197–200. https://doi.org/10.1016/j.jvolgeores.2012.10.012.

Rowan, L. C.; Hook, S. J.; Abrams, M. J.; Mars, J. C. Mapping Hydrothermally Altered Rocks at Cuprite, Nevada, Using the Advanced Spaceborne Thermal Emission and Reflection Radiometer (ASTER), a New Satellite-Imaging System. *Econ. Geol.* **2003**, *98* (5), 1019–1027. https://doi.org/10.2113/gsecongeo.98.5.1019.

Rowan, L. C.; Schmidt, R. G.; Mars, J. C. Distribution of Hydrothermally Altered Rocks in the Reko Diq, Pakistan Mineralized Area Based on Spectral Analysis of ASTER Data. *Remote Sens. Environ.* **2006**, *104* (1), 74–87. https://doi.org/https://doi.org/10.1016/j.rse.2006.05.014.

Rybin, A.; Chibisova, M.; Webley, P.; Steensen, T.; Izbekov, P.; Neal, C.; Realmuto, V. Satellite and Ground Observations of the June 2009 Eruption of Sarychev Peak Volcano, Matua Island, Central Kuriles. *Bull. Volcanol.* **2011**, *73* (9), 1377–1392. https://doi.org/10.1007/s00445-011-0481-0.

Saepuloh, A.; Koike, K.; Sumintadireja, P.; Nugraha, A. Digital Geological Mapping Using ASTER Level-1B in Relation with Heat Source of Geothermal System in an Active Volcano. *ISTECS* **2008**, *X*, 1–11.

Saepuloh, A.; Urai, M.; Widiwijayanti, C.; Aisyah, N. Observing 2006-2010 Ground Deformations of Merapi Volcano (Indonesia) Using ALOS/PALSAR and ASTER TIR Data. In *International Geoscience and Remote Sensing Symposium (IGARSS)*; 2011; pp 1634–1637. https://doi.org/10.1109/IGARSS.2011.6049545.

Saepuloh, A.; Urai, M.; Aisyah, N.; Sunarta; Widiwijayanti, C.; Subandriyo; Jousset, P. Interpretation of Ground Surface Changes Prior to the 2010 Large Eruption of Merapi Volcano Using ALOS/PALSAR, ASTER TIR and Gas Emission Data. *J. Volcanol. Geotherm. Res.* **2013**, *261*, 130–143. https://doi.org/10.1016/j.jvolgeores.2013.05.001.

Sasai, Y.; Hapada, M.; Sabit, J. P.; Zlotnickl, J.; Tanaka, Y.; Cordon Jr., J. M.; Uyeda, S.; Nagao, T.; Sincioco, J. S. Geomagnetic and Topographic Survey of the Main Crater Lake in Taal Volcano (Philippines): Preliminary Report. *Chigaku Zasshi (Jounal Geogr.* **2008**, *117* (5), 894–900. https://doi.org/10.5026/jgeography.117.894.

Schneider, D.; Delgado Granados, H.; Huggel, C.; Kääb, A. Assessing Lahars from Ice-Capped Volcanoes Using ASTER Satellite Data, the SRTM DTM and Two Different Flow Models: Case Study on Iztaccíhuatl (Central Mexico). *Nat. Hazards Earth Syst. Sci.* **2008**, *8* (3), 559–571. https://doi.org/10.5194/nhess-8-559-2008.

Scholte, K. H.; Hommels, A.; Meer, F. D.; Kroonenberg, S. B.; Hanssen, R. G.; Aliyeva, E. Preliminary Aster and InSAR Imagery Combination for Mud Volcano Dynamics, Azerbaijan. In *International Geoscience and Remote Sensing Symposium*; 2003; Vol. 3, pp 13–16.

Sekertekin, A.; Arslan, N. Monitoring Thermal Anomaly and Radiative Heat Flux Using Thermal Infrared Satellite Imagery – A Case Study at Tuzla Geothermal Region. *Geothermics.* **2019**, *78*, 243–254. https://doi.org/10.1016/j.geothermics.2018.12.014.

Selles, A.; Deffontaines, B.; Hendrayana, H.; Violette, S. The Eastern Flank of the Merapi Volcano (Central Java, Indonesia): Architecture and Implications of Volcaniclastic Deposits. *J. Asian Earth Sci.* **2015**, *108*, 33–47. https://doi.org/10.1016/j.jseaes.2015.04.026.

Silvestri, M.; Cardellini, C.; Chiodini, G.; Buongiorno, M. F. Satellite-Derived Surface Temperature and in Situ Measurement at Solfatara of Pozzuoli (Naples, Italy). *Geochemistry, Geophys. Geosystems.* **2016**, *17* (6), 2095–2109. https://doi.org/10.1002/2015GC006195.

Silvestri, M.; Rabuffi, F.; Pisciotta, A.; Musacchio, M.; Diliberto, I. S.; Spinetti, C.; Lombardo, V.; Colini, L.; Buongiorno, M. F. Analysis of Thermal Anomalies in Volcanic Areas Using Multiscale and Multitemporal Monitoring: Vulcano Island Test Case. *Remote Sens.* **2019**, *11* (2), 2. https://doi.org/10.3390/rs11020134.

Smets, B.; D'Oreye, N.; Kervyn, F.; Kervyn, M.; Albino, F.; Arellano, S. R.; Bagalwa, M.; Balagizi, C.; Carn, S. A.; Darrah, T. H.; et al. Detailed Multidisciplinary Monitoring Reveals Pre- and Co-Eruptive Signals at Nyamulagira Volcano (North Kivu, Democratic Republic of Congo). *Bull. Volcanol.* **2014**, *76* (1), 1–35. https://doi.org/10.1007/s00445-013-0787-1.

Solikhin, A.; Thouret, J. C.; Gupta, A.; Harris, A. J. L.; Liew, S. C. Geology, Tectonics, and the 2002-2003 Eruption of the Semeru Volcano, Indonesia: Interpreted from High-Spatial Resolution Satellite Imagery. *Geomorphology* **2012**, *138* (1), 364–379. https://doi.org/10.1016/j.geomorph.2011.10.001.

Sosio, R.; Crosta, G. B.; Hungr, O. Numerical Modeling of Debris Avalanche Propagation from Collapse of Volcanic Edifices. *Landslides* **2012**, *9* (3), 315–334. https://doi.org/10.1007/s10346-011-0302-8.

Spinetti, C.; Buongiorno, M. F.; Silvestri, M.; Zoffoli, S. Mt. Etna Volcanic Plume from ASTER and HYPERION Data by ASI-SRV Modules. *Int. Geosci. Remote Sens. Symp.* **2011**, No. July, 4018–4021. https://doi.org/10.1109/IGARSS.2011.6050113.

Spinetti, C.; Barsotti, S.; Neri, A.; Buongiorno, M. F.; Doumaz, F.; Nannipieri, L. Investigation of the Complex Dynamics and Structure of the 2010 Eyjafjallajökull Volcanic Ash Cloud Using Multispectral Images and Numerical Simulations. *J. Geophys. Res. Atmos.* **2013**, *118* (10), 4729–4747. https://doi.org/10.1002/jgrd.50328.

Stebel, K.; Amigo, A.; Thomas, H.; Prata, A. J. First Estimates of Fumarolic SO2 Fluxes from Putana Volcano, Chile, Using an Ultraviolet Imaging Camera. *J. Volcanol. Geotherm. Res.* **2015**, *300*, 112–120. https://doi.org/10.1016/j.jvolgeores.2014.12.021.

Stevens, N. F.; Garbeil, H.; Mouginis-Mark, P. J. NASA EOS Terra ASTER: Volcanic Topographic Mapping and Capability. *Remote Sens. Environ.* **2004**, *90* (3), 405–414. https://doi.org/10.1016/j.rse.2004.01.012.

Suminar, W.; Saepuloh, A.; Meilano, I. Identifying Hazard Parameter to Develop Quantitative and Dynamic Hazard Map of an Active Volcano in Indonesia. *AIP Conf. Proc.* **2016**, *1730*, 1. https://doi.org/10.1063/1.4947403.

Takarada, S. The Volcanic Hazards Assessment Support System for the Online Hazard Assessment and Risk Mitigation of Quaternary Volcanoes in the World. *Front. Earth Sci.* **2017**, *5*, 14. https://doi.org/10.3389/feart.2017.00102.

Tayebi, M. H.; Tangestani, M. H.; Vincent, R. K.; Neal, D. Spectral Properties and ASTER-Based Alteration Mapping of Masahim Volcano Facies, SE Iran. *J. Volcanol. Geotherm. Res.* **2014**, *287*, 40–50. https://doi.org/10.1016/j.jvolgeores.2014.09.013.

Thomas, H. E.; Watson, I. M. Observations of Volcanic Emissions from Space: Current and Future Perspectives. *Nat. Hazards* **2010**, *54* (2), 323–354. https://doi.org/10.1007/s11069-009-9471-3.

Thompson, J. O.; Ramsey, M. S.; Hall, J. L. MMT-Cam: A New Miniature Multispectral Thermal Infrared Camera System for Capturing Dynamic Earth Processes. *IEEE Trans. Geosci. Remote Sens.* **2019**, *57* (10), 7438–7446. https://doi.org/10.1109/tgrs.2019.2913344.

Torres, R.; Mouginis-Mark, P.; Self, S.; Garbeil, H.; Kallianpur, K.; Quiambao, R. Monitoring the Evolution of the Pasig-Potrero Alluvial Fan, Pinatubo Volcano, Using a Decade of Remote Sensing Data. *J. Volcanol. Geotherm. Res.* **2004**, *138* (3–4), 371–392. https://doi.org/10.1016/j.jvolgeores.2004.08.005.

Tralli, D. M.; Blom, R. G.; Zlotnicki, V.; Donnellan, A.; Evans, D. L. Satellite Remote Sensing of Earthquake, Volcano, Flood, Landslide and Coastal Inundation Hazards. *ISPRS J. Photogramm. Remote Sens.* **2005**, *59* (4), 185–198. https://doi.org/10.1016/j.isprsjprs.2005.02.002.

Troncoso, L.; Bustillos, J.; Romero, J. E.; Guevara, A.; Carrillo, J.; Montalvo, E.; Izquierdo, T. Hydrovolcanic Ash Emission between August 14 and 24, 2015 at Cotopaxi Volcano (Ecuador): Characterization and Eruption Mechanisms. *J. Volcanol. Geotherm. Res.* **2017**, *341*, 228–241. https://doi.org/10.1016/j.jvolgeores.2017.05.032.

Trunk, L.; Bernard, A. Investigating Crater Lake Warming Using ASTER Thermal Imagery: Case Studies at Ruapehu, Poás, Kawah Ijen, and Copahué Volcanoes. *J. Volcanol. Geotherm. Res.* **2008**, *178* (2), 259–270. https://doi.org/10.1016/j.jvolgeores.2008.06.020.

Tsu, H.; Yamaguchi, Y.; Fujisada, H.; Kahle, A.; Sato, I.; Kato, M.; Watanabe, H.; Kudoh, M.; Pniel, M. ASTER Early Science Outcome and Operation Status. *Sensors, Syst. Next-Generation Satell.* **2001**, *4169* (IV), 1–8.

Ulusoy, İ. Temporal Radiative Heat Flux Estimation and Alteration Mapping of Tendürek Volcano (Eastern Turkey) Using ASTER Imagery. *J. Volcanol. Geotherm. Res.* **2016**, *327*, 40–54. https://doi.org/10.1016/j.jvolgeores.2016.06.027.

Urai, M. Observation of SO2 in Miyakejima Island by ASTER, MODIS; *Occasional Papers.Research Center for the Pacific Island, Kagoshima University* 2003; Vol. 37, 50-57.

Urai, M. Sulfur Dioxide Flux Estimation from Volcanoes Using Advanced Spaceborne Thermal Emission and Reflection Radiometer - a Case Study of Miyakejima Volcano, Japan. *J. Volcanol. Geotherm. Res.* **2004**, *134* (1–2), 1–13. https://doi.org/10.1016/j.jvolgeores.2003.11.008.

Urai, M. Volcano Observations with ASTER and ASTER Image Database for Volcanoes. In *International Geoscience and Remote Sensing Symposium (IGARSS)*; 2011; pp 3661–3663. https://doi.org/10.1109/IGARSS.2011.6050018.

Urai, M.; Pieri, D. ASTER Applications in Volcanology. In *Remote Sensing and Digital Image Processing*; Ramachandran, B., Justice, C., Abrams, M., Eds.; NASAs Earth Observing System and the Science of ASTER and MODIS. Springer, New York, Land Remote Sensing and Global Environmental Change, 2011; Vol. 11, pp 245–272. https://doi.org/10.1007/978-1-4419-6749-7_12.

Urai, M.; Fukui, K.; Yamaguchi, Y.; Pieri, D. Volcano Observation Potential and Global Volcano Monitoring Plan with ASTER. *Bull. Volcanol. Soc. Japan.* **1999**, *44* (3), 131–141. https://doi.org/10.18940/kazan.44.3_131.

Urai, M.; Kawanabe, Y.; Itoh, J.; Takada, A.; Kato, M. Ash Fall Areas Associated with the Usu 2000 Eruption Observed by ASTER. *Bull. Geol. Surv. Japan.* **2001**, *52* (4–5), 189–197. https://doi.org/10.9795/bullgsj.52.189.

Urai, M.; Geshi, N.; Staudacher, T. Size and Volume Evaluation of the Caldera Collapse on Piton de La Fournaise Volcano during the April 2007 Eruption Using ASTER Stereo Imagery. *Geophys. Res. Lett.* **2007**, *34* (22), L22318–L22318. https://doi.org/10.1029/2007GL031551.

Urai, M.; Ishizuka, Y. Advantages and Challenges of Space-Borne Remote Sensing for Volcanic Explosivity Index (VEI): The 2009 Eruption of Sarychev Peak on Matua Island, Kuril Islands, Russia. *J. Volcanol. Geotherm. Res.* **2011**, *208* (3–4), 163–168. https://doi.org/10.1016/j.jvolgeores.2011.07.010.

Vaughan, R. G.; Hook, S. J. Using Satellite Data to Characterize the Temporal Thermal Behavior of an Active Volcano: Mount St. Helens, WA. *Geophys. Res. Lett.* **2006**, *33* (20), L20303–L20303. https://doi.org/10.1029/2006GL027957.

Vaughan, R. G.; Abrams, M. J.; Hook, S. J.; Pieri, D. C. Satellite Observations of New Volcanic Island in Tonga. *Eos (Washington. DC).* **2007**, *88* (4), 37–41. https://doi.org/10.1029/2007EO040002.

Vaughan, R. G.; Kervyn, M.; Realmuto, V.; Abrams, M.; Hook, S. J. Satellite Measurements of Recent Volcanic Activity at Oldoinyo Lengai, Tanzania. *J. Volcanol. Geotherm. Res.* **2008**, *173* (3–4), 196–206. https://doi.org/10.1016/j.jvolgeores.2008.01.028.

Vaughan, R. G.; Keszthelyi, L. P.; Davies, A. G.; Schneider, D. J.; Jaworowski, C.; Heasler, H. Exploring the Limits of Identifying Sub-Pixel Thermal Features Using ASTER TIR Data. *J. Volcanol. Geotherm. Res.* **2010**, *189* (3–4), 225–237. https://doi.org/10.1016/j.jvolgeores.2009.11.010.

Vaughan, R. G.; Keszthelyi, L. P.; Lowenstern, J. B.; Jaworowski, C.; Heasler, H. Use of ASTER and MODIS Thermal Infrared Data to Quantify Heat Flow and Hydrothermal Change at Yellowstone National Park. *J. Volcanol. Geotherm. Res.* **2012a**, *233–234*, 72–89. https://doi.org/10.1016/j.jvolgeores.2012.04.022.

Vaughan, R. G.; Lowenstern, J. B.; Keszthelyi, L. P.; Jaworowski, C.; Heasler, H. Mapping Temperature and Radiant Geothermal Heat Flux Anomalies in the Yellowstone Geothermal System Using ASTER Thermal Infrared Data. *Trans. - Geotherm. Resour. Counc.* **2012b**, *36* (2), 1403–1409.

Viramonte, J.; Godoy, S.; Arnosio, M.; Becchio, R.; Poodts, M. El Campo Geotermal de La Caldera Del Cerro Blanco: Utilización de Imágenes Aster. *Buenos Aires, Asoc. Geológica Argentina* **2005**, *2*, 505–512.

Völker, D.; Kutterolf, S.; Wehrmann, H. Comparative Mass Balance of Volcanic Edifices at the Southern Volcanic Zone of the Andes between 33°S and 46°S. *J. Volcanol. Geotherm. Res.* **2011**, *205* (3–4), 114–129. https://doi.org/10.1016/j.jvolgeores.2011.03.011.

Volynets, A. O.; Edwards, B. R.; Melnikov, D.; Yakushev, A.; Griboedova, I. Monitoring of the Volcanic Rock Compositions during the 2012-2013 Fissure Eruption at Tolbachik Volcano, Kamchatka. *J. Volcanol. Geotherm. Res.* **2015**, *307*, 120–132. https://doi.org/10.1016/j.jvolgeores.2015.07.014.

Wadge, G.; Burt, L. Stress Field Control of Eruption Dynamics at a Rift Volcano: Nyamuragira, D.R.Congo. *J. Volcanol. Geotherm. Res.* **2011**, *207* (1–2), 1–15. https://doi.org/10.1016/j.jvolgeores.2011.06.012.

Wadge, G.; Saunders, S.; Itikarai, I. Pulsatory Andesite Lava Flow at Bagana Volcano. *Geochemistry, Geophys. Geosystems.* **2012**, *13* (11), 11. https://doi.org/10.1029/2012GC004336.

Wadge, G.; McCormick Kilbride, B. T.; Edmonds, M.; Johnson, R. W. Persistent Growth of a Young Andesite Lava Cone: Bagana Volcano, Papua New Guinea. *J. Volcanol. Geotherm. Res.* **2018**, *356*, 304–315. https://doi.org/10.1016/j.jvolgeores.2018.03.012.

Walter, T. R.; Subandriyo, J.; Kirbani, S.; Bathke, H.; Suryanto, W.; Aisyah, N.; Darmawan, H.; Jousset, P.; Luehr, B. G.; Dahm, T. Volcano-Tectonic Control of Merapi's Lava Dome Splitting: The November 2013 Fracture Observed from High Resolution TerraSAR-X Data. *Tectonophysics.* **2015**, *639*, 23–33. https://doi.org/10.1016/j.tecto.2014.11.007.

Wantim, M. N.; Suh, C. E.; Ernst, G. G. J.; Kervyn, M.; Jacobs, P. Characteristics of the 2000 Fissure Eruption and Lava Flow Fields at Mount Cameroon Volcano, West Africa: A Combined Field Mapping and Remote Sensing Approach. *Geol. J.* **2011**, *46* (4), 344–363. https://doi.org/10.1002/gj.1277.

Wantim, M. N.; Kervyn, M.; Ernst, G. G. J.; del Marmol, M. A.; Suh, C. E.; Jacobs, P. Numerical Experiments on the Dynamics of Channelised Lava Flows at Mount Cameroon Volcano with the FLOWGO Thermo-Rheological Model. *J. Volcanol. Geotherm. Res.* **2013**, *253*, 35–53. https://doi.org/10.1016/j.jvolgeores.2012.12.003.

Watanabe, H.; Matsuo, K. Rock Type Classification by Multi-Band TIR of ASTER. *Geosci. J.* **2003**, *7* (4), 347–358. https://doi.org/10.1007/bf02919567.

Watt, S. F. L.; Pyle, D. M.; Mather, T. A. Evidence of Mid- to Late-Holocene Explosive Rhyolitic Eruptions from Chaitén Volcano, Chile. *Andean Geol.* **2013**, *40* (2), 216–226. https://doi.org/10.5027/andgeov40n2-a02.

Wessels, R. L.; Coombs, M. L.; Schneider, D. J.; Dehn, J.; Ramsey, M. S. High-Resolution Satellite and Airborne Thermal Infrared Imaging of the 2006 Eruption of Augustine Volcano. *US Geol. Surv. Prof. Pap.* **2010**, *1769* (1769), 527–552. https://doi.org/10.3133/pp176922.

Wessels, R. L.; Vaughan, R. G.; Patrick, M. R.; Coombs, M. L. High-Resolution Satellite and Airborne Thermal Infrared Imaging of Precursory Unrest and 2009 Eruption at Redoubt Volcano, Alaska. *J. Volcanol. Geotherm. Res.* **2013**, *259*, 248–269. https://doi.org/10.1016/j.jvolgeores.2012.04.014.

West, M. E. Recent Eruptions at Bezymianny Volcano-A Seismological Comparison. *J. Volcanol. Geotherm. Res.* **2013**, *263*, 42–57. https://doi.org/10.1016/j.jvolgeores.2012.12.015.

Whelley, P. L.; Newhall, C. G.; Bradley, K. E. The Frequency of Explosive Volcanic Eruptions in Southeast Asia. *Bull. Volcanol.* **2015**, *77* (1), 1. https://doi.org/10.1007/s00445-014-0893-8.

Williams, D. B.; Ramsey, M. S. On the Applicability of Laboratory Thermal Infrared Emissivity Spectra for Deconvolving Satellite Data of Opaque Volcanic Ash Plumes. *Remote Sens.* **2019**, *11* (19), 2318. https://doi.org/10.3390/rs11192318.

Williams, D. B.; Ramsey, M. S.; Wickens, D. J.; Karimi, B. Identifying Eruptive Sources of Drifting Volcanic Ash Clouds Using Back-Trajectory Modeling of Spaceborne Thermal Infrared Data. *Bull. Volcanol.* **2019**, *81* (9), 53. https://doi.org/10.1007/s00445-019-1312-y.

Worden, A.; Dehn, J.; Ripepe, M.; Donne, D. D. Frequency Based Detection and Monitoring of Small Scale Explosive Activity by Comparing Satellite and Ground Based Infrared Observations at Stromboli Volcano, Italy. *J. Volcanol. Geotherm. Res.* **2014**, *283*, 159–171. https://doi.org/10.1016/j.jvolgeores.2014.07.007.

Worni, R.; Huggel, C.; Stoffel, M.; Pulgarín, B. Challenges of Modeling Current Very Large Lahars at Nevado Del Huila Volcano, Colombia. *Bull. Volcanol.* **2012**, *74* (2), 309–324. https://doi.org/10.1007/s00445-011-0522-8.

Wright, R.; Rothery, D. A.; Blake, S.; Harris, A. J. L.; Pieri, D. C. Simulating the Response of the EOS Terra ASTER Sensor to High-Temperature Volcanic Targets. *Geophys. Res. Lett.* **1999**, *26* (12), 1773–1776. https://doi.org/10.1029/1999GL900360.

Wright, R.; Rothery, D. A.; Blake, S.; Pieri, D. C. Improved Remote Sensing Estimates of Lava Flow Cooling: A Case Study of the 1991-1993 Mount Etna Eruption. *J. Geophys. Res. Solid Earth.* **2000**, *105* (B10), 23681–23694. https://doi.org/10.1029/2000jb900225.

Wright, R.; Carn, S. A.; Flynn, L. P. A Satellite Chronology of the May-June 2003 Eruption of Anatahan Volcano. *J. Volcanol. Geotherm. Res.* **2005**, *146* (1-3 SPEC. ISS.), 102–116. https://doi.org/10.1016/j.jvolgeores.2004.10.021.

Wright, R.; Garbeil, H.; Davies, A. G. Cooling Rate of Some Active Lavas Determined Using an Orbital Imaging Spectrometer. *J. Geophys. Res. Solid Earth.* **2010**, *115* (6). https://doi.org/10.1029/2009JB006536.

Xi, X.; Johnson, M. S.; Jeong, S.; Fladeland, M.; Pieri, D.; Diaz, J. A.; Bland, G. L. Constraining the Sulfur Dioxide Degassing Flux from Turrialba Volcano, Costa Rica Using Unmanned Aerial System Measurements. *J. Volcanol. Geotherm. Res.* **2016**, *325*, 110–118. https://doi.org/10.1016/j.jvolgeores.2016.06.023.

Yamaguchi, Y.; Kahle, A. B.; Tsu, H.; Kawakami, T.; Pniel, M. Overview of Advanced Spaceborne Thermal Emission and Reflection Radiometer (ASTER). *IEEE Trans. Geosci. Remote Sens.* **1998**, *36* (4), 1062–1071. https://doi.org/10.1109/36.700991.

Yulianto, F.; Suwarsono; Sofan, P. The Utilization of Remotely Sensed Data to Analyze the Estimated Volume of Pyroclastic Deposits and Morphological Changes Caused by the 2010–2015 Eruption of Sinabung Volcano, North Sumatra, Indonesia. *Pure Appl. Geophys.* **2016**, *173* (8), 2711–2725. https://doi.org/10.1007/s00024-016-1342-8.

Zlotnicki, J.; Sasai, Y.; Toutain, J. P.; Villacorte, E. U.; Bernard, A.; Sabit, J. P.; Gordon, J. M.; Corpuz, E. G.; Harada, M.; Punongbayan, J. T.; et al. Combined Electromagnetic, Geochemical and Thermal Surveys of Taal Volcano (Philippines) during the Period 2005-2006. *Bull. Volcanol.* **2009**, *71* (1), 29–47. https://doi.org/10.1007/s00445-008-0205-2.

Zouzias, D.; Miliaresis, G. C.; Seymour, K. S. Interpretation of Nisyros Volcanic Terrain Using Land Surface Parameters Generated from the ASTER Global Digital Elevation Model. *J. Volcanol. Geotherm. Res.* **2011**, *200* (3–4), 159–170. https://doi.org/10.1016/j.jvolgeores.2010.12.012.

References

1. Kahle, A.B.; Palluconi, F.D.; Hook, S.J.; Realmuto, V.J.; Bothwell, G. The Advanced Spaceborne Thermal Emission and Reflectance Radiometer (Aster). *Int. J. Imaging Syst. Technol.* **1991**, *3*, 144–156. [CrossRef]

2. Ramsey, M.S. Synergistic Use of Satellite Thermal Detection and Science: A Decadal Perspective Using ASTER. *Geol. Soc. Spec. Publ.* **2016**, *426*, 115–136. [CrossRef]

3. Pieri, D.; Abrams, M. ASTER Watches the World's Volcanoes: A New Paradigm for Volcanological Observations from Orbit. *J. Volcanol. Geotherm. Res.* **2004**, *135*, 13–28. [CrossRef]

4. Urai, M. Sulfur Dioxide Flux Estimation from Volcanoes Using Advanced Spaceborne Thermal Emission and Reflection Radiometer—A Case Study of Miyakejima Volcano, Japan. *J. Volcanol. Geotherm. Res.* **2004**, *134*, 1–13. [CrossRef]

5. Carter, A.; Ramsey, M. Long-Term Volcanic Activity at Shiveluch Volcano: Nine Years of ASTER Spaceborne Thermal Infrared Observations. *Remote Sens.* **2010**, *2*, 2571–2583. [CrossRef]

6. Ramsey, M.S.; Wessels, R.L.; Anderson, S.W. Surface Textures and Dynamics of the 2005 Lava Dome at Shiveluch Volcano, Kamchatka. *Bull. Geol. Soc. Am.* **2012**, *124*, 678–689. [CrossRef]

7. Yamaguchi, Y.; Kahle, A.B.; Tsu, H.; Kawakami, T.; Pniel, M. Overview of Advanced Spaceborne Thermal Emission and Reflection Radiometer (ASTER). *IEEE Trans. Geosci. Remote Sens.* **1998**, *36*, 1062–1071. [CrossRef]

8. Gillespie, A.; Rokugawa, S.; Matsunaga, T.; Cothern, J.S.; Hook, S.; Kahle, A.B. A temperature and emissivity separation algorithm for Advanced Spaceborne Thermal Emission and Reflection Radiometer (ASTER) images. *IEEE Trans. Geosci. Remote Sens.* **1998**, *36*, 1113–1126. [CrossRef]

9. Hirano, A.; Welch, R.; Lang, H. Mapping from ASTER Stereo Image Data: DEM Validation and Accuracy Assessment. *ISPRS J. Photogramm. Remote Sens.* **2003**, *57*, 356–370. [CrossRef]

10. ASTER Validation Team (AVT). *ASTER Global DEM Validation Summary Report*; report no. 28; The Ministry of Economy, Trade and Industry (METI) and the National Aeronautics and Space Administration (NASA): Washington, DC, USA, 2009.

11. Tachikawa, T.; Hato, M.; Kaku, M.; Iwasaki, A. Characteristics of ASTER GDEM version 2. In Proceedings of the 2011 IEEE International Geoscience and Remote Sensing Symposium, Vancouver, BC, Canada, 24–29 July 2011; pp. 3657–3660.

12. National Research Council (NRC). *Thriving on Our Changing Planet: A Decadal Strategy for Earth Observation from Space*; The Natl. Acad. Press: Washington, DC, USA, 2018; 560p, ISBN 978-0-309-46757-5.

13. Gawarecki, S.J.; Lyon, R.J.P.; Nordberg, W. Infrared spectral returns and imagery of the Earth from space and their application to geological problems: Scientific experiments for manned orbital flight. *Am. Astronaut. Soc. Sci. Technol.* **1965**, *4*, 13–133.

14. Williams, R.S., Jr.; Friedman, J.D. Satellite Observation of Effusive Volcanism. *Br. Interplanet. Soc. J.* **1970**, *23*, 441–450.

15. Scorer, R.S. Etna: The Eruption of Christmas 1985 As Seen By Meteorological Satellite. *Weather* **1986**, *41*, 378–384. [CrossRef]

16. Harris, A. *Thermal Remote Sensing of Active Volcanoes: A User's Manual*; Cambridge University Press: Cambridge, UK, 2013.

17. Ramsey, M.S.; Harris, A.J.L. Volcanology 2020: How Will Thermal Remote Sensing of Volcanic Surface Activity Evolve over the next Decade? *J. Volcanol. Geotherm. Res.* **2013**, *249*, 217–233. [CrossRef]

18. Harris, A.J.L.; Butterworth, A.L.; Carlton, R.W.; Downey, I.; Miller, P.; Navarro, P.; Rothery, D.A. Low-Cost Volcano Surveillance from Space: Case Studies from Etna, Krafla, Cerro Negro, Fogo, Lascar and Erebus. *Bull. Volcanol.* **1997**, *59*, 49–64. [CrossRef]

19. Wright, R.; Flynn, L.; Garbeil, H.; Harris, A.; Pilger, E. Automated Volcanic Eruption Detection Using MODIS. *Remote Sens. Environ.* **2002**, *82*, 135–155. [CrossRef]

20. Hirn, B.R.; Di Bartola, C.; Laneve, G.; Cadau, E.; Ferrucci, F. SEVIRI onboard Meteosat Second Generation, and the quantitative monitoring of effusive volcanoes in Europe and Africa. In Proceedings of the IEEE International Geoscience and Remote Sensing Symposium (IGARSS 2008), Boston, MA, USA, 7–11 July 2008; Inst. of Electr. and Electron. Eng.: New York, NY, USA, 2008; pp. 374–377.

21. Ramsey, M.S.; Chevrel, M.O.; Coppola, D.; Harris, A.J.L. The Influence of Emissivity on the Thermo-Rheologic Al Modeling of the Channelized Lava Flows at Tolbachik Volcano. *Ann. Geophys.* **2019**, *61*. [CrossRef]

22. Reath, K.A.; Ramsey, M.S.; Dehn, J.; Webley, P.W. Predicting Eruptions from Precursory Activity Using Remote Sensing Data Hybridization. *J. Volcanol. Geotherm. Res.* **2016**, *321*, 18–30. [CrossRef]

23. National Research Council (NRC). *Earth Science and Applications from Space: National Imperatives for the Next Decade and Beyond*; The Natl. Acad. Press: Washington, DC, USA, 2007; 454p, ISBN 978-0-309-10387-9.

24. Raup, B.; Racoviteanu, A.; Khalsa, S.J.S.; Helm, C.; Armstrong, R.; Arnaud, Y. The GLIMS Geospatial Glacier Database: A New Tool for Studying Glacier Change. *Glob. Planet. Chang.* **2007**, *56*, 101–110. [CrossRef]

25. Ramsey, M.S. Mapping the City Landscape from Space: The Advanced Spaceborne Thermal Emission and Reflectance Radiometer (ASTER) Urban Environmental Monitoring Program. *Earth Sci. City* **2003**, *1*, 337–361. [CrossRef]

26. Urai, M.; Fukui, K.; Yamaguchi, Y.; Pieri, D. Volcano Observation Potential and Global Volcano Monitoring Plan with ASTER. *Bull. Volcanol. Soc. Japan* **1999**, *44*, 131–141. [CrossRef]

27. Urai, M.; Pieri, D. ASTER Applications in Volcanology. In *Remote Sensing and Digital Image Processing*; Ramachandran, B., Justice, C., Abrams, M., Eds.; NASAs Earth Observing System and the Science of ASTER and MODIS; Land Remote Sensing and Global Environmental Change; Springer: New York, NY, USA, 2011; Volume 11, pp. 245–272. [CrossRef]

28. Abrams, M.; Tsu, H.; Hulley, G.; Iwao, K.; Pieri, D.; Cudahy, T.; Kargel, J. The Advanced Spaceborne Thermal Emission and Reflection Radiometer (ASTER) after Fifteen Years: Review of Global Products. *Int. J. Appl. Earth Obs. Geoinf.* **2015**, *38*, 292–301. [CrossRef]

29. Duba, K.; Ramsey, M.; Wessels, R.; Dehn, J. Optical Satellite Volcano Monitoring: A Multi-Sensor Rapid Response System. *Geosci. Remote Sens.* **2009**, 473–496. [CrossRef]

30. Davies, A.G.; Chien, S.; Wright, R.; Miklius, A.; Kyle, P.R.; Welsh, M.; Johnson, J.B.; Tran, D.; Schaffer, S.R.; Sherwood, R. Sensor Web Enables Rapid Response to Volcanic Activity. *Eos (Wash. DC)* **2006**, *87*, 2–4. [CrossRef]

31. Wright, R.; Flynn, L.P.; Garbeil, H.; Harris, A.J.L.; Pilger, E. MODVOLC: Near-Real-Time Thermal Monitoring of Global Volcanism. *J. Volcanol. Geotherm. Res.* **2004**, *135*, 29–49. [CrossRef]

32. Coppola, D.; Laiolo, M.; Cigolini, C.; Donne, D.D.; Ripepe, M. Enhanced Volcanic Hot-Spot Detection Using MODIS IR Data: Results from the MIROVA System. *Geol. Soc. Lond. Spec. Publ.* **2016**, *426*, 181–205. [CrossRef]

33. Dehn, J.; Dean, K.; Engle, K. Thermal Monitoring of North Pacific Volcanoes from Space. *Geology* **2000**, *28*, 755–758. [CrossRef]

34. Urai, M. Volcano Observations with ASTER and ASTER Image Database for Volcanoes. *Int. Geosci. Remote Sens. Symp.* **2011**, *2011*, 3661–3663. [CrossRef]

35. Amici, S.; Piscini, A.; Buongiorno, M.F.; Pieri, D. Geological Classification of Volcano Teide by Hyperspectral and Multispectral Satellite Data. *Int. J. Remote Sens.* **2013**, *34*, 3356–3375. [CrossRef]

36. Ramsey, M.S.; Flynn, L.P. Strategies, Insights, and the Recent Advances in Volcanic Monitoring and Mapping with Data from NASA's Earth Observing System. *J. Volcanol. Geotherm. Res.* **2004**, *135*, 1–11. [CrossRef]

37. Scheidt, S.; Lancaster, N.; Ramsey, M.S. Eolian Dynamics and Sediment Mixing in the Gran Desierto, Mexico, Determined from Thermal Infrared Spectroscopy and Remote-Sensing Data. *Bull. Geol. Soc. Am.* **2011**, *123*, 1628–1644. [CrossRef]

38. Rose, S.R.; Watson, I.M.; Ramsey, M.S.; Hughes, C.G. Thermal Deconvolution: Accurate Retrieval of Multispectral Infrared Emissivity from Thermally-Mixed Volcanic Surfaces. *Remote Sens. Environ.* **2014**, *140*, 690–703. [CrossRef]

39. Flynn, I.T.W.; Ramsey, M.S. Uncertainties in Pyroclastic Density Current Hazard Assessments of Fuego Volcano, Guatemala. *Remote Sens.* **2020**, in review.

40. Henney, L.A.; Rodríguez, L.A.; Watson, I.M. A Comparison of SO 2 Retrieval Techniques Using Mini-UV Spectrometers and ASTER Imagery at Lascar Volcano, Chile. *Bull. Volcanol.* **2012**, *74*, 589–594. [CrossRef]

41. Realmuto, V.J. The potential use of the earth observing system data to monitor the passive emissions of sulfur dioxide from volcanoes. In *Remote Sensing of Active Volcanism, Geophysical Monograph 116*; AGU: Washington, DC, USA, 2000; pp. 101–115.

42. Patrick, M.R.; Witzke, C.-N. Thermal Mapping of Hawaiian Volcanoes with ASTER Satellite Data. *US Geol. Surv. Sci. Investig. Rep.* **2011**, *5110*, 22.

43. Rose, S.R.; Ramsey, M.S. The 2005 and 2007 Eruptions of Klyuchevskoy Volcano, Russia: Behavior and Effusion Mechanisms. In *Monitoring Volcanoes in the North Pacific: Observations from Space*; Dean, K.G., Dehn, J., Dehn, J., Eds.; -540-24125-6, 389 pp: DVD; Springer-Praxis Books: New York, NY, USA, 2015; pp. 973–978.

44. Harris, A.; Chevrel, M.; Coppola, D.; Ramsey, M.; Hrysiewicz, A.; Thivet, S.; Villeneuve, N.; Favalli, M.; Peltier, A.; Kowalski, P.; et al. Validation of an Integrated Satellite-Data-Driven Response to an Effusive Crisis: The April–May 2018 Eruption of Piton de La Fournaise. *Ann. Geophys. Geophys.* **2019**, *61*. [CrossRef]

45. Stevens, N.F.; Garbeil, H.; Mouginis-Mark, P.J. NASA EOS Terra ASTER: Volcanic Topographic Mapping and Capability. *Remote Sens. Environ.* **2004**, *90*, 405–414. [CrossRef]

46. Huggel, C.; Schneider, D.; Miranda, P.J.; Delgado Granados, H.; Kääb, A. Evaluation of ASTER and SRTM DEM Data for Lahar Modeling: A Case Study on Lahars from Popocatépetl Volcano, Mexico. *J. Volcanol. Geotherm. Res.* **2008**, *170*, 99–110. [CrossRef]

47. Ramsey, M.; Dehn, J. Spaceborne Observations of the 2000 Bezymianny, Kamchatka Eruption: The Integration of High-Resolution ASTER Data into near Real-Time Monitoring Using AVHRR. *J. Volcanol. Geotherm. Res.* **2004**, *135*, 127–146. [CrossRef]

48. Tralli, D.M.; Blom, R.G.; Zlotnicki, V.; Donnellan, A.; Evans, D.L. Satellite Remote Sensing of Earthquake, Volcano, Flood, Landslide and Coastal Inundation Hazards. *ISPRS J. Photogramm. Remote Sens.* **2005**, *59*, 185–198. [CrossRef]

Article

Integration and Visualization of Mineralogical and Topographical Information Derived from ASTER and DEM Data

Kana Kurata [1,2,*] **and Yasushi Yamaguchi** [1]

[1] Graduate School of Environmental Studies, Nagoya University, D2-1 (510) Furo-cho, Chikusa-ku, Nagoya 464-8601, Japan; yasushi@nagoya-u.jp
[2] NTT Media Intelligence Laboratories, Nippon Telegraph and Telephone Corporation, 1-1 Hikarinooka, Yokosuka 239-0847, Japan
* Correspondence: kana.kurata.cb@hco.ntt.co.jp

Received: 7 December 2018; Accepted: 14 January 2019; Published: 16 January 2019

Abstract: This paper proposes a method of combining and visualizing multiple lithological indices derived from Advanced Spaceborne Thermal Emission and Reflection Radiometer (ASTER) data and topographical information derived from digital elevation model (DEM) data in a single color image that can be easily interpreted from a geological point of view. For the purposes of mapping silicate rocks, carbonate rocks, and clay minerals in hydrothermal alteration zones, two new indices derived from ASTER thermal infrared emissivity data were developed to identify silicate rocks, and existing indices were adopted to indicate the distribution of carbonate rocks and the species and amounts of clay mineral. In addition, another new method was developed to visualize the topography from DEM data. The lithological indices and topographical information were integrated using the hue–saturation–value (HSV) color model. The resultant integrated image was evaluated by field survey and through comparison with the results of previous studies in the Cuprite and Goldfield areas, Nevada, USA. It was confirmed that the proposed method can be used to visualize geological information and that the resulting images can easily be interpreted from a geological point of view.

Keywords: ASTER; DEM; lithological mapping; thermal infrared

1. Introduction

In geological remote sensing, surface materials such as rocks and minerals are characterized and/or discriminated based on their spectral features, which are captured by a multi- or hyperspectral sensor mounted on an aircraft or spacecraft (e.g., [1–4]). The Advanced Spaceborne Thermal Emission and Reflection Radiometer (ASTER) is a multispectral imaging sensor on NASA's Terra spacecraft and has already imaged almost all land areas worldwide except for those in the polar regions. ASTER has three optical subsystems: the visible and near-infrared (VNIR), shortwave-infrared (SWIR), and thermal infrared (TIR) radiometers, as described in Table 1 [5]. The VNIR and SWIR bands were designed to capture the diagnostic absorption features of clay, carbonate, and iron oxide minerals, as well as vegetation, whereas the TIR band was designed to measure the surface temperature and detect the emissivity patterns of silicate rocks.

Several methods have been developed and applied to express or quantify the diagnostic spectral features of rocks and minerals using ASTER data. The band ratio technique (e.g., [6]) has been widely used to enhance spectral patterns because of its simple calculation. It can suppress the effects of topography, but can examine only one absorption feature with one band ratio. The spectral angle mapper (SAM) [7], spectral indices [8], and conventional classification methods such as supervised classification (e.g., [9]) are also popularly used to map the distributions of different rocks and minerals.

However, these methods cannot express gradual changes and/or mixtures of multiple constituents. Moreover, when a color composite image is generated by assigning each index a color primitive of the conventional red–green–blue (RGB) color model, there is a limitation in that the RGB color composite can deal with only up to three indices at a time. Additionally, it is necessary to interpret the meanings of the colors on a case-by-case basis, and this is one of the major difficulties facing general geologists when interpreting color images generated by remote sensing techniques. Moreover, the spectral features of rocks and minerals are generally analyzed and expressed separately in each spectral region, and it is thus difficult to integrate results derived from spectral bands of different wavelength regions into one image. One reason for this is that the spatial resolutions of spectral bands in different wavelength regions are different; in the case of ASTER, the resolutions are 15, 30, and 90 m for the VNIR, SWIR, and TIR bands, respectively.

Table 1. Spectral ranges and spatial resolutions of the Advanced Spaceborne Thermal Emission and Reflection Radiometer (ASTER) spectral bands.

Subsystem.	Band No.	Spectral Range	Spatial Resolution
VNIR	1	0.52–0.60	15 m
	2	0.63–0.69	
	3N	0.78–0.86	
	3B	0.78–0.86	
SWIR	4	1.600–1.700	30 m
	5	2.145–2.185	
	6	2.185–2.225	
	7	2.235–2.285	
	8	2.295–2.365	
	9	2.360–2.430	
TIR	10	8.125–8.475	90 m
	11	8.475–8.825	
	12	8.925–9.275	
	13	10.25–10.95	
	14	10.95–11.65	

Another problem is that topographical information is not available in many of the resultant color images generated by the methods described above, because these methods tend to suppress the effects of topography, which often hampers the discrimination of rocks and minerals. However, the topography often serves as an important indicator of geology and is also important in the identification of the locations of target features (e.g., [10,11]). There have been very few methods combining lithological indices and topography, necessitating the development of an effective method that can combine them. As ASTER has along-track stereo capabilities, it is always possible to obtain the digital elevation model (DEM) whenever the surface image is captured by the other spectral bands. Moreover, high-spatial-resolution global DEM datasets have recently become available (e.g., [12,13]), and it is beneficial for geologists to integrate spectral information with topographical information derived from DEM data. LiDAR is another important source of topographical information, and surface roughness derived from LiDAR can be used to map geological features (e.g., [14]).

Kurata and Yamaguchi [15] proposed the use of the HSV color model [16] to combine the spectral indices derived from the ASTER data in different wavelength regions and generate a single integrated color image. In their method, spectral indices indicating the mineral species and amount are allocated to the hue (H) and saturation (S) elements, respectively, whereas topographical information is included in the image as the value (V) element. In this particular case, ASTER SWIR band 4 with 30 m spatial resolution was pansharpened using the VNIR bands with 15 m spatial resolution and was assigned to the V element by assuming that the pansharpened image represented topography, because most of rocks and minerals except gypsum have no absorptions in band 4 and generally have the highest reflectance in band 4 if the surface is dry. However, the pansharpened image was often affected by the albedo of surface materials and could not clearly indicate topography. Moreover, silicate rocks could not be readily distinguished in the resultant color image.

In the present study, the concept proposed by Kurata and Yamaguchi [15] was used to employ the HSV color model to integrate the spectral indices in the different wavelength regions. Clay, carbonate, and silicate minerals and rocks were selected for discrimination because they are important for mineralogical mapping and mineral exploration and they have characteristic spectral features in the SWIR and TIR regions. Iron oxide minerals are also important for mineral exploration. Distribution of iron oxide minerals is independent of distribution of clay and carbonate minerals, but they often overlap each other. As a result, it would be difficult to display all these minerals simultaneously in a single image, and thus we did not include iron oxide minerals in this study. New spectral indices to distinguish silicate rocks are proposed in this paper and are allocated to the H and S elements in a different manner from that proposed by Kurata and Yamaguchi [15]. In addition, the colors allocated to the clay, carbonate, and silicate minerals are also different from those of Kurata and Yamaguchi [15]. Moreover, the topographical information to be allocated to the V element was derived from a DEM, not the pansharpening of the SWIR band data. Thus, the proposed method is a substantial improvement on that developed by Kurata and Yamaguchi [15]. The goal of the present work was to develop an effective technique that enables the combination and visualization of multiple lithological indices derived from ASTER data and topographical information derived from DEM data in a single-color image that can be easily interpreted from a geological point of view.

2. Methods and Materials

2.1. Fundamental Concept

In this study, a method of integrating and visualizing lithological and topographical information derived from ASTER and DEM data in a single-color image was developed. The HSV color model [16] was employed to generate a color image by allocating lithological and topographical indices to the three HSV color model elements; the HSV model was used as an alternative to the RGB color model, which has been conventionally used to produce color composite images in remote sensing. The HSV color model uses the H, S, and V elements to express colors, whereas the RGB model uses the intensities of the three-color primitives red (R), green (G), and blue (B). The values of the HSV elements and RGB primitives used to express a particular color are mutually convertible. The H element represents the type of color and is typically expressed as an angle between $0°$ and $360°$; for example, the values 0, 120, and 240 correspond to red, green, and blue, respectively. The S and V elements have values ranging from 0 to 1 that determine the vividness and brightness of the color, respectively.

For the purpose of mapping the distribution and gradual change of clay minerals in hydrothermal alteration zones, the orthogonal transformation and band ratios were used for the SWIR surface reflectance product (AST_07XT), and the species and amounts of clay minerals were allocated to the H and S elements, respectively. Additionally, two new spectral indices were proposed to map silicate rocks using the TIR bands, and an existing spectral index [17] was employed to map carbonate rocks using the TIR bands as well. The silica content and its mineral form (crystal or amorphous) were allocated to the H and S elements, respectively, for each target pixel corresponding to a silicate rock. Topographical information derived from DEM data was allocated to the V element.

2.2. Data and Test Sites

2.2.1. Data

The ASTER standard data products of TIR surface emissivity (AST_05) and SWIR surface reflectance (AST_07XT) were used for lithological mapping in this study. The surface emissivity product was generated using the emissivity temperature separation algorithm developed by Gillespie et al. [18].

It is important to mention that data selection was very important, particularly for the TIR region. In this study, only summer night data were considered, as they have a relatively small dependence on slope orientations and high signal-to-noise (S/N) ratios and thus, a high accuracy of separation between kinetic temperature and spectral emissivity.

The acquisition dates for the ASTER data used in this study are as follows: SWIR surface reflectance of Cuprite: 2007.04.28; TIR surface emissivity of Cuprite: 2005.07.03; Sierra Nevada including SN_obs: 2011.08.03; Sierra Nevada including SN_gr: 2016.08.16; Navajo Sandstone: 2017.06.29; Kilauea: 2017.02.21; Sahara: 2009.06.12; and Great Dyke: 2005.11.13.

The DEM dataset of the 3D Elevation Program (3DEP) were also used to visualize the topography. 3DEP has been operated by the United States Geological Survey (USGS) and provides seamless DEM datasets with a spatial resolution of 1/3″ (in the 48 conterminous states, Hawaii, territories of the United States, and part of Alaska), 1″ (in the conterminous states and part of Alaska), or 2″ (only in Alaska). The 3DEP datasets are available for free online from the website of USGS.

2.2.2. Test Sites

The Cuprite and Goldfield areas in Nevada, USA, were selected as test sites in this study because a variety of rocks and minerals including silicate rocks, carbonate rocks, and hydrothermal alteration zones are widely distributed in these areas. In addition, these areas are easily accessible and have been well studied as famous geological remote sensing test sites and reference area for remote sensing sensors and application, allowing the present results to be compared with those of previous studies (e.g., [19–22]).

2.3. Lithological Indices

2.3.1. TIR Spectral Indices for Mapping Silicate Rocks

The emissivity patterns for rocks show a systematic relation to reflect the silica content and the types of mineral present [23]. Figure 1a shows the spectral shapes derived from ASTER TIR emissivity data at 10 locations around the world with distributions of different types of silicate rocks (Table 2). Several dozen pixels were sampled at each location and were averaged to represent the typical emissivity spectrum of each rock type. On the basis of the results in Figure 1a, new indices are proposed for silicate rocks: The T-depth, which represents the silica content, and the T-angle, which represents the mineral form. The T-depth represents a decrease in the emissivity in bands 10, 11, and 12, relative to those in bands 13 and 14, and it is defined as:

$$\text{T-depth} = \frac{B13 + B14}{2} - \frac{B10 + B11 + B12}{3} \tag{1}$$

here B10, B11, B12, B13, and B14 are the emissivities of ASTER TIR bands 10, 11, 12, 13, and 14, respectively.

The emissivities of felsic rocks with a higher silica content in bands 10, 11, and 12 are lower than those of mafic rocks (Figure 1a). Thus, felsic rocks have larger T-depths, and mafic rocks have smaller T-depths. A larger T-depth corresponds to a high silica content of the target rock. For example, the emissivity average of bands 10–12 was lowest for the quartzose sands in the Sahara Desert, followed by the Navajo Sandstone in Utah with almost 100% quartz grains; the basalt lava in Kilauea Volcano, Hawaii, with a thin silica coating [24]; and the quartzite in Cuprite, Nevada. The T-depth values of these targets were high, as shown in Table 3.

Table 2. Characteristics of the sampling areas for thermal infrared (TIR) emissivity spectra.

Sampling Area and Approximate Location	SiO$_2$ (%)	Lithology	Minerals	Texture
Felsic Samples				
1. Cup_qz (37°37′35″ N/117°16′58″ W) Quartzite in the northwest of Cuprite, Nevada, USA	≒100	Quartzite	Quartz	Well-sorted quartz grains
2. Navajo (37°42′05″ N/111°22′35″ W) Navajo Sandstone in Utah, USA	≒100	Quartzose sandstone	Quartz	Well-sorted quartz grains
3. Sahara (25°47′14″ N/25°21′07″ E) Sand dunes in western New Valley, Egypt	≒100	Quartzose sediments	Quartz	Unconsolidated Well-sorted quartz grains
4. SN_obs (37°54′40″ N, 119°01′20″ W) Obsidian on the hills in south side of Mono Lake, California, USA	≒100	Obsidian	Glass	Amorphous
5. Kilauea (19°16′55″ N, 155°21′08″ W) Basalt lava with thin silica coating in Kilauea Volcano, Hawaii, USA	?	Amorphous silica on pahoehoe lava	Glass	Amorphous
6. SN_gr1 (37°52′47″ N, 119°20′58″ W) Granite of Lembert dome in Yosemite, California, USA	66~70	Granite	Quartz Feldspar	Crystalline
7. SN_gr2 (37°28′ 48″ N, 119°29′ 09″ W) Granite near Olmsted Point in Yosemite, California, USA	66~70	Granite	Quart Feldspar Biotite	Crystalline
8. SN_gr3 (37°50′43″ N, 119°26′40″ W) Granite in the North of Tenaya Lake in Yosemite, California, USA	66~70	Granite	Quartz Feldspar Biotite	Crystalline
Mafic Samples				
9. Dyke_gb (20°19′07″ S, 29°48′27″ E) Gabbro in southern in the southern Great Dyke, Midlands, Zimbabwe	45~52	Gabbro	Amphibole Pyroxene Plagioclase	Crystalline
10. Dyke_sp (20°26′23″ S, 29°44′35″ E) Serpentinite in the southern Great Dyke, Midlands, Zimbabwe	32~43	Serpentinite	Serpentine	Crystalline

Table 3. Average T-depth values in each sampling area.

Sampling Area.	Average of T-Depth
Felsic Samples	
Cup_qz	10.058
Navajo	12.464
Sahara	31.009
SN_obs	4.804
Kilauea	10.388
SN_gr1	4.944
SN_gr2	5.440
SN_gr3	4.542
Mafic Samples	
Dyke_gb	2.394
Dyke_sp	1.387

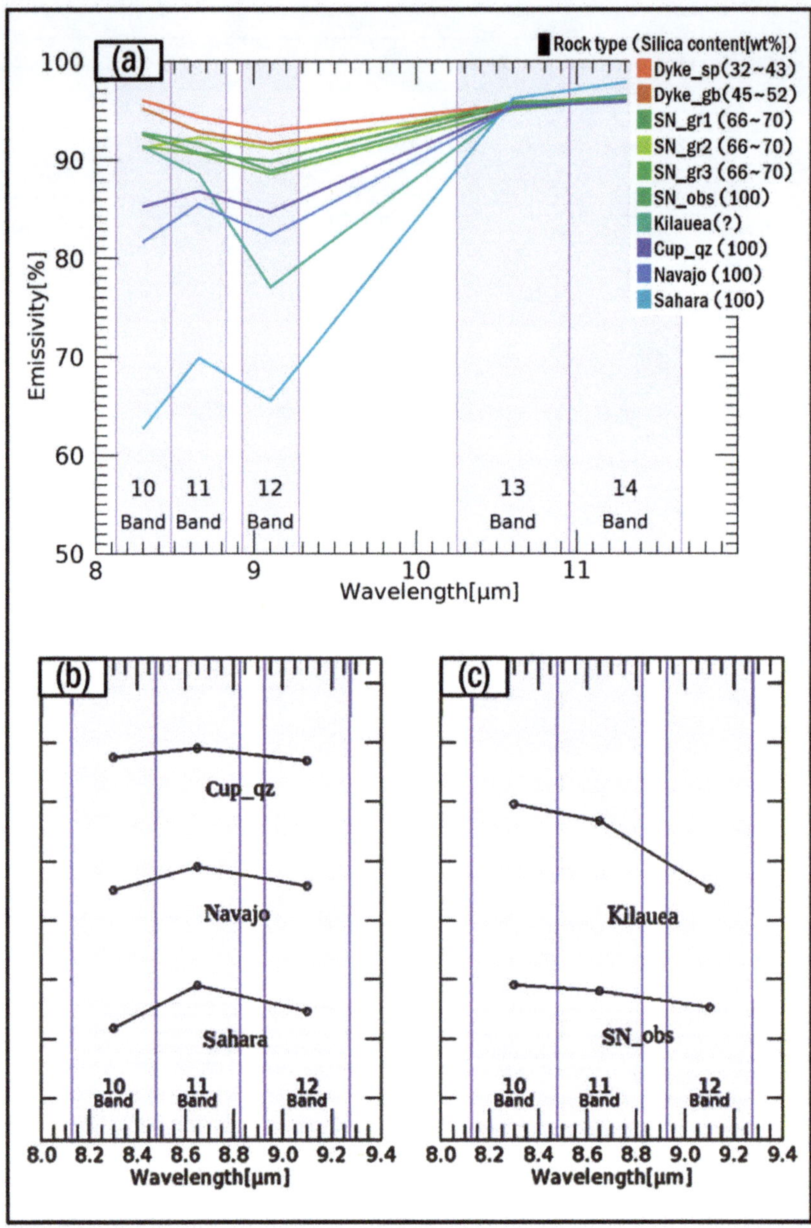

Figure 1. (**a**) Typical emissivity spectra in the TIR region in the sampling areas; (**b**) typical emissivity spectra of bands 10, 11, and 12 in the three sampling areas, where quartzose materials are distributed; (**c**) typical emissivity spectra of SN_obs (obsidian) and Kilauea, where a thin amorphous silica layer covers the surface. Please note that the spectra in (**b**,**c**) are offset for clarity. The vertical scale resolution is 10% emissivity.

The T-angle quantifies the emissivity patterns in ASTER bands 10, 11, and 12 to detect the mineral form—in this particular case, quartz or amorphous silica—by using the orthogonal transformation.

This index is defined as the angle between two vectors expressed in the three-dimensional space of the ASTER bands 10, 11, and 12 emissivities. The two vectors are a reference vector and a variable (or target) vector (Figure 2). In this case, the reference vector represents a spectral pattern linearly increasing from band 10 to 12 and is set to have an orientation of 0°. The transform coefficients were calculated according to the method by Jackson [25]. The T-angle is given by

$$
\begin{aligned}
\cos \text{T-angle} &= -\tfrac{\sqrt{2}}{2}\text{B10} + \tfrac{\sqrt{2}}{2}\text{B12} \\
\sin \text{T-angle} &= \tfrac{\sqrt{6}}{6}\text{B10} - \tfrac{\sqrt{6}}{3}\text{B11} + \tfrac{\sqrt{6}}{6}\text{B12} \\
\text{T-angle} &= \arctan\!\left(\tfrac{\sin \text{T-angle}}{\cos \text{T-angle}}\right) \quad (0° \leq \text{T-angle} \leq 360°).
\end{aligned}
\tag{2}
$$

The spectral patterns can be described by the T-angle, which ranges from 0° to 360°. Quartz-rich materials have a diagnostic spectral pattern in this wavelength region (Figure 1b), whereas amorphous-silica-rich materials have a different spectral pattern (Figure 1c), which is represented by T-angles of approximately 210° and 300°, respectively. Quartz has relatively low emissivity in ASTER bands 10 and 12, and a relatively high emissivity in band 11 due to its fundamental asymmetric Si-O-Si stretching vibration [17,23]. In contrast, amorphous-silica-rich materials such as glass, opal, and obsidian show emissivity patterns that decrease from band 10 to band 12 due to their broad emissivity depression at around 9.1 microns [24,26]. As the two proposed indices, T-depth and T-angle, have fixed value ranges, the index values derived from different scenes or different areas can easily be compared.

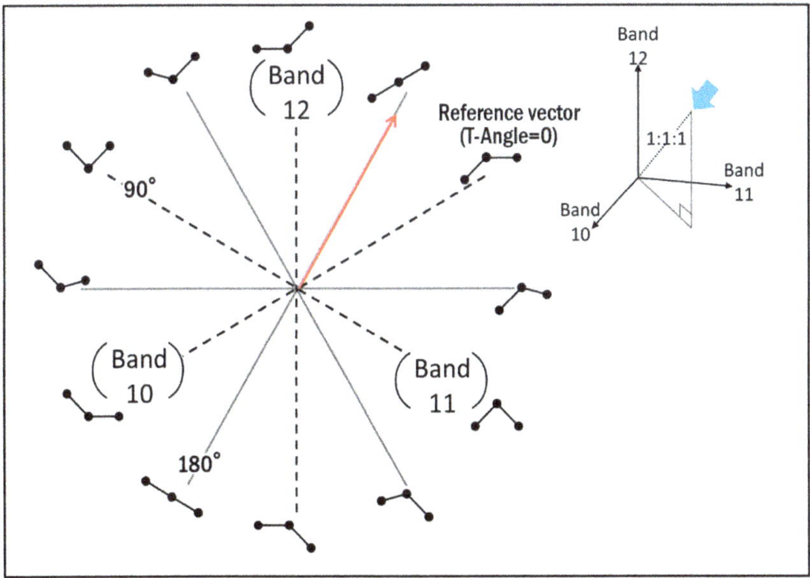

Figure 2. Concept of orthogonal transformation in the three-dimensional space of the reflectance of ASTER bands 10, 11, and 12.

2.3.2. Delineating Carbonate Rocks

Carbonate rocks are generally detected using SWIR spectral data. However, in this study, a carbonate index for ASTER TIR data [17] was used to delineate carbonate rocks because ASTER SWIR band 9 data show some problems in the surface reflectance product. It may be possible to replace this index of 90 m spatial resolution with the relative band depth (RBD) of bands 7, 8, and 9 of 30 m spatial resolution when reliable surface reflectance data including SWIR band 9 are available. The carbonate index used here is defined as:

$$\text{carbonate index} = B13/B14. \tag{3}$$

2.3.3. Delineating Hydrothermal Alteration Zones

To map hydrothermal alteration areas, two clay mineral indices developed by Kurata and Yamaguchi [15] were adopted. The clay index quantifies the SWIR spectral patterns by using the orthogonal transformation, and the spectral pattern changes depending on type of clay mineral. This index is defined as the angle between two vectors expressed in the three-dimensional space of the reflectances of ASTER SWIR bands 5, 6, and 7 in a manner similar to the definition of the T-angle. The two vectors are a reference vector and a variable (or target) vector. The angle between the two vectors is calculated from the inner product of the vectors with elements representing the reflectances of ASTER SWIR bands 5, 6, and 7 and two orthogonal unit vectors. The transform coefficients were calculated according to the method by Jackson [25]. The clay index assigned to the H element is calculated using the following formula:

$$
\begin{aligned}
\cos(\text{clay index}) &= -\tfrac{\sqrt{2}}{2}B5 + \tfrac{\sqrt{2}}{2}B7 \\
\sin(\text{clay index}) &= \tfrac{\sqrt{6}}{6}B5 - \tfrac{\sqrt{6}}{3}B6 + \tfrac{\sqrt{6}}{6}B7 \\
\text{clay index} &= \arctan\!\left(\tfrac{\sin(\text{clay index})}{\cos(\text{clay index})}\right) \quad (0° \leq \text{clay index} \leq 360°)
\end{aligned}
\tag{4}
$$

where B5, B6, and B7 are the reflectances of ASTER SWIR bands 5, 6, and 7, respectively. We used the ASTER standard data product of SWIR surface reflectance (AST_07XT), including the crosstalk correction. The spectral pattern of alunite takes a clay index of 10°, and the clay indices of kaolinite and montmorillonite would be approximately 45 and 90.

The index describing the clay mineral amount is defined as

$$\text{SWIR depth} = (B4 \times 3)/(B5 + B6 + B7). \tag{5}$$

This index indicates the depth of the absorption caused by the –OH radicals of clay minerals in the SWIR region corresponding to ASTER bands 5, 6, and 7.

2.3.4. Allocation of Spectral Indices to the H and S Elements

The lithological indices were integrated by allocating each of them to the H or S element of the HSV color model. Figure 3 shows the meanings of the colors determined by the following procedure. In the integration stage, the following minerals were considered in order of descending priority: clay minerals, carbonate minerals, and silicate minerals. This order was selected because it reflects the general relative importance of these minerals in metal explorations. Namely, the indices of silicate rocks were first allocated to all pixels in the image as a background, then the carbonate and clay mineral indices were progressively overwritten to the pixels where these indices exceeded the corresponding threshold.

The T-depth values ranging from 1.16 to 9.85 were linearly rescaled to the range 210–315 and assigned to the H element of the HSV color model; this means that the silica content is represented by the pixel color. The two thresholds 1.16 and 9.85 were determined empirically by comparing the T-depth values (Table 3) and rock types of target pixels on actual ASTER images. For example, the lower threshold of 1.16 corresponds to the average T-depth value of the mafic rocks that include the basalt in Cuprite and the mafic intrusions of the Great Dyke. Please note that the T-depth values in Table 3 are just typical examples. As we used approximately three times as many pixels to obtain the thresholds, the T-depth average in Table 3 does not coincide with the threshold. Pixels whose T-depth is lower than 1.16 are regarded as mafic rock and assigned an H value of 210°. Similarly, the higher threshold of 9.85 was determined using the T-depth values of the felsic rocks on actual ASTER images.

Pixels whose T-depth is higher than 9.85 are regarded as felsic rock and assigned an H value of 300°. The thresholds for the clay and carbonate indices were similarly determined.

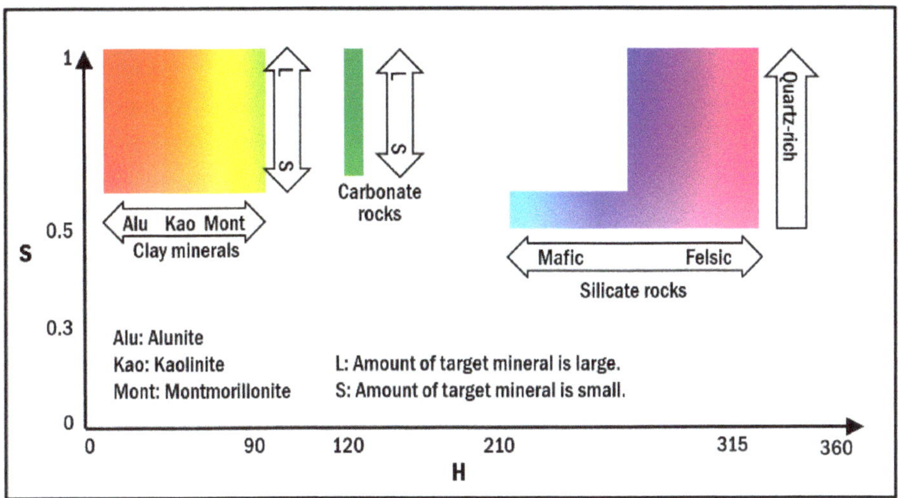

Figure 3. Meanings of the colors determined by the procedures to allocate lithological indices to the hue (H) and saturation (S) elements of the hue–saturation–value (HSV) color model.

Silica-rich materials (e.g., quartzose sand and obsidian) are thus represented by H values close to 315, corresponding to pink, whereas silica-poor materials (e.g., basalt) are represented by H values close to 210, corresponding to blue. In this case, the S element was set to $(0.5 + \alpha)$, where α, which ranges from 0.0 to 0.5, was determined from the value of the T-angle, which represents the mineral form of the silicates, whether the target felsic rock is composed of crystalline quartz or amorphous silica. T-angles ranging from 210.0 to 310.0 were linearly rescaled to the given range of α values. This assignment means that the parts of the images in which the T-angle was between 210.0 and 310.0 are represented by a more vivid color for high T-angles, whereas other areas are displayed in a dull color. As a result of this process, quartz and amorphous silica could be distinguished from each other, as shown in Figure 4a–d, respectively.

Then, to integrate the carbonate index into the image, the carbonate index values were linearly rescaled to the range between 0.0 and 1.0 by excluding the top 2% and bottom 2% as unexpected values. This procedure is scene-dependent. In this case, pixels with rescaled indices exceeding the lower threshold of 0.65 were selected as areas with carbonate rocks. The H element of these pixels was then set to 120, corresponding to light green, and the S element was replaced by the rescaled index in the target area; this means that areas containing carbonate rocks are shown in light green with a vividness representing the index value.

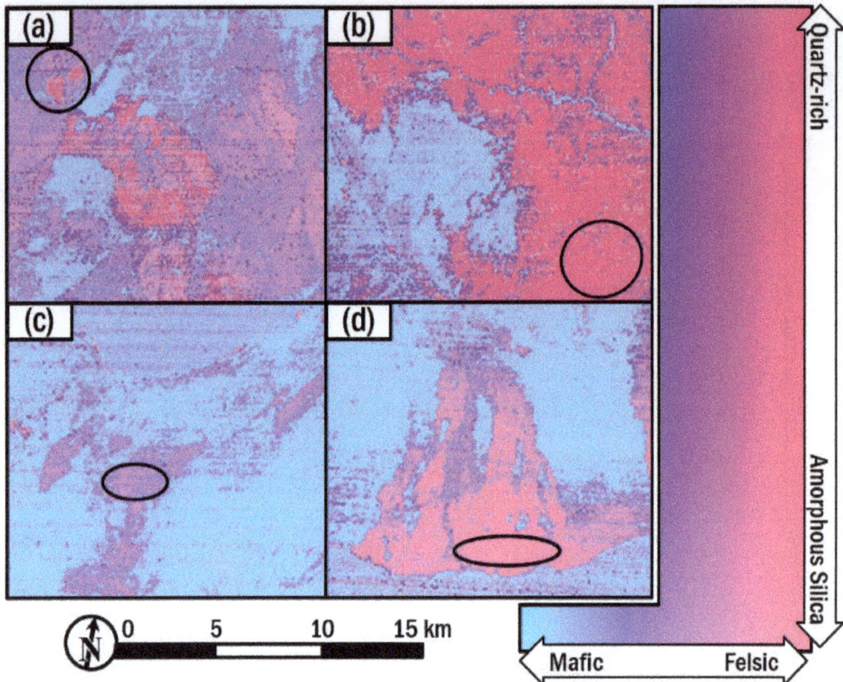

Figure 4. Images generated by integrating the T-depth and T-angle, which can distinguish between quartz (**a**,**b**), and amorphous silica (**c**,**d**): (**a**) quartzite in Cuprite, Nevada, USA; (**b**) Navajo Sandstone in Utah, USA; (**c**) Obsidian in the hills in the southern part of Mono Lake, California, USA; (**d**) thin silica coating on pahoehoe lava in Kilauea, Hawaii, USA. The ellipses indicate representative areas where these target rocks are distributed. Coordinates of the ellipse centers; (**a**) 37°37′17.14″ N/117°17′26.87″ W, (**b**) 37°42′05.32″ N/111°22′35.22″ W, (**c**) 37°53′39.43″ N, 119°00′29.60″ W, (**d**) 19°16′34.93″ N, 155°09′21.74″ W.

Lastly, indices representing the clay mineral species (clay index) and amount (SWIR depth) were allocated to the H and S elements, respectively. For the purpose of visualizing gradual changes in the clay minerals in hydrothermal alteration zones, clay index values between 10.0 and 110.0, which includes the spectral patterns of alunite, kaolinite, and montmorillonite (or sericite)—minerals commonly distributed in hydrothermal alteration zones—were assigned to the H element after rescaling to the range of 0, corresponding to red, to 90, corresponding to greenish yellow, using the following formula:

$$H_{clay} = (90 - 0)\left(\frac{h - 10}{110 - 10}\right)^{1/1.2} \tag{6}$$

where h is the clay index. The SWIR depth and carbonate index were rescaled to the range of 0.0 to 1.0 and were allocated to the S element. The pixels satisfying the following conditions were selected as hydrothermal alteration zones: (1) the clay index is between 10.0 and 110.0, and (2) the stretched SWIR-depth value exceeds 0.6. The H and S elements were modified for the pixels that fulfill these conditions.

2.4. DEM Visualization by the V Element

A new method of visualizing the topography as a grayscale relief map (GRM) obtained from DEM data was developed in this study. This method combines two existing methods: the overground openness by Yokoyama et al. [27] and the inverted slope by Yajima and Yamaguchi [28]. The overground openness describes a sky extent measured by a solid angle (steradian) over the point within a certain distance. In this particular case, it was measured at each pixel for an area within 30 m. The overground openness of a target pixel is defined as the average of the slope of the ground in eight radial directions, and the slope in each direction is determined by taking the maximum slope between the target pixel and the pixels within a certain distance in each direction. However, the actual topography cannot easily be understood from the overground openness because the angles of slopes are not shown, even though mountain ridges and valleys are accurately indicated. However, the inverted slope can be used to visualize the angles of slopes by using the Guth hybrid slope algorithm [29], which can estimate the slope at a target pixel by taking the maximum slope measured in eight radial directions from the target pixel to adjacent pixels. A steeper slope is expressed as darker in an inverted slope image, whereas a gentle slope is expressed as brighter; flat areas are expressed as white. As a result, characteristic topographical features (e.g., conus or tables) corresponding to different lithological units can be visually interpreted. However, both ridges and valleys are represented as white, and it is difficult to distinguish them in an inverted slope image, as shown in Figure 5a.

The proposed GRM method utilizes the advantages of these two existing methods and combines them in the following equation:

$$GRM = (\text{Overground openness} \times \gamma) + \text{Inverted slope}. \tag{7}$$

In this study, the value of γ was set to 3.0 based on visual comparisons of images with several different values. The GRM is expressed in grayscale, and it can thus be allocated to the V element of the HSV color model. In addition, as this method is independent of the illumination direction, meaning there is no selective enhancement by illumination. Therefore, as shown in Figure 5b, mountain ridges (white) and valleys (black) are easily distinguishable in images generated using this new method.

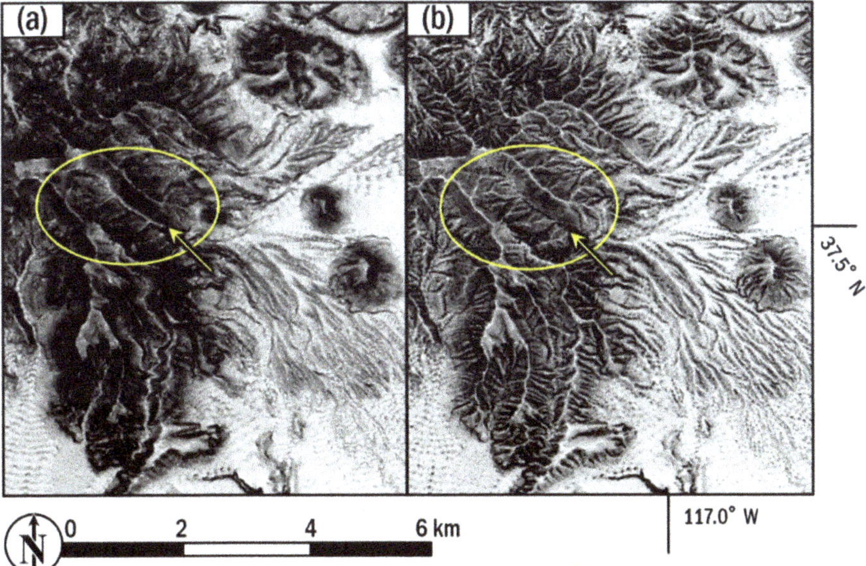

Figure 5. (a) Inverted slope image of Stonewall Mountain located to the southeast of Cuprite; (b) Grayscale relief map (GRM) image of the same area. In (a), the valley trending northwest-southeast shown by the arrow is white and looks similar to the parallel white ridges in the ellipse. In (b), this valley is black and is easily distinguishable from the ridges in the ellipse.

3. Results

Figure 6 shows the results of applying the image processing method proposed in this study. All of the lithological indices and topographical information were combined to generate a single image that clearly exhibits the distribution of silicate rocks, carbonate rocks, and hydrothermal alteration zones, as well as the topography, using the HSV color model. The V elements, which correspond to the brightness of the colors, express the topography. Because dark colors indicate steep slopes, the topography is depicted as the surface ruggedness overlaid on the integrated image. Figure 7 shows GRM images along with photographs of the actual topographical features represented in the images.

Colors ranging from blue to pink represent the silica content based on the T-depth. The silica content is low in blue areas as "basalt" in the image, high in pink areas, and intermediate when the color is between blue and pink. In addition, areas in which quartz is abundant on the surface are represented by vivid pink as "quartzite" in the image. Areas rich in carbonate rocks are shown in light green with a high vividness in the color image because the only carbonate rock area has higher carbonate index values than the other regions in the image. A hydrothermal alteration zone depicted in the integrated image exhibits a concentric pattern of hues; the color changes from red at the center to yellow and finally greenish yellow in the surrounding marginal areas. This characteristic pattern of colors indicates a typical spatial distribution of clay minerals in a hydrothermal alteration zone, from alunite in the central acidic alteration zone through kaolinite to montmorillonite in the peripheral alkali alteration zone (e.g., [20,30]). There are 3 zones having the characteristic pattern in the integrated image, Cuprite, Goldfield and an unconfirmed area.

As shown in Section 2.3.4, it was confirmed that quartzite in Cuprite (Figure 4a), quartzose in Navajo Sandstone (Figure 4b), and obsidian near Mono Lake (Figure 4c) and amorphous silica on pahoehoe lava in Kilauea (Figure 4d) are displayed in a pinkish color because of their high T-depth values. Additionally, the former two, which consist of almost 100% quartz, are exhibited in a vivid color according to a particular T-angle value, whereas the latter two, which consists of amorphous silica, is exhibited in a pale color. A comparison of the present image with the results obtained in previous studies (e.g., [20,22]) reveals that chalcedony previously identified at the center of the eastern Cuprite hill (silicified area in Figure 8a) is clearly depicted in the image obtained using the proposed method, as shown in Figure 8b. In addition, a flat plateau of basalt in the southwestern part of the Goldfield area confirmed by Ashley [31] is also clearly depicted in pale blue, indicating mafic rock. Therefore, the T-depth can be considered to accurately represent the silica content of chemical compositions, and the T-angle can be used to successfully discriminate between quartz and amorphous silica, which have different mineralogical forms.

The results of the field verification are shown in Figure 9. The reflectance spectrum of the silicified rock with chalcedony at the center of the hill (Figure 9a) shows a weak absorption at 2.25 microns due to hydroxyl [32]. Alunite and kaolinite were identified in the hydrothermally altered rock samples collected south of the hill (Figure 9b,c). These results are consistent with the integrated image shown in Figure 8 as well as previous studies, indicating a typical zonation of hydrothermal alteration (e.g., [20]).

The integrated image was also compared with the mineral map generated using the AVIRIS data [22]. Reddish areas in Figure 8b coincided well with the alunite areas, orange areas with the alunite + kaolinite or kaolinite areas, and yellowish areas with the kaolinite + white mica ± alunite or halloysite areas in Swayze et al. [22]. Calcite distributions in these two studies also agreed well.

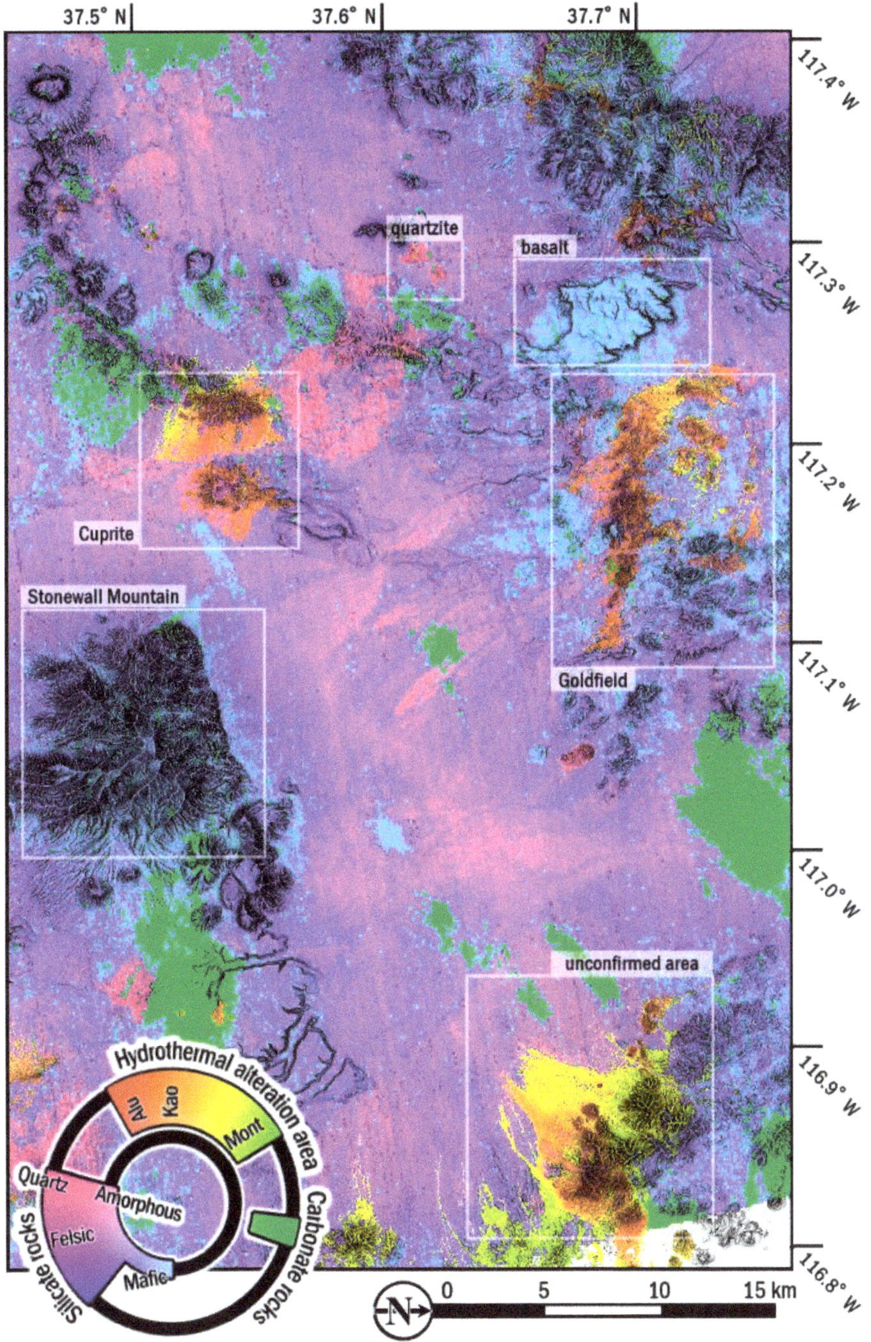

Figure 6. Integrated image simultaneously exhibiting multiple lithological indices and the topography.

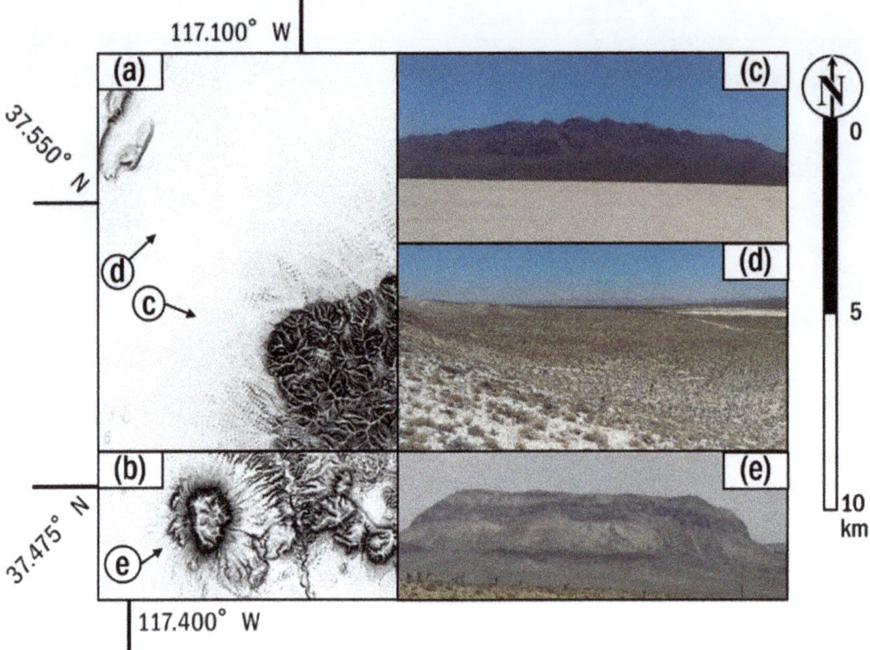

Figure 7. GRM images and photographs the topographical features contained in the images: (**a**,**b**) GRM image; (**c**) Stonewall Mountains; (**d**) Alluvial plain; (**e**) Table plateau.

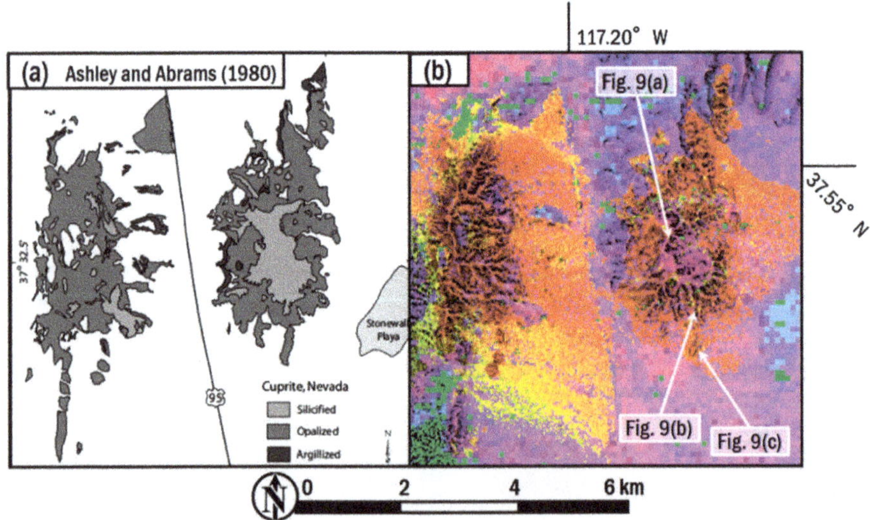

Figure 8. Comparison between (**a**) Zonation of the hydrothermal alteration in Cuprite, Nevada by Ashley and Abrams [20] and (**b**) integrated image.

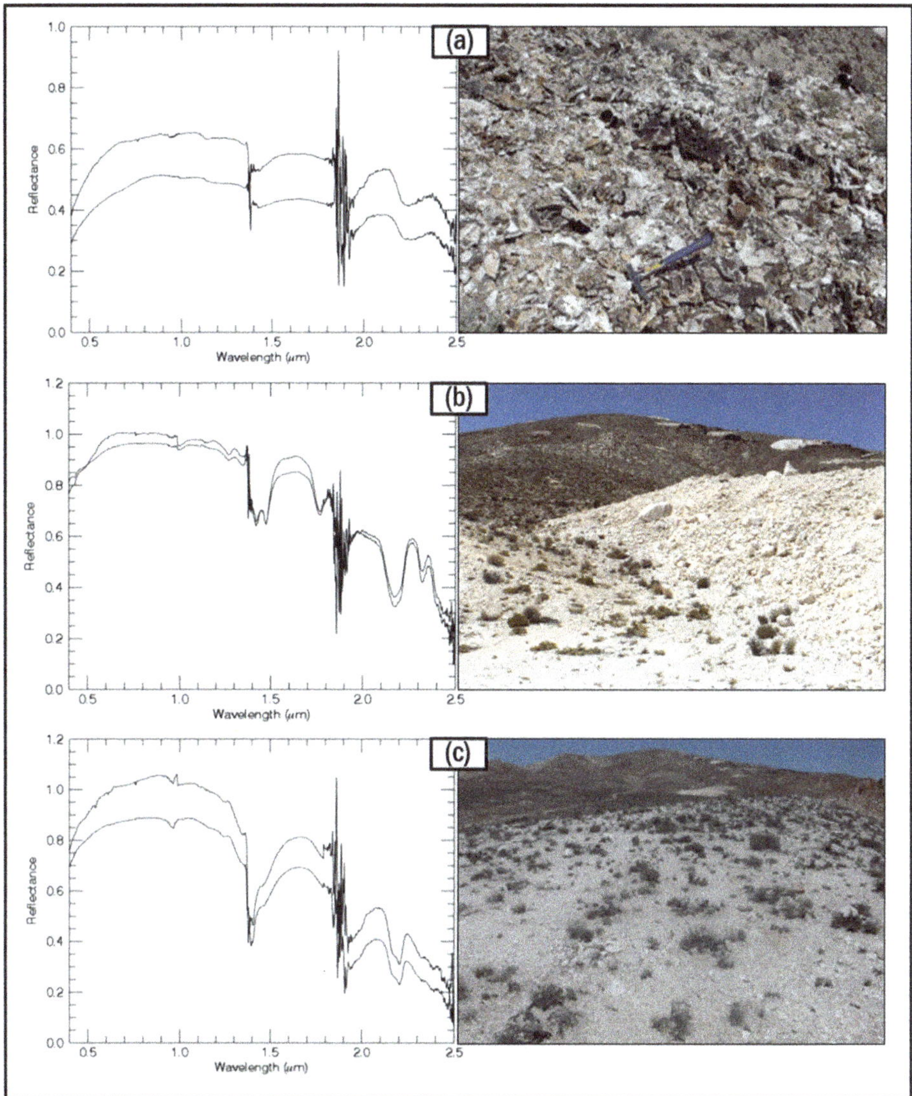

Figure 9. Reflectance spectra and outcrop photographs of the hydrothermal alteration zone in Cuprite. See the locations in Figure 8. The upper and lower spectra for each sample indicate measurements for the fresh surface and the weathered surface, respectively. Absolute reflectance values are not meaningful due to imprecise measurement conditions. The noisy patterns at around 1.4 and 1.9 microns are due to the atmospheric water vapor as these spectra were obtained outdoor by using sunlight as an illumination source. (**a**) Silicified rock with chalcedony at the center of the alteration zone, (**b**) hydrothermally altered rock that includes alunite, and (**c**) hydrothermally altered rock that includes kaolinite.

4. Discussion

4.1. Integration Method

The method of integrating lithological and topological information in an HSV image was evaluated on the basis of two considerations: whether the HSV model is more appropriate than the conventional RGB color composite for data integration and whether the method shows site or target dependence. The RGB color composite has been widely used and explored; for example, it has been used to represent combinations of band ratios for various mineralogical features in [33]. An integrated image generated by the HSV color model (Figure 10a) and the corresponding RGB color composite image (Figure 10b) were compared based on how well the zonal structure of the hydrothermal alteration in the image could be visualized. The proposed integration method has several advantages over the conventional method. First, in the RGB color composite, up to three indices of the surface materials are directly allocated to the intensities of the RGB color primitives, whereas more than three appropriate indices can be arbitrarily selected for allocation to the HSV elements. Additionally, in the RGB color composite images, the mapping target must be assessed by interpreting the meanings of colors on a case-by-case basis, whereas the meanings of the colors in the HSV images generated with the proposed method are determined in advance, making them easily interpretable, and important targets can be readily enhanced by varying the S element value.

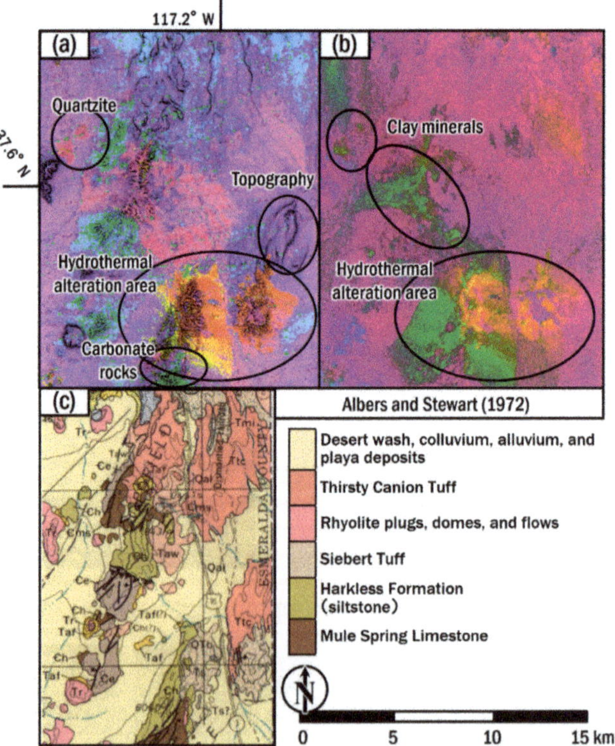

Figure 10. Images showing the zonal structures of the hydrothermal alteration generated by: (**a**) The method proposed in this paper; (**b**) The RGB color composite method with the R, G, and B primitives representing the kaolinite group index, the AlOH group content, and the FeOH group content, respectively, as developed by Cudahy [34]. (**c**) Geological map of this area [35].

Regarding the second consideration, it is important to analyze different regions or targets because the proposed method may show site and/or target dependence. This study focused on the geological mapping of multiple types of rocks and minerals and selected a large target area (approximately 4,000 km^2) containing distributions of a variety of rock types. Apart from the method having target dependence in that appropriate indices were selected for lithological mapping, there is additional site dependence in terms of the allocation of indices to HSV elements and rescaling them to a specific value range. The T-depth, T-angle, and clay index originally had definite value ranges, but the carbonate index and SWIR depth did not, and thus the resultant ranges could vary according to the data used in the analysis. This inconsistency forces the index range to be adjusted on a case-by-case basis, which causes site dependence. This means that the processes of allocating indices to the HSV elements is fundamentally dependent on both the target and the site, similarly to the problem of the RGB color composite technique.

However, the fundamental concept of this method can be applied to other targets and areas by selecting appropriate indices to detect the targets according to their spectral features. As a next step, it might be necessary to develop a method that can allocate index values to fixed ranges or select indices that have definite value ranges.

4.2. Lithological Indices and Geological Interpretation

As discussed, the integrated image could be used to easily interpret various geological situations that cannot be shown by separate lithological indices and topographical images. Some examples illustrating this are discussed here. First, the method for mapping hydrothermal alterations was successfully used to clearly visualize the different geological situations of the two hydrothermal alteration zones in Cuprite and Goldfield. Concentric zonation patterns were observed in both areas but with opposing patterns of changing colors: from red at the center of the concentric zonal structure to greenish yellow at the margin in Cuprite and from greenish yellow at the center to red at the margin in Goldfield. This implies that a region that underwent hydrothermal alteration with an acidic environment was distributed on the edge of the circular structure in Goldfield [31,36]. As highly acidic environments generally form at the center of hydrothermal alteration zones, this phenomenon could have produced a characteristic pattern in Goldfield in which the center of the hydrothermal alteration zone was not coincident with the center of the concentric zonal structure, unlike Cuprite, where the two centers were coincident. This inference is consistent with the suggestion in a previous study [31] that in Goldfield, a series of hydrothermal alteration zones is arranged in a circle along the circular structure formed by volcanic activity. This discussion indicates that this method could provide important geological information on hydrothermal alteration zones and the integrated image is a powerful tool to effectively and efficiently exhibit the geological characteristics.

Second, in the eastern part of Goldfield, the characteristic spatial distribution of the colors indicates the presence of a hydrothermal alteration zone ("unconfirmed area" in Figure 6). Clay minerals are distributed in the hills, and some linear distributions of greenish yellow and mixed distributions of red, yellow, and greenish yellow colors are observable in the flatlands around the hills. A geological interpretation of these patterns indicates that outcrops of a hydrothermal alteration zone are present on the hills and sediments derived from those outcrops were transported and spread onto the flat alluvial plains. This information was obtained from image interpretation only. Unfortunately access to the area is restricted due to military use. In a similar way, the topographical information enables the outcrops to be distinguished from the alluvia, which includes the sediments that flowed from the outcrops, as in the area around the flat plateau of basalt in the western part of Goldfield [35] ("basalt" in Figure 6). Furthermore, there are two different topographies in this area that were colored by vivid pink representing quartzose rocks but showed different morphologies: rugged and flat surfaces. These differences could have been caused by the different rock types, such as quartzite at the rugged hills and quartzose sedimentary rock or sediments in the flat surfaces. It can thus be concluded that the integrated image generated by the method proposed in this study is easily geologically interpretable.

Further studies are required to investigate the effects of vegetation cover. In this paper, we chose areas with sparse vegetation. It is likely that the H and S values for geological components will change if these components are mixed with a vegetation component. Removal of the vegetation component might be required prior to the application of the proposed integration method.

5. Conclusions

The goal of this study was to develop a new method of combining and visualizing multiple lithological indices derived from ASTER data in different wavelength regions and topography derived from DEM data in a single-color image that can be easily interpreted from a geological point of view. The proposed method consists of three parts: a technique to combine multiple indices using the HSV color model, the derivation of appropriate spectral indices to discriminate among different lithologies, and a method of visualizing the topography derived from DEM data. Consequently, a method that can be used to visualize geological information was successfully developed, and the resulting images were demonstrated to be more easily interpretable than those generated using conventional methods. It was confirmed that integrated images generated by combining lithological indices and topographical information enable the interpretation of detailed geological situations. The integration method can also be applied to various targets by adjusting the spectral indices and their allocation to the elements in the HSV color model. The newly developed lithological indices, the T-depth and T-angle, have definite value ranges and thus enable scene-independent analysis of silicate rocks, although the new method still has some shortcomings and target dependence. For example, the quantitative correspondence between the index values of T-depth and silica content of the actual lithology is not clear. Furthermore, the parameters used in the allocation process of the carbonate index and SWIR depth as well as the ranges of index values and thresholds are site-dependent. These points must be improved when the method is applied to other regions or geological targets. Regarding the visualization of the topography, it was confirmed that the new GRM method can clearly exhibit topographical differences.

The integration of multiple indices was effective in the geological mapping performed in this study. By adopting the proposed method of integrating multiple indices, even people who are not familiar with remote sensing can use geological remote sensing more easily to detect or map a particular geological unit and better understand its characteristics. This would facilitate the introduction of geologic remote sensing to geological studies and would contribute to bridging the gap between the fields of geology and geological remote sensing.

Author Contributions: Conceptualization, methodology, field validation and writing, K.K. and Y.Y.; programming and data curation, K.K.; funding acquisition, Y.Y.

Funding: This work was funded by JSPS KAKENHI (Grant no. 15H04225).

Acknowledgments: We would like to thank NASA and LPDAAC for providing the ASTER data used in this study, and USGS for providing the 3DEP DEM dataset used in this study.

Conflicts of Interest: The authors declare no conflict of interest.

References

1. Van der Meer, F.D.; van der Werff, H.M.A.; van Ruitenbeek, F.J.A.; Hecker, C.A.; Bakker, W.H.; Noomen, M.F.; van der Meijde, M.; Carranza, E.J.M.; Smeth, J.B.D.; Woldai, T. Multi- and hyperspectral geologic remote sensing: A review. *Int. J. Appl. Earth Obs. Geoinf.* **2012**, *14*, 112–128. [CrossRef]
2. Kereszturi, G.; Schaefer, L.N.; Schleiffarth, W.K.; Procter, J.; Pullanagari, R.R.; Mead, S.; Kennedy, B. Integrating airborne hyperspectral imagery and LiDAR for volcano mapping and monitoring through image classification. *Int. J. Appl. Earth Obs. Geoinf.* **2018**, *73*, 323–339. [CrossRef]
3. Rogge, D.; Rivard, B.; Segl, K.; Grant, B.; Feng, J. Mapping of NiCu–PGE ore hosting ultramafic rocks using airborne and simulated EnMAP hyperspectral imagery, Nunavik, Canada. *Remote Sens. Environ.* **2014**, *152*, 302–317. [CrossRef]

4. Kruse, F.A. Mapping surface mineralogy using imaging spectrometry. *Geomorphology* **2012**, *137*, 41–56. [CrossRef]

5. Yamaguchi, Y.; Kahle, A.B.; Tsu, H.; Kawakami, T.; Pniel, M. Overview of Advanced Spaceborne Thermal Emission and Reflection Radiometer (ASTER). *IEEE Trans. Geosci. Remote Sens.* **1998**, *36*, 1062–1071. [CrossRef]

6. Rowan, L.C.; Wetlaufer, P.H.; Goetz, A.F.H.; Billingsley, F.C.; Stewart, J.H. *Discrimination of Rock Types and Detection of Hydrothermally Altered Areas in South-Central Nevada by the Use of Computer-Enhanced ERTS Images*; Geological Survey Professional Paper; U.S. Department of Energy Office of Scientific and Technical Information: Oak Ridge, TN, USA, 1976; Volume 883, p. 35.

7. Kruse, F.A.; Lefkoff, A.B.; Boardman, J.W.; Heidebrecht, K.B.; Shapiro, A.T.; Barloon, P.J.; Goetz, A.F.H. The spectral image-processing system (SIPS)—Interactive visualization and analysis of imaging spectrometer data. *Remote Sens. Environ.* **1993**, *44*, 145–163. [CrossRef]

8. Yamaguchi, Y.; Naito, C. Spectral indices for lithologic discrimination and mapping by using the ASTER SWIR bands. *Int. J. Remote Sens.* **2003**, *24*, 4311–4323. [CrossRef]

9. Fatima, K.; Khattak, M.U.K.; Kausar, A.B.; Toqeer, M.; Haider, N.; Rehman, A.U. Minerals identification and mapping using ASTER satellite image. *J. Appl. Remote Sens.* **2017**, *11*, 46006. [CrossRef]

10. Jakob, S.; Gloaguen, R.; Laukamp, C. Remote sensing-based exploration of structurally-related mineralizations around Mount Isa, Queensland, Australia. *Remote Sens.* **2016**, *8*, 358. [CrossRef]

11. Zimmermann, R.; Brandmeier, M.; Andreani, L.; Mhopjeni, K.; Gloaguen, R. Remote sensing exploration of Nb-Ta-LREE-enriched carbonatite (Epembe/Namibia). *Remote Sens.* **2016**, *8*, 620. [CrossRef]

12. Fujisada, H.; Urai, M.; Iwasaki, A. Technical methodology for ASTER global DEM. *IEEE Trans. Geosci. Remote Sens.* **2012**, *50*, 3725–3736. [CrossRef]

13. Takaku, I.; Tadono, T.; Tsutsui, K. Generation of high resolution global DSM from ALOS PRISM. *Int. Arch. Photogramm. Remote Sens. Spat. Inf. Sci.* **2014**, *XL-4*, 243–248. [CrossRef]

14. Whelley, P.L.; Glaze, L.S.; Calder, E.S.; Harding, D.J. LiDAR-derived surface roughness texture mapping: Application to Mount St. Helens pumice plain deposit analysis. *IEEE Trans. Geosci. Remote Sens.* **2014**, *52*, 426–438. [CrossRef]

15. Kurata, K.; Yamaguchi, Y. Combining multiple lithological indices derived from ASTER data by using the HSV color model. *J. Remote Sens. Soc. Jpn.* **2018**, *38*, 163–173, (In Japanese with an English abstract).

16. Smith, R.A. Color gamut transform pairs. *ACM Siggraph Computer Graphics* **1978**, *3*, 12–19. [CrossRef]

17. Ninomiya, Y.; Fu, B. Quartz index, carbonate index and SiO_2 content index defined for ASTER TIR data. *J. Remote Sens. Soc. Jpn.* **2002**, *22*, 50–61, (In Japanese with an English abstract).

18. Gillespie, A.R.; Matsunaga, T.; Rokugawa, S.; Hook, S.J. Temperature and emissivity from Advanced Thermal Emission and Reflection Radiometer (ASTER) images. *IEEE Trans. Geosci. Remote Sens.* **1998**, *36*, 1113–1126. [CrossRef]

19. Abrams, M.J.; Ashley, R.P.; Rowan, L.C.; Goetz, A.F.H.; Kahle, A.B. Mapping of hydrothermal alteration in the Cuprite mining district, Nevada using aircraft scanner images for the spectral region 0.46–2.36 µm. *Geology* **1977**, *5*, 713–718. [CrossRef]

20. Ashley, R.P.; Abrams, M.J. *Alteration Mapping Using Multispectral Images-Cuprite Mining District, Esmeralda County, Nevada*; U.S. Geological Survey Open File Report; U.S. Geological Survey: Reston, VA, USA, 1980; pp. 80–367.

21. Kahle, A.B.; Goetz, A.F.H. Mineralogic information from a new airborne thermal infrared multispectral scanner. *Science* **1983**, *222*, 24–27. [CrossRef]

22. Swayze, G.A.; Clark, R.N.; Goetz, A.F.H.; Livo, K.E.; Breit, G.N.; Kruse, F.A.; Sutley, S.J.; Snee, L.W.; Lowers, H.A.; Post, J.L. Mapping advanced argillic alteration at Cuprite, Nevada using imaging spectroscopy. *Econ. Geol.* **2014**, *109*, 1179–1221. [CrossRef]

23. Lyon, R.J.P. Analysis of rocks by spectral infrared emission (8 to 25 microns). *Econ. Geol.* **1965**, *60*, 715–736. [CrossRef]

24. Kahle, A.B.; Gillespie, A.R.; Abbott, E.A.; Abrams, M.J.; Walter, R.E.; Hoover, G. Relative dating of Hawaiian lava flows using multispectral thermal infrared images: A new tool for geologic mapping of young volcanic terranes. *J. Geophys. Res.* **1988**, *93*, 15239–15251. [CrossRef]

25. Jackson, R.D. Spectral indices in n-space. *Remote Sens. Environ.* **1983**, *13*, 409–421. [CrossRef]

26. Balick, L.; Gillespie, A.; French, A.; Danilina, I.; Allard, J.-P.; Mushkin, A. Longwave thermal infrared spectral variability in individual rocks. *IEEE Geosci. Remote Sens. Lett.* **2009**, *6*, 52–56. [CrossRef]

27. Yokoyama, R.; Shirasawa, M.; Kikuchi, Y. Representation of topographical features by openness. *J. Jpn. Soc. Photogramm. Remote Sens.* **1999**, *38*, 26–34. (In Japanese with an English abstract)

28. Yajima, T.; Yamaguchi, Y. Application of inverted slope images generated from ASTER GDEM to geological mapping. In Proceedings of the 56th Spring Conference of the Remote Sensing Society of Japan, Ibaraki, Japan, 15–16 May 2014; pp. 57–58. (In Japanese with an English abstract)

29. Guth, P.L. Slope and aspect calculations on gridded digital elevation models: Examples from a geomorphometric toolbox for personal computers. *Z. Geomorphol. Suppl.* **1995**, *101*, 31–52.

30. Bedini, E.; van der Meer, F.; van Ruitenbeek, F. Use of HyMap imaging spectrometer data to map mineralogy in the Rodalquilar caldera, southeast Spain. *Int. J. Remote Sens.* **2009**, *30*, 327–348. [CrossRef]

31. Ashley, R.P. *Relation between Volcanism and Ore Deposition at Goldfield, Nevada*; Nevada Bureau of Mines and Geology Report; Nevada Bureau of Mines and Geology: Reno, NV, USA, 1979; Volume 33, pp. 77–86.

32. Hunt, G.R.; Salisbury, J.W.; Lenhoff, C.J. Visible and near-infrared spectra of minerals and rocks: VII. Acidic igneous rocks. *Mod. Geol.* **1973**, *4*, 217–224.

33. Kalinowski, A.A.; Oliver, S. *ASTER Processing Manual. Remote Sensing Applications*; Internal Report; Geoscience Australia: Canberra, Australia, 2004.

34. Cudahy, T. *Australian Satellite ASTER Geoscience Product Notes, Version 1*; No. EP-30-07-12-44; CSIRO ePublish: Canberra, Australia, 2012.

35. Albers, P.J.; Stewart, H.J. *Geology and Mineral Deposits of Esmeralda County, Nevada*; Nevada Bureau of Mines and Geology: Reno, NV, USA, 1972; Volume 78, p. 88.

36. Blakely, R.J.; John, D.A.; Box, S.; Berger, B.R.; Fleck, R.J.; Ashley, R.P.; Newport, G.R.; Heinemeyer, G.R. Crustal controls on magmatic-hydrothermal systems: A geophysical comparison of White River, Washington, with Goldfield, Nevada. *Geosphere* **2007**, *3*, 91–107. [CrossRef]

Article

Global 15-Meter Mosaic Derived from Simulated True-Color ASTER Imagery

Louis Gonzalez [1],*, Valérie Vallet [2] and Hirokazu Yamamoto [3],*

[1] Univ. Lille, CNRS, UMR 8518 - LOA - Laboratoire d'Optique Atmosphérique, F-59000 Lille, France
[2] Univ. Lille, CNRS, UMR 8523 - PhLAM - Physique des Lasers Atomes et Molécules, F-59000 Lille, France; valerie.vallet@univ-lille.fr
[3] Geological Survey of Japan, National Institute of Advanced Industrial Science and Technology, 1-1-1-C7, Higashi, Tsukuba 305-8567, Japan
* Correspondence: louis.gonzalez@univ-lille.fr (L.G.); hirokazu.yamamoto@aist.go.jp (H.Y.)

Received: 31 January 2019; Accepted: 17 February 2019; Published: 20 February 2019

Abstract: This work proposes a new methodology to build an Earth-wide mosaic using high-spatial resolution (15 m) Advanced Spaceborne Thermal Emission and Reflection Radiometer (ASTER) images in pseudo-true color. As ASTER originally misses a blue visible band, we have designed a cloud of artificial neural networks to estimate the ASTER blue reflectance from Level-1 data acquired by the Moderate Resolution Imaging Spectroradiometer (MODIS) on the same satellite Terra platform. Next, the granules are radiometrically harmonized with a novel color-balancing method and seamlessly blended into a mosaic. We demonstrate that the proposed algorithms are robust enough to process several thousands of scenes acquired under very different temporal, spatial, and atmospheric conditions. Furthermore, the created mosaic fully preserves the ASTER fine structures across the various building steps. The proposed methodology and protocol are modular so that they can easily be adapted to similar sensors with enormous image libraries.

Keywords: Terra ASTER; Terra MODIS; True Color imagery; Mosaic; atmospheric correction; Artificial Neural Network

---

1. Introduction

The Terra satellite was launched on 18 December 1999. This satellite platform has five instruments which include the Advanced Spaceborne Thermal Emission and Reflection Radiometer (ASTER) and the Moderate Resolution Imaging Spectroradiometer (MODIS). ASTER was built by the Japanese Ministry of Economy, Trade and Industry (METI) [1], while MODIS was designed by National Aeronautics and Space Administration (NASA), Goddard Space Flight Center (GSFC).

ASTER is a 15 m resolution, 14 bands multispectral instrument. It has been used for change detection, calibration, validation, and land surface studies from individual granules analysis [1]. However, the global monitoring of the Earth and ocean surfaces will be greatly helped by the integration of the satellite granule database from 2003 to 2012 into a unique natural color global mosaic, referred to as CLAMS (Color-Land ASTER MosaicS). The distributed ASTER granules cannot produce natural-color images, since ASTER sensors lack a blue visible band as illustrated by Figure 1. Generally, false color RGB composites are created by assigning red, green and blue to visible near infrared bands 3N, 2, and 1, respectively. However, ice/snow areas appear grey, desert areas yellow, and vegetation red, as illustrated on the left hand side of Figure 2.

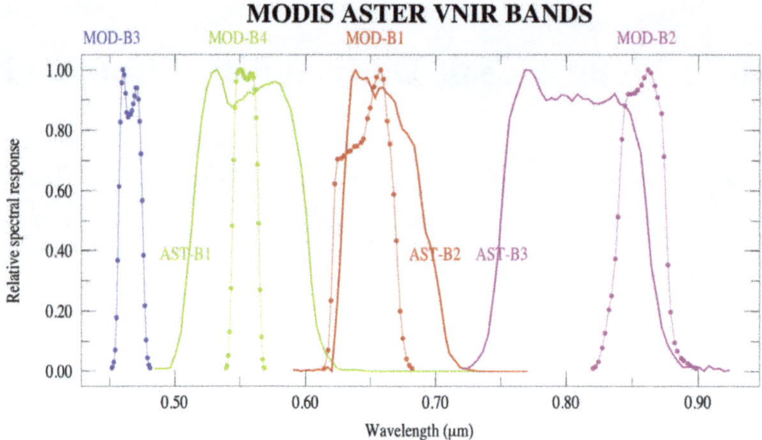

Figure 1. Normalized spectral response functions of ASTER (solid line) and MODIS (dotted lines) VNIR bands.

Figure 2. Examples of ASTER granules over different areas (ice/snow, desert, vegetation) rendered either with the false color composite (R:G:B = Bands 3N:2:1) on the left hand side, or with the simulated true color composite (R:G:B = Bands 3N:2:simulated blue) on the right hand side.

In this work, we aim at constructing a global mosaic at 15 m of ground resolution, which blends our true-color visible ASTER images.

Several studies have addressed various ways of generating the simulated true color imagery for optical satellite sensors, which do not have a part of visible bands. Chen and Tsai [2] proposed spectral transformation techniques between Système Probatoire de l'Observation de la Terre (SPOT) false color image and Landsat-5 Thematic Mapper (TM) true color imagery. They used an unsupervised fuzzy c-means classifier and spectral control points. Knudsen [3] proposed a pseudo-natural color methodology for aerial imageries by using a simple least-squares adjusted linear model for the relationship between the blue band and the green, red and near-infrared (NIR) bands. This paper showed the possibilities of cost-efficiency by using the color-infrared (CIR) traditional photogrammetric products recorded by a traditional aerial camera. Patra et al. [4] developed spectral transformation techniques with spectral control points to IKONOS false-color imagery and natural color (simulated true color) imagery, and they evaluated their developed transformation method by using Quickbird, MODIS, Indian remote sensing satellite (IRS-P6) LISS-4, LISS-3, and AWiFS sensors. Huixi and Yunhao [5] applied the atmospheric correction algorithm, ATCOR to Landsat-7 ETM+, SPOT, and Terra ASTER imageries. They used the spectral similarity scale (SSS) method with a spectral library for generating pseudo natural color composites, and could obtain excellent results. Zhu et al. [6] developed a non-linear model based on a spectrum machine learning (SML) method with the spectral library, and they applied it to Landsat-7 ETM+, SPOT, and ASTER imageries. The atmospheric correction algorithm FLAASH from ENVI software was first applied, followed by the SML method to establish an implicit non-linear relationship between the blue band and other bands. The next-generation GOES-R advanced baseline imager (ABI) does not have a green band, and high-resolution atmospheric model simulations have been used to produce the ABI reflective band imagery required for true-color imagery [7,8]. Most researchers suggest that true-color algorithms are affected by sun-target-sensor geometry and atmospheric conditions.

There have been several efforts to construct cloud-free true color base-maps from moderate or high spatial resolution (less than 100 m) satellite data. ASTER has been operated over 19 years since Terra satellite launch in 1999. Thus, the huge available acquisitions database makes it possible to generate cloud-free global mosaics. The large number of ASTER image granules (about 780,000) used in this study were collected over many years, different seasons, and under varying vegetation and illumination conditions. Without the appropriate corrections, the resulting mosaic can appear as a patchwork of individual images. To avoid this, it deems necessary to apply atmospheric corrections and to smoothen seasonal effects. There have been several attempts of true-color mosaic constructions for various satellite optical sensors, mostly using adjustments of radiometric characteristics. Guindon [9] proposed radiometric adjustment for seamlessness mosaic of Landsat-5 MSS for northwestern Ontario area. Liew et al. [10] calculated solar zenith angle corrected radiances, and have used a brightness thresholding method to identify the best cloud-free and non-shadow pixels among the pixels from the multiple images at a given region. They successfully tested their mosaic technique with SPOT images acquired over the South East Asia region. Du et al. [11] applied a radiometric equalization techniques for representative pixel pairs in each overlap area, selected by means of a principal component analysis and calculation of linear correlation coefficients. They proved the methodology by mosaicking 6–7 Landsat-5 TM granules over the Boreal Ecosystem-Atmosphere Study (BOREAS) transect. Bindschadler et al. [12] generated a seamless cloud-free Landsat-7 ETM+ mosaic of Antarctica by radiometric adjustment. Roy et al. [13] produced a mosaic of the conterminous United States (CONUS) using 6521 Landsat-7 ETM+ imageries from December 2007 to November 2008. Choi et al. [14] developed the mosaic algorithm for high resolution images captured by Kompsat-2 sensor. This algorithm can be applied to different images affected by seasonal change, and is applicable to other high resolution optical sensor images. There exist two global mosaics, GeoCover2000 [15] (Landsat-7) and Landsat-8 VNIR maps [16], that have processed 30 m Landsat images and pansharpened them to 15 m. In summary, most attempts of generating mosaics of high resolution optical imageries were mostly carried out over regional areas, as the construction of a global cloud-free mosaic of high resolution imageries is extremely difficult due to the large amount of spatially and temporarily varying acquisitions.

We propose a protocol to assemble a true-color global mosaic from high-resolution ASTER imagery. The protocol first implies the construction of the missing ASTER blue band making use of the Moderate Resolution Imaging Spectroradiometer (MODIS) Terra instrument as it sits on board of the same satellite platform acquires time and space synchronized images with all visible bands, though at a lower spatial resolution (250–500 m). In 2012, we developed a first algorithm in which, after atmospheric corrections, the ASTER pseudo-blue band was constructed from a single artificial neural network (ANN), making it possible to generate ASTER granules in true color composite. These processed ASTER images are currently distributed by the AIST MADAS system as one of ASTER-VA products [17,18]. This first algorithm was not completely satisfactory and a couple of years ago, we decided to improve the blue color retrieval in particular over the ocean, as described in Section 2. The new retrieval algorithm this time uses a cloud of ANN's, which preserve the finest details of the atmospheric components, such as dust, smoke and thin clouds.

The second key feature of the constructed CLAMS mosaic is that it is worldwide color-balanced, in the sense that the radiometric differences between the adjacent images introduced by the solar incident angle, atmosphere, and illumination condition are equalized. This is achieved by a novel color-balancing method for ASTER based on a MODIS reflectance reference library (FondsDeSol, FDS) proposed by Gonzalez et al. [19]. By automatically selecting appropriate color reference information from the FDS library according to the geographical scope and acquisition season information of the target images, the proposed approach provides effective solutions for eliminating color error propagations between adjacent granules over the globe. We will illustrate the rendered color quality of the mosaic in Section 3.1 and on the website http://newtec.univ-lille1.fr/LARGEMOSAICSFR/.

The third feature of the CLAMS mosaic is that it preserves the ASTER fine 15 m structures across the various building steps, as demonstrated in Section 3.2.

2. Methodology for ASTER Blue Reflectance Reconstruction

In the entire document, all presented data correspond to reflectance values from MODIS L1B or ASTER L3A products to which minimum atmospheric corrections were applied to compensate for the Rayleigh scattering contribution, and correct for ozone according to the 6S radiative transfer code [20], with the same equations as in Ref. [21]. These corrected reflectances are denoted hereafter as ρ. In all the figures, we present ASTER granules geometrically corrected to Plate Carrée projection (North oriented) within a bounding box of 75 km $\times$ 75 km, and we specify both the dates and geolocalization of their central points.

The ASTER granules suffer from two main defects. The first is linked to signal saturations in the 0.56 μm and 0.66 μm bands over bright surfaces as explained in Section 2.1, while the second corresponds to the missing 0.4 μm band which is reconstructed with ANNs using MODIS reflectance data as detailed in Section 2.2. To train the ANN and to reconstruct the saturated ASTER pixels missing values we use as input data the 0.555, 0.645, and 0.858 μm bands from MODIS and the 0.469 μm blue band as target output data. To retrieve the ASTER blue band we use as input data the 0.56, 0.66, 0.81 μm ASTER bands.

2.1. Reconstruction of Saturated ASTER Level1 Pixels

In the process of correcting the bright surface saturation we do not expect to retrieve the real reflectances but simply to create reasonable values giving true color composites close to reality. Saturated values only appear for the 0.56 and 0.66 μm bands (noted ρ_1 and ρ_2). A correction model was empirically determined using MODIS reflectances over the ASTER saturated areas. It turned out that simple first and second-order polynomial functions of the 0.81 μm (ρ_{3N}) reflectances could be used to assign unsaturated reflectance values ρ_1 and ρ_2, as detailed in Table 1.

Table 1. Polynomial expansions for the saturation correction of bands ρ_1 (0.56 μm), ρ_2 (0.66 μm) using ρ_{3N} (0.81 μm).

	Ocean Colors
$\rho_{3N} < 0.2$	$\rho_1 = -1.627869 \times \rho_{3N}^2 + 0.714701 \times \rho_{3N} + 0.370053$
	$\rho_2 = -17.420099 \times \rho_{3N}^2 + 6.117370 \times \rho_{3N} - 0.180111$
$\rho_{3N} > 0.2$	$\rho_1 = 2.298600 \times \rho_{3N}^2 - 1.202424 \times \rho_{3N} + 0.591962$
	$\rho_2 = 0.725354 \times \rho_{3N} + 0.202700$
	Desert
	$\rho_1 = 0.564969 \times \rho_{3N} - 0.002576$
	$\rho_2 = 0.808582 \times \rho_{3N} - 0.001048$
	Clouds
	$\rho_1 = 1.176044 \times \rho_{3N} - 0.081713$
$\rho_1 > 0.15$	$\rho_1 = 1.096000 \times \rho_{3N} - 0.020926$
	$\rho_2 = 1.220622 \times \rho_{3N} - 0.086758$
$\rho_2 > 0.15$	$\rho_2 = 1.186130 \times \rho_{3N} - 0.090857$

Figures 3–5 demonstrate that the desaturation algorithm visually corrects the missing values over desert, ocean and cloudy areas.

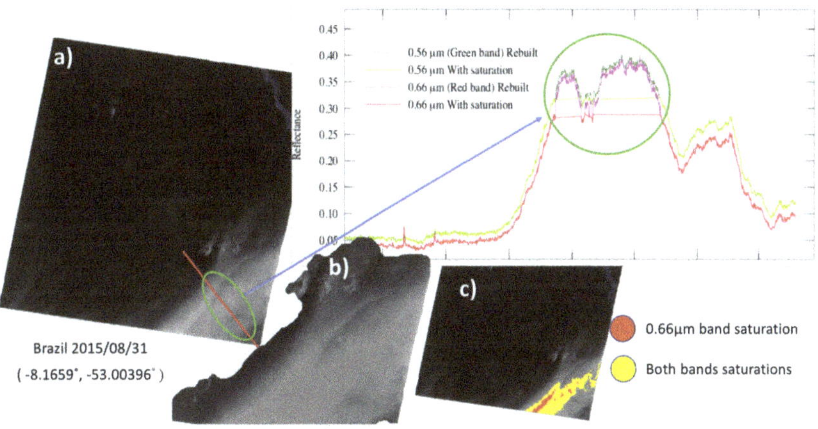

Figure 3. Visualization of the saturation corrections over cloudy areas: (**a**) final true color composite; (**b**) original saturated granule; (**c**) mask of the saturated areas (in red for the 0.66 μm band and in yellow for both 0.56 and 0.66 μm bands).

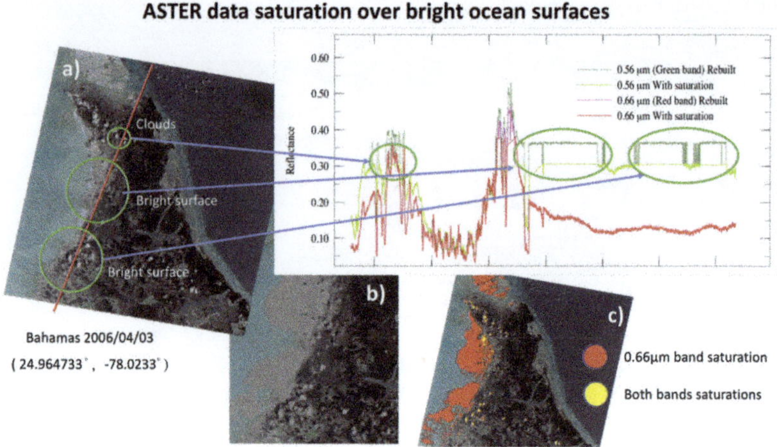

Figure 4. Visualization of the saturation corrections over the ocean: (**a**) final true color composite; (**b**) original saturated granule; (**c**) mask of the saturated areas (in red for the 0.66 µm band and in yellow for both 0.56 and 0.66 µm bands).

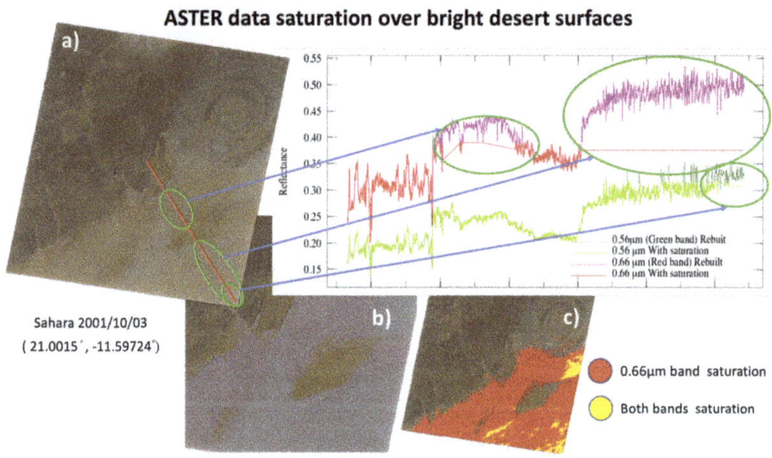

Figure 5. Visualization of the saturation corrections over desert areas: (**a**) final true color composite; (**b**) original saturated granule; (**c**) mask of the saturated areas (in red for the 0.66 µm band and in yellow for both 0.56 and 0.66 µm bands).

2.2. ANN to Reconstruct ASTER Blue Reflectances

Fine-tuning the parameters of an ANN (Stuttgart Neural Network Simulator (SNNS) [22]) to faithfully reconstruct ASTER blue reflectances turned out to be a tricky process. In 2012, we designed an ANN that used for the training inputs the 3 MODIS 0.555, 0.645, 0.858 µm bands (spectrally close to the VNIR ASTER bands, 0.56, 0.66, 0.81 µm), and the blue MODIS band at 0.469 µmas target values. This ANN is a fully connected feedforward network [23]. The ANN topology we present yields the best blue retrieval out of many tests we carried out. It consists of two hidden layers with 10 nodes each (See Figure 6). The training data set used a selection of about fifty MODIS granules for which we hand-selected representative areas such as water, ocean colors, deserts, rocks, white and bright surfaces, etc. Once trained, the network was tested on various MODIS granules and the training ANN data set was enlarged with those showing inaccurate (poor correlation with respect to MODIS real blue reflectances) reconstructed blue values. However, the blue retrieval was not fully accurate, yielding yellow colors over hazy areas and bright reefs as illustrated by the examples on the left hand side of Figure 7.

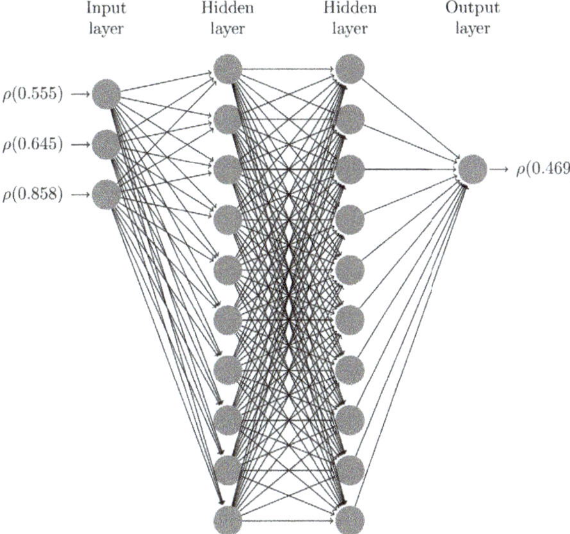

Figure 6. Topology of the ANN with one input layer taking 0.555, 0.645, and 0.858 µm reflectance values, the 0.465 µm.

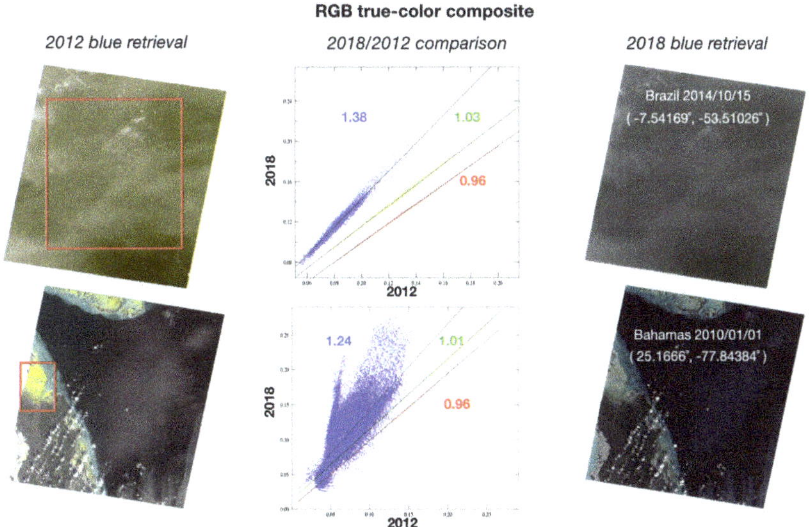

Figure 7. Comparisons of ASTER granules obtained with the blue retrieval algorithms of 2012 (**left**) and 2018 (**right**). The center panel shows the correlations between the reflectance values collected within the red boxes, obtained with the 2012 (x-axis) and 2018 (y-axis) algorithms.

Based on our recent experience with ANNs design for aerosol optical depth retrievals [21], we found out that ANNs training and accuracy are improved if applied to specific classes of reflectance values. Thus, our new algorithm uses MODIS values that are classified into three groups according to the values $x = 0.2 \times \rho_{MODIS}(0.645) + 0.8 \times \rho_{MODIS}(0.555)$, defining two main areas, the first one corresponding to water, dark areas and vegetation ($x < 0.2$), the second one to desert areas ($x > 0.2$). The third group encompasses the brightest cloudy pixels selected by according to the following three-fold tests:

$$\{\rho_{MODIS}(0.555) > 0.2\} \ \& \ \{\rho_{MODIS}(0.858) > 0.2\} \ \& \ \left\{ \frac{\rho_{MODIS}(0.555)}{\rho_{MODIS}(0.645)} > 0.9 \right\}. \tag{1}$$

The histograms of the three groups are drawn in Figure 8. Each of the three histograms are split to define subclasses of pixel values defining the inputs of ANNs. The determination of the number of data by classes (bins) was carried out in successive stages by monitoring both the convergence of the networks and the quality of the blue retrieval over about one hundred MODIS granules that compose the training data sets. It soon became clear that the first group requires a more precise division (800 networks) and converges with an average number of 4000 pixels per bin. In group 2, the subdivision is wider with 1900 networks and an average number of 2200 pixels per bin was necessary. The third group requires 1300 networks with an average number of 3200 pixels per bin. Figure 9 demonstrates how the histogram classification is used to split the MODIS data. For each bin, a dedicated ANN with a feedforward topology, with the same architecture as before (3 inputs, 2 hidden layers with 10 nodes each, and 1 output value, see Figure 6), is designed. Altogether an ensemble of 4000 ANNs is trained. The training dataset was enlarged to about one hundred MODIS granules.

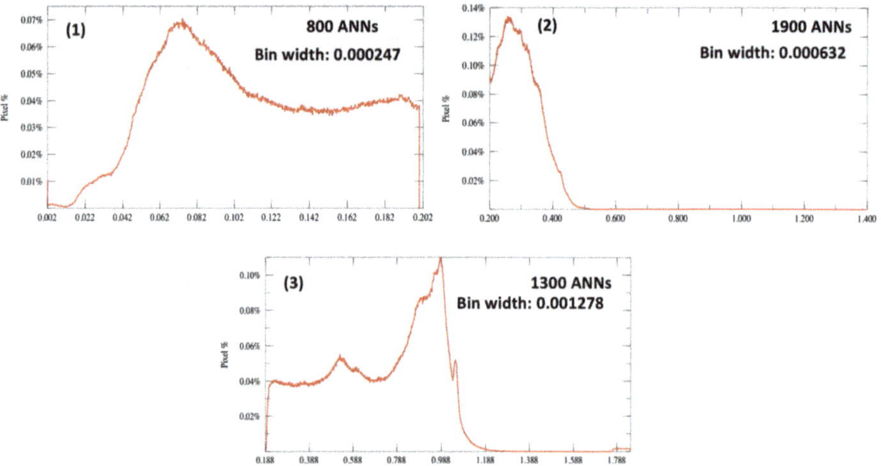

Figure 8. Histograms of the three groups of MODIS data to define the ANNs inputs.

For the retrieval, the three ASTER bands (0.56, 0.66, 0.81 µm) are scaled to match the MODIS ones (0.555, 0.645, 0.858 µm) as:

$$\{\rho_1 = \rho_{ASTER}(0.56) \times 0.9140\} \ ; \ \{\rho_2 = \rho_{ASTER}(0.66) \times 0.8918\} \ ; \ \{\rho_3 = \rho_{ASTER}(0.81) \times 0.8857\}, \tag{2}$$

the coefficients being determined from linear regressions.

The central panel of Figure 7 demonstrates that the 2018 blue retrieval algorithm yields enhanced blue reflectances, improving the final image rendering. Furthermore, the absolute quality of the 2018 blue reconstruction is proven by Figure 10 showing the excellent correlations (at least 0.98 with low standard deviations, max of 0.03) between all three bands of ASTER and MODIS. This is further highlighted by Figure 11 where a comparable agreement is found with Landsat-8 OLI L1T products which holds a real blue channel. This Landsat-8 L1T were obtained via the USGS EarthExplorer.

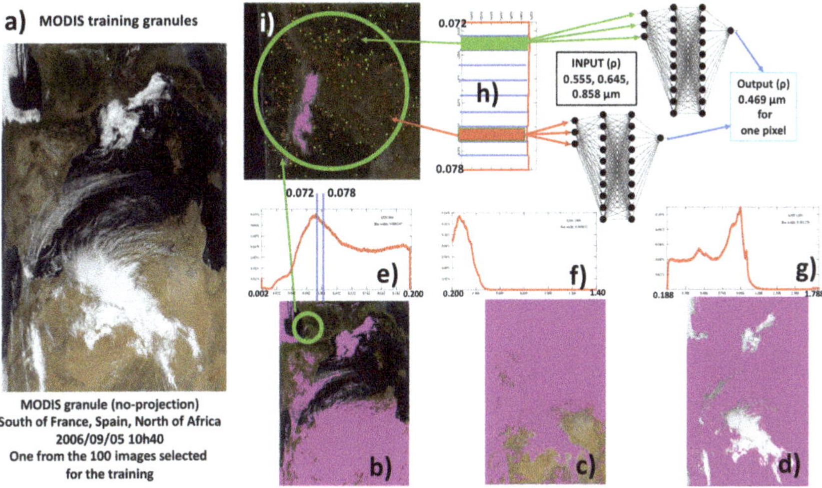

Figure 9. Scheme illustrating the pixel classification of the training (**a**) MODIS data. Subfigures (**b–d**) display the selected pixels (unselected pixels in magenta) sorted into classes (bins) by the (**e–g**) histograms. In histogram (**e**), we select 10 bins from $0.072 < x < 0.078$ (**h**), and display the spatial location of two bins as green and red dots in the zoom (**i**). The MODIS reflectance values within a bin are used as inputs of the dedicated ANN (feedforward ANN with 3 inputs, 2 hidden layers with 10 nodes each, and 1 output) to retrieve the 0.469 μm blue band.

In order to preserve the spectral response of ASTER instrument, the red and green bands of the final visible composite are the ASTER original measurements, while the blue is reconstructed from the cloud of ANNs. As the MODIS and ASTER bands response are slightly shifted (See Figure 1), one can observe in Figure 12 subtle differences for red and green areas in the final color rendering of the natural ASTER composite with respect to MODIS.

To further illustrate the accuracy of the blue reconstruction, we present in Figure 13, examples of heterogeneous natural-color reconstructions. The final results preserve the 15 m ASTER sensitivity, as we notably recognize the signatures of thin clouds on both dark (Figure 13a,b) and bright backgrounds (Figure 13c) with their shadows, as well as the fine patterns of bright sand dunes in deserts (Figure 13d).

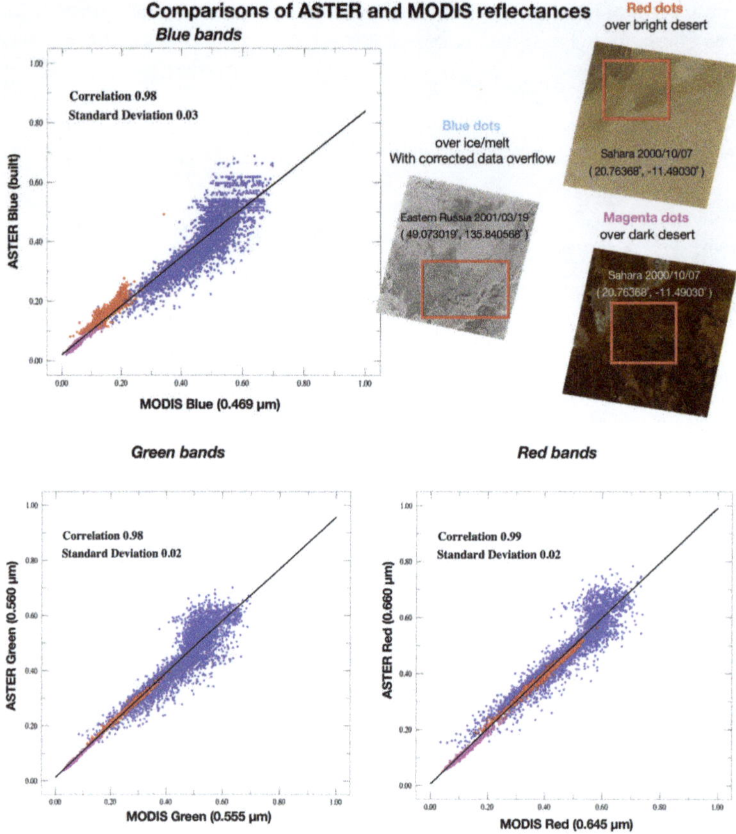

Figure 10. Correlations between ASTER and MODIS reflectance values for the red, green and blue bands, across several surfaces (selected within the red boxes), as ice/melt areas (blue dots), bright desert areas (red dots), and dark desert areas (magenta dots).

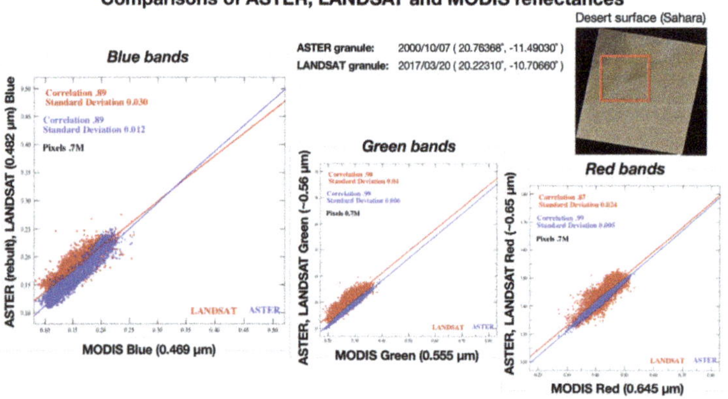

Figure 11. Correlations between ASTER (blue), OLI (red) and MODIS reflectance values (selected within the red box) for the red, green and blue bands over a desert area.

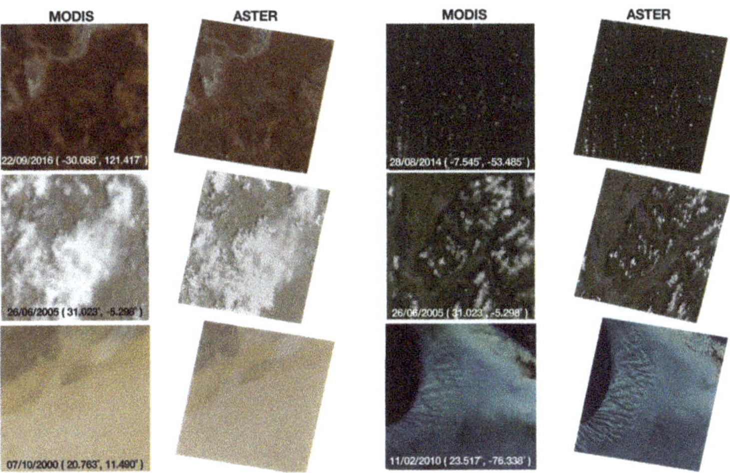

Figure 12. Comparison between the color rendering by MODIS on the left and ASTER (natural color simulated) on the right.

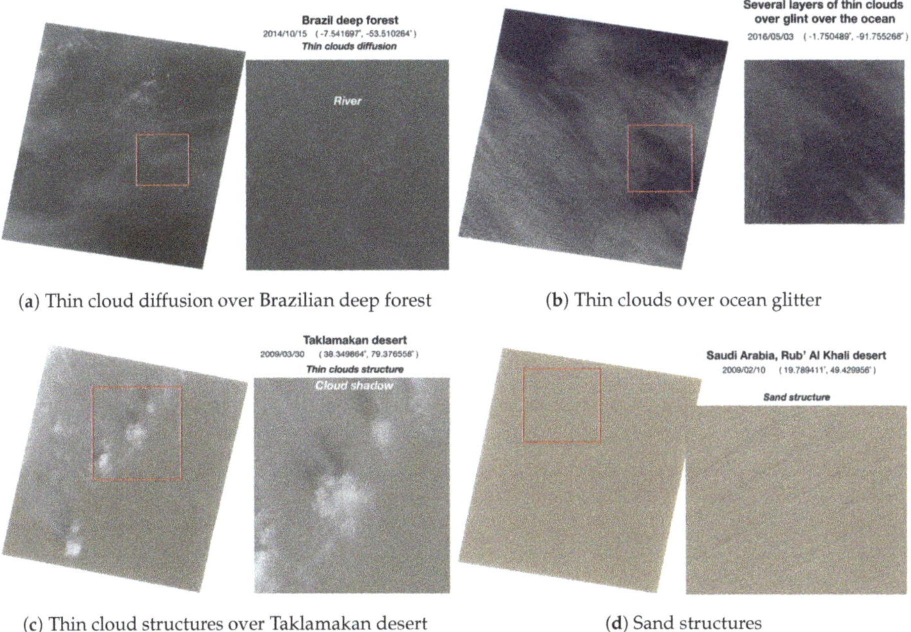

(**a**) Thin cloud diffusion over Brazilian deep forest

(**b**) Thin clouds over ocean glitter

(**c**) Thin cloud structures over Taklamakan desert

(**d**) Sand structures

Figure 13. Examples of heterogeneous reconstructions of natural-color ASTER granules in different areas. The left figure is the ASTER granule and the right inset a zoom into the red box.

3. Building a Global Mosaic

The protocol to construct the CLAMS global mosaic released in 2012 has involved the scanning of the whole distributed ASTER portfolio from 2003 onwards to define a classification algorithm that sorts images with respect to their temporal distance to a seasonal reference date, but also with increasing degree of cloudiness as to minimize cloudy granules. About 780,000 granules where analyzed from where 70% were downloaded and colorized, while 40% contribute to the final mosaic.

3.1. Color-Matching of ASTER Granules to FondsDeSol Reference

In order to construct a seamless global mosaic image, radiometric and phenomenological scene differences have to be minimized across space and time. Helmerand and Ruefenacht [24] used histogram matching of adjacent scenes to build small-size regional mosaics, but this approach leads to increasingly growing bias when applied to large-scale areas. The availability of a world-wide cloud-free surface reflectance database FondsDeSol [19] centered at the end of august is used as reference to equalize the ASTER granules over the whole Earth.

To achieve the color equalization of the ASTER RGB bands and remove as far as possible bidirectional effects and seasonal changes, we have used the iterative distribution transfer (IDT) algorithm proposed by Pitié et al. [25], which allows within few iterations, typically less than 10, an excellent color match to the FondsDeSol reference. It is noteworthy that this colorization process allows to partially dehaze ASTER granules that are blurred by the atmospheric aerosols. An example is displayed in Figure 14, which presents in subset Figure 14 (1) patched natural color ASTER granules tainted by the atmospheric components. Figure 14 (2) shows the same patch colorized by the IDT process, the only visible transitions being that of the granules boundaries. Indeed, seasonal effects are smoothened but may be still be slightly visible when the land surface is significantly altered across various seasons (field structures, forest, rivers, etc.).

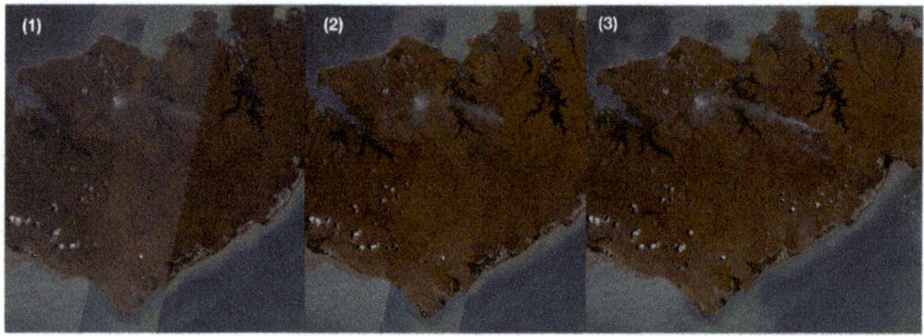

Figure 14. Illustrations of the ASTER mosaic quality over Van Diemen Gulf (Australia): (1) patch of the natural color ASTER granules; (2) patch the IDT colorized granules; (3) seamless blending of IDT colorized granules.

3.2. Seamless ASTER Granules Blending

To create seamless mosaics, we relied on an effective methodology using the Laplacian pyramid-based algorithm introduced by Burt and Adelson [26]. Multi-scale Gaussian pyramid representations of the overlapping areas of the patched granules are used to merge all the spatial structural details at the 15 m resolution, the color matching being already achieved by the FondsDeSol color equalization preceding step. Figure 15 illustrates the reflectance values over transects of two different surface types in the Nusa Tenggara (Indonesia) and Van Diemen Gulf (Australia). The black curves represent the original ASTER natural color granules, the magenta or purple lines refer to the values across IDT colorized granules, while the R,G,B curves correspond to the final results after pyramid blending. Note that the different steps shift the reflectance values but preserve the fine structures of the original data. In Figure 15a, the colorization and blending effects are minor across the granules boundaries (vertical black lines). In Figure 15b we have highlighted with red circles large changes in the reflectance values between the two last steps of the mosaic process (IDT and blending). In this example the blending step (displayed in the (3) inset) brings in thin smoke plumes from the background image (https://earthexplorer.usgs.gov/).

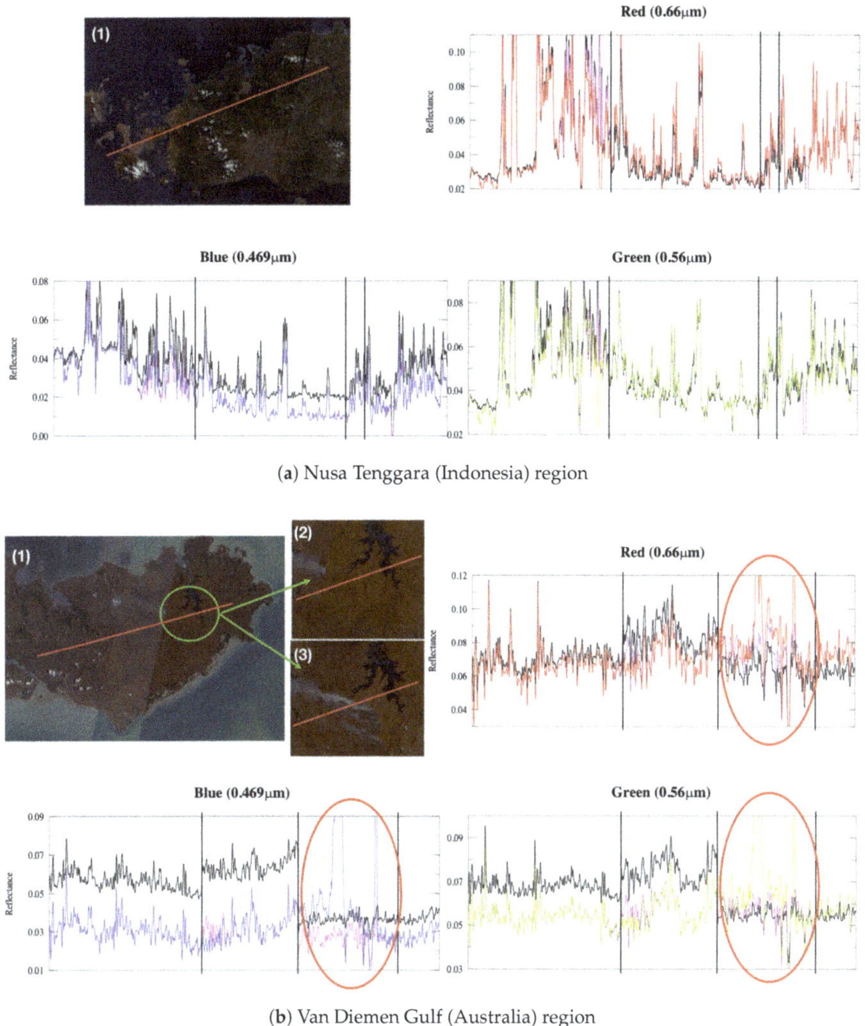

(**a**) Nusa Tenggara (Indonesia) region

(**b**) Van Diemen Gulf (Australia) region

Figure 15. RGB reflectance values across the red marked transects; black lines refer to the patch of the natural color ASTER granules (1); magenta lines to the patch of IDT colorized granules (2); red, green, and blue lines correspond to the final reflectance values in the seamless CLAMS mosaic (3). Black vertical lines mark the granules boundaries.

3.3. Filling the Gaps

For limited numbers of very small areas, where we did not find any suitable ASTER granules, we patched the mosaic with less than 100 Landsat-7 ETM+ granules processed by NASA/JPL CalTech, which is derived from GeoCover2000 [15]. These were color-equalized with the IDT algorithm and merged into the mosaic with the pyramid blending method.

3.4. Distribution of the Mosaic

Partial mosaics samples can be seen on http://newtec.univ-lille1.fr/LARGEMOSAICSFR/ with a web interface that allows to visualize and zoom into several subsets of the global mosaic, with the 2012 blue algorithm (France, Saudi Arabia), and the 2018 blue algorithm (Australia, Indonesia, South Africa, Bahamas, Antarctic). The global mosaic with the 2012 blue algorithm will be available on the https://gbank.gsj.jp/madas/?lang=en web-platform in 2019, while the latest higher quality one shall be also available in the near future.

4. Conclusions and Future Work

This work is the first world-wide mosaic of ASTER images in natural colors. Two main problems of the distributed data have been tackled, namely the saturation over bright surfaces, and most importantly the lack of a blue band. The latter was reconstructed using a large cloud of ANNs trained on MODIS reflectance values, improving the blue rendering with respect to the 2012 version. The granules with the 2012 version are readily available on the https://gbank.gsj.jp/madas/?lang=en web-platform, while the latest higher quality ones shall be available in the near future. The availability of true-color ASTER granules is by itself relevant to help comparisons with other huge high-resolution database as Landsat, in particular to monitor surface changes.

The construction of the mosaic implies to harmonize the colors of spatially and temporarily separated granules and to fusion them with seamless blending algorithms. We have demonstrated that the various steps preserve the 15 m spatial structures of the original ASTER images. The created global mosaic is a real 15 m resolution mosaic that can serve as a reference to validate the quality of other global maps [15,16]. The presented mosaic protocol with the improved blue retrieval will shortly be applied to create a revised version of the ASTER mosaic. The proposed algorithmic protocol has processed several hundred thousands ASTER images and its robustness and modularity can be adapted to similar sensors with enormous image libraries, such as SPOT.

Author Contributions: Conceptualization, L.G.; Methodology, L.G. and V.V.; Software, L.G. and V.V.; Supervision, H.Y.; Validation, L.G. and H.Y.; Writing—original draft, L.G., V.V. and H.Y.

Funding: This research received no external funding.

Acknowledgments: The research for the ASTER in this paper was partially supported by the Ministry of Economy, Trade and Industry in Japan. Original ASTER imagery is provided by "NASA/METI/AIST/Japan Spacesystems, and U.S./Japan ASTER Science Team". All MODIS products are courtesy of the online Data Pool at the NASA Land Processes Distributed Active Archive Center (LP DAAC), USGS/Earth Resources Observation and Science (EROS) Center, Sioux Falls, South Dakota.

Conflicts of Interest: The authors declare no conflict of interest.

References

1. Abrams, M.; Hook, S.; Ramachandran, B. *ASTER User Handbook, Version 2*; Jet Propulsion Laboratory: Pasadena, CA, USA, 2002. Available online: http://asterweb.jpl.nasa.gov/content/03_data/04_Documents/aster_user_guide_v2.pdf (accessed on 3 January 2019).
2. Chen, C.F.; Tsai, H. A spectral transformation technique for generating SPOT natural color image. In Proceedings of the 19th Asian Conference on Remote Sensing, Manila, Philippines, 16–20 November 1998.
3. Knudsen, T. Technical Note: Pseudo natural color aerial imagery for urban and suburban mapping. *Int. J. Remote Sens.* **2005**, *26*, 2689–2698. [CrossRef]
4. Patra, S.K.; Shekher, M.; Solanki, S.S.; Ramachandran, R.; Krishnan, R. A technique for generating natural colour images from false colour composite images. *Int. J. Remote Sens.* **2006**, *27*, 2977–2989. [CrossRef]
5. Huixi, X.; Yunhao, C. A technique for simulating pseudo natural color images based on spectral similarity scales. *IEEE Geosci. Remote Sens.* **2012**, *9*, 70–74. [CrossRef]
6. Zhu, C.; Luo, J.; Ming, D.; Shen, Z.; Li, J. Method for generating SPOT natural-colour composite images based on spectrum machine learning. *Int. J. Remote Sens.* **2012**, *33*, 1309–1324. [CrossRef]
7. Hillger, D.W.; Grasso, L.; Miller, S.D.; Brummer, R.; DeMaria, R.J. Synthetic advanced baseline imager true-color imagery. *J. Appl. Remote Sens.* **2011**, *5*, 053520. [CrossRef]
8. Miller, S.D.; Schmidt, C.C.; Schmit, T.J.; Hillger, D.W. A case for natural colour imagery from geostationary satellites, and an approximation for the GOES-R ABI. *Int. J. Remote Sens.* **2012**, *33*, 3999–4028. [CrossRef]

9.	Guindon, B. Assessing the radiometric fidelity of high resolution satellite image mosaics. *ISPRS J. Photogramm. Remote Sens.* **1997**, *52*, 229–243. [CrossRef]

10.	Liew, S.C.; Li, M.; Kwoh, L.K.; Chen, P.; Lim, H. "Cloud-free" multi-scene mosaics of SPOT images. In Proceedings of the IGARSS '98—1998 IEEE International Geoscience and Remote Sensing, Seattle, WA, USA, 6–10 July 1998; Volume 2, pp. 1083–1085. [CrossRef]

11.	Du, Y.; Cihlar, J.; Beaubien, J.; Latifovic, R. Radiometric normalization, compositing, and quality control for satellite high resolution image mosaics over large areas. *IEEE T. Geosci. Remote Sens.* **2001**, *39*, 623–634. [CrossRef]

12.	Bindschadler, R.; Vornberger, P.; Fleming, A.; Fox, A.; Mullins, J.; Binnie, D.; Paulsen, S.J.; Granneman, B.; Gorodetzky, D. The Landsat image mosaic of Antarctica. *Remote Sens. Environ.* **2008**, *112*, 4214–4226. [CrossRef]

13.	Roy, D.P.; Ju, J.; Kline, K.; Scaramuzza, P.L.; Kovalskyy, V.; Hansen, M.; Loveland, T.R.; Vermote, E.; Zhang, C. Web-enabled Landsat Data (WELD): Landsat ETM+ composited mosaics of the conterminous United States. *Remote Sens. Environ.* **2010**, *114*, 35–49. [CrossRef]

14.	Choi, J.; Jung, H.S.; Yun, S.H. An efficient mosaic algorithm considering seasonal variation: Application to KOMPSAT-2 satellite images. *Sensors* **2015**, *15*, 5649–5665. [CrossRef] [PubMed]

15.	Plesea, L. Remote access to very large image repositories, a high performance computing perspective. In Proceedings of the Earth-Sun System Technology Conference, Adelphi, MD, USA, 28–30 June 2005.

16.	Descartes Maps. Available online: http://www.descarteslabs.com/maps.html (accessed on 3 January 2019).

17.	AIST "ASTER-VA", Value-Added Satellite Observation Data, Provided at No Cost. Available online: https://gbank.gsj.jp/madas/ (accessed on 3 January 2019).

18.	Obata, K.; Tsuchida, S.; Yamamoto, H.; Thome, K. Cross-calibration between ASTER and MODIS visible to near-infrared bands for improvement of ASTER radiometric calibration. *Sensors* **2017**, *17*, 1793. [CrossRef] [PubMed]

19.	Gonzalez, L.; Bréon, F.M.; Caillault, K.; Briottet, X. A sub km resolution global database of surface reflectance and emissivity based on 10-years of MODIS data. *ISPRS J. Photogramm. Remote Sens.* **2016**, *122*, 222–235. [CrossRef]

20.	Vermote, E.; Kotchenova, S.; Tanré, D.; Deuzé, J.; Herman, M.; Roger, J.C.; Morcrette, J. 6SV Code. 2015. Available online: http://6s.ltdri.org (accessed on 20 February 2019).

21.	Gonzalez, L.; Briottet, X. North Africa and Saudi Arabia Day/Night Sandstorm Survey (NASCube). *Remote Sens.* **2017**, *9*, 896. [CrossRef]

22.	Zell, A.; Mache, N.; Hübner, R.; Mamier, G.; Vogt, M.; Schmalzl, M.; Herrmann, K.U. SNNS (Stuttgart Neural Network Simulator). In *Neural Network Simulation Environments*; Springer: Boston, MA, USA, 1994; pp. 165–186.

23.	Miikkulainen, R. Topology of a Neural Network. In *Encyclopedia of Machine Learning*; Sammut, C., Webb, G.I., Eds.; Springer: Boston, MA, USA, 2010; pp. 988–989. [CrossRef]

24.	Helmer, E.H.; Ruefenacht, B. Cloud-Free Satellite Image Mosaics with Regression Trees and Histogram Matching. *Photogramm. Eng. Remote Sens.* **2005**, *71*, 1079–1089. [CrossRef]

25.	Pitié, F.; Kokaram, A.C.; Dahyot, R. Automated colour grading using colour distribution transfer. *Comput. Vis. Image Underst.* **2007**, *107*, 123–137. [CrossRef]

26.	Burt, P.J.; Adelson, E.H. A Multiresolution Spline with Application to Image Mosaics. *ACM Trans. Graph.* **1983**, *2*, 217–236. [CrossRef]

Sample Availability: Examples of the global natural colored mosaic are available on the http://newtec.univ-lille1.fr/LARGEMOSAICSFR/ website.

Article

New Insights of Geomorphologic and Lithologic Features on Wudalianchi Volcanoes in the Northeastern China from the ASTER Multispectral Data

Han Fu [1,2], Bihong Fu [1,*], Yoshiki Ninomiya [3] and Pilong Shi [1]

[1] Aerospace Information Research Institute, Chinese Academy of Sciences, Beijing 100094, China; fuhan2017@radi.ac.cn (H.F.); shipl@radi.ac.cn (P.S.)

[2] University of Chinese Academy of Sciences, Beijing 100094, China

[3] Geological Survey of Japan, National Institute of Advanced Industrial Science and Technology, Higashi 1-1-1, Tsukuba 305-8567, Japan; yoshiki.ninomiya@aist.go.jp

* Correspondence: fubh@radi.ac.cn; Tel.: +86-10-8217-8096

Received: 17 October 2019; Accepted: 12 November 2019; Published: 14 November 2019

Abstract: Advanced Spaceborne Thermal Emission and Reflection Radiometer (ASTER) imaging system onboard NASA's (National Aeronautics and Space Administration's) Terra satellite is capable of measuring multispectral reflectance of the earth's surface targets in visible and infrared (VNIR) to shortwave infrared (SWIR) (until 2006) as well as multispectral thermal infrared (TIR) regions. ASTER VNIR stereo imaging technique can provide high-resolution digital elevation models (DEMs) data. The DEMs data, three-dimensional (3D) perspective, and ratio images produced from the ASTER multispectral data are employed to analyze the geomorphologic and lithologic features of Wudalianchi volcanoes in the northeastern China. Our results indicate that the 14 major conical volcanic craters of Wudalianchi volcanoes are arranged as three sub-parallel zones, extending in a NE (Northeast) direction, which is similar to the direction of regional fault system based on the ASTER DEMs data. Among the 14 volcanic craters in Wudalianchi, the Laoheishan, and Huoshaoshan lavas flows, after the historic eruptions, pouring down from the crater, partially blocked the Baihe River, which forms the Five Large Connected Pools, known as the Wudalianchi Lake. Lithologic mapping shows that ASTER multispectral ratio imagery, particularly, the Lava Flow Index (LFI) (LFI = B10/B12) imagery, can clearly distinguish different lava flow units, and at least four stages of volcanic eruptions are revealed in the Wudalianchi Quaternary volcano cluster. Thus, ASTER multispectral TIR data can be used to determine relative dating of Quaternary volcanoes in the semi-arid region. Moreover, ASTER 3D perspective image can present an excellent view for tracking the flow directions of different lavas of Wudalianchi Holocene volcanoes.

Keywords: 3D perspective view; morphology; lithology; Wudalianchi volcano; ASTER multispectral data

---

1. Introduction

Remote sensing can provide a certain resolution of spectral, spatial, and temporal coverage based on the type of sensor for geologic mapping and monitoring at numerous volcanoes throughout the world [1,2]. The use of spaceborne and airborne remote sensing data to monitor the volcanoes and map the products of eruptions has been ongoing for decades [3]. Satellite remote sensing data have been widely used to detect or monitor the eruption of volcanoes [4–6]. However, how to document the products of volcanic eruptions and morphology using satellite remote sensing data is another challenge for geoscientists [6,7]. The weathering of active lava flows in arid and semi-arid environments is accompanied by changes in their thermal infrared emittance spectra [8,9]. The spectral differences caused by the weathering can be measured and mapped with multispectral imaging system [8].

The Advanced Spaceborne Thermal Emission and Reflection Radiometer (ASTER) sensor was launched in 18 December, 1999, onboard the first NASA's Earth Observation System (EOS) series of satellites, Terra [10,11]. The ASTER covered a wide spectral region with 14 bands from visible to thermal infrared with high spatial, spectral and radiometric resolution as shown in Table 1. Three visible and near infrared (VNIR) bands, six shortwave infrared (SWIR) bands (until 2006), and five thermal infrared (TIR) bands with the spatial resolution of 15 m, 30 m, and 90 m, respectively. In addition, the bands 3N and 3B in near infrared bands have a stereoscopic capability, which can be used to generate DEM [12]. ASTER-TIR is the first satellite-borne multispectral TIR remote sensing system with spectral, spatial, and radiometric resolutions adequate to be used for geologic applications, such as determining the relative age dating of lavas [13,14]. Compared with two bands of Landsat TM or ETM in the SWIR region (between 1.6 to 2.5 microns), the ASTER SWIR sensor has six bands in this region and has the capability to identify mineral component of surface rocks in the semi-arid to arid region [15,16]. The ASTER multispectral SWIR and TIR sensors can provide an important tool for monitoring heat flow related to volcanic activities [4,5]. Therefore, ASTER can provide a potential tool for mapping the products from active volcanoes from regional to global scales.

Table 1. Wavelength Range of The Advanced Spaceborne Thermal Emission and Reflection Radiometer Data [10,12].

Band	Wavelength Range (μm)	Band Type	Spatial Resolution (m)
B1	0.52–0.60	VNIR (visible and near infrared)	15
B2	0.63–0.69	VNIR	15
B3N	0.76–0.86	VNIR	15
B3B	0.76–0.86	VNIR	15
B4	1.60–1.70	SWIR (shortwave infrared)	30
B5	2.145–2.185	SWIR	30
B6	2.185–2.225	SWIR	30
B7	2.235–2.285	SWIR	30
B8	2.295–2.365	SWIR	30
B9	2.36–2.43	SWIR	30
B10	8.125–8.475	TIR (thermal infrared)	90
B11	8.475–8.825	TIR	90
B12	8.925–9.275	TIR	90
B13	10.25–10.95	TIR	90
B14	10.95–11.65	TIR	90

There are two Quaternary volcanoes with historic eruptions in northeastern China (Figure 1a, WD, Wudalianchi Volcano; TC, Changbaishan Tianchi Volcano). Petrology, chronology, and geochemistry of these Quaternary volcanoes have been widely addressed by researchers [17–20]. However, general research on the topographic and geomorphologic features of these Quaternary volcanoes is still lacks. Furthermore, no detailed map shows the distribution of lava flows from these different stages.

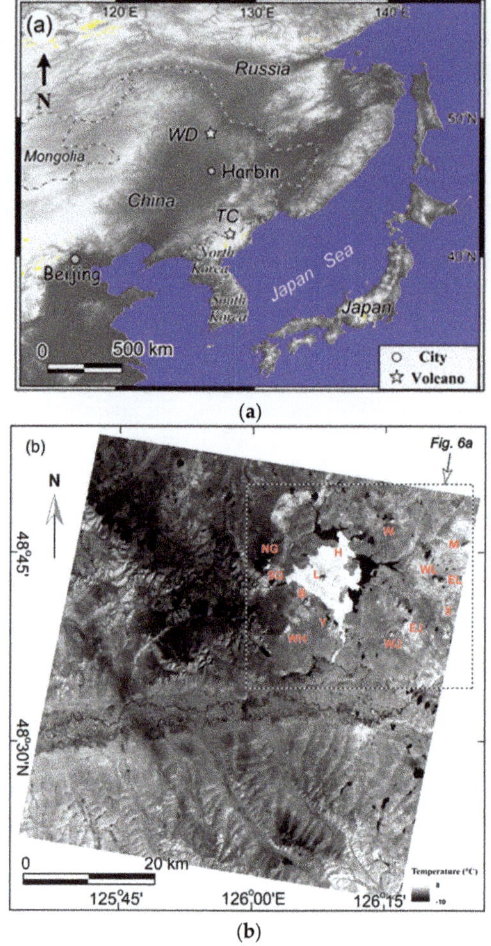

Figure 1. (**a**) Topographic map generated from the USGS (United States Geological Survey) SRTM (Shuttle Radar Topography Mission) DEM (Digital Elevation Model) data, showing the locations of Wudalianchi (WD) and Tianchi (TC) volcanoes in northeast China. (**b**) ASTER thermal infrared (band 13) image observed at 6 April, 2004 in the Wudalianchi volcanic region. WH: Wohushan; SG: South Gelaqiushan; NG: North Gelaqiushan; B: Bijiashan; L: Laoheishan; H: Huoshaoshan; W: Weishan; WJ: West Jiaodebushan; EJ: East Jiaodebushan; X: Xiaogushan; WL: West Longmenshan; EL: East Longmenshan; M: Molabushan; Y: Yaoquanshan.

In this study, we documented the products of eruptions and morphology of Quaternary volcanoes in northeast China using satellite remote sensing techniques. The main goal of this study is to verify how to map geomorphologic features and lava flows of Quaternary volcanoes using ASTER multispectral data. A field investigation was conducted around the Wudalianchi volcanoes from late July to early August, 2005 as part of a collaborative project between the Institute of Geology and Geophysics, Chinese Academy of Sciences and Geological Survey of Japan, The National Institute of Advanced Industrial Science and Technology. During the field investigation, we observed the difference of vegetation coverage in the Wudalianchi volcanoes, which can provide useful information to understand the spectral difference of the lava flows in the Laoheishan and Huoshaoshan lava flow regions.

2. Geologic Setting

The Wudalianchi and Changbaishan volcanoes in northeastern China have been widely studied [18–22]. Previous studies indicated that they are intraplate volcanoes far from the West Pacific Plate subduction zone [22–24].

The Wudalianchi volcanic cluster is located on Wudalianchi city, about 350 km north of Harbin, capital of Heilongjiang Province (Figure 1a). The volcanic group covers an area of about 800 km^2 and is composed of 14 major volcanic cones (Figure 1b). These lava rocks in the Wudalianchi region are strong alkaline potassium-rich volcanic rock with an average K (Potassium) content of 5.28% and average SiO$_2$ content of 50.46%. [17,25].

The K-Ar (Potassium-Argon) dating of Wudalianchi monogenetic volcanic products suggested three stages of eruptions: Early-middle Pleistocene (circa 1.33 ± 0.08 to 0.8 ± 0.02 Ma), Late Pleistocene (circa 0.63–0.3 Ma), and historic periods [18,26]. Among the 14 volcanoes in the Wudalianchi volcanic cluster, the lava flows distributed around Xiaogushan (X), Yaoquanshan (Y), Wohushan (WH), and West Jiaodebushan (WJ) are alkaline basalt that belongs to the Early-middle Pleistocene; the lava flows from Molabushan (M), Bijiashan (B), South Gelaqiushan (SG), North Gelaqiushan (NG), Weishan (W), and East Jiaodebushan (EJ) are alkaline basalt that belongs to the Late Pleistocene [27]. However, it is still a big debate on the historic volcanic activity. Some researchers indicated that the most recent major explosive eruption occurred between 1719–1721 AD and formed craters of the Laoheishan (L) and Huoshaoshan (H) volcanoes [17,28]. Historical local documents suggest that the lava flows from Laoheishan and Huoshaoshan craters were mainly occurred in 1719–1721 AD and 1776 AD [26,27,29]. Tectonically, the Wudalianchi volcanic cluster is located in the triangular area formed by the three structural units of the Greater Khingan uplift, the Lesser Khingan uplift and the Songliao faulted basin. Previous studies suggested that the distribution of 14 volcanoes in the Wudalianchi area is associated with the Northeast (NE)-striking faults, Northwest (NW)-striking faults and near Eastwest (EW)-striking faults or fracture zones [30,31]. However, the geological interpretation of ASTER DEM image (Figure 2) shows that the NE and NW-striking faults are major structure in the Wudalianchi volcanic field. The most of volcanoes are extending as NE-striking direction (such as L, B, M and WH in Figure 2). We inferred that these conjugate faults might provide the pathway for magma migration during the volcanic eruption.

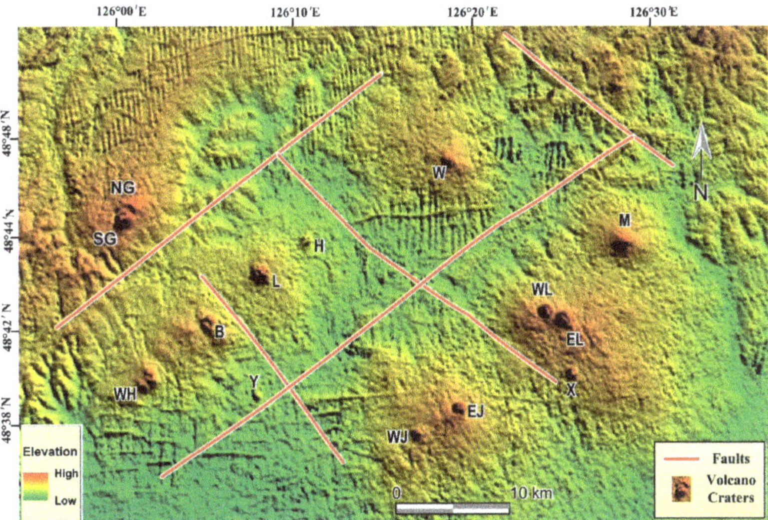

Figure 2. Geological sketch showing the relation between tectonic context and volcanism in the Wudalianchi region. Letters refer to name of volcanic craters are same as Figure 1b.

Concerning the geodynamics of the formation of the Wudalianchi volcanoes, some studies suggested that they are the intraplate volcano associated with the westward extending subduction slab of the West Pacific Plate [22–24]. Based on analysis of the topographic image as well as the previous data given by seismic evidence [32,33], we have drawn a three-dimensional geodynamic sketch map using Artificial Intelligence (AI) software, which shows the relationship between the active intraplate volcanoes in NE China and the deep subduction of the Pacific slab as shown in Figure 3. This geodynamic sketch map implies that the intraplate volcanoes in the interior of the Asian continent are not the back-arc volcanoes related to the subducting Pacific slab such as the Japanese Islands, but the continental volcanoes are likely induced by the deep subduction and dehydration of the west Pacific stagnant slab, possibly through hot and wet upwelling in the big mantle wedge under the NE China as suggested by the geophysical studies [34–37].

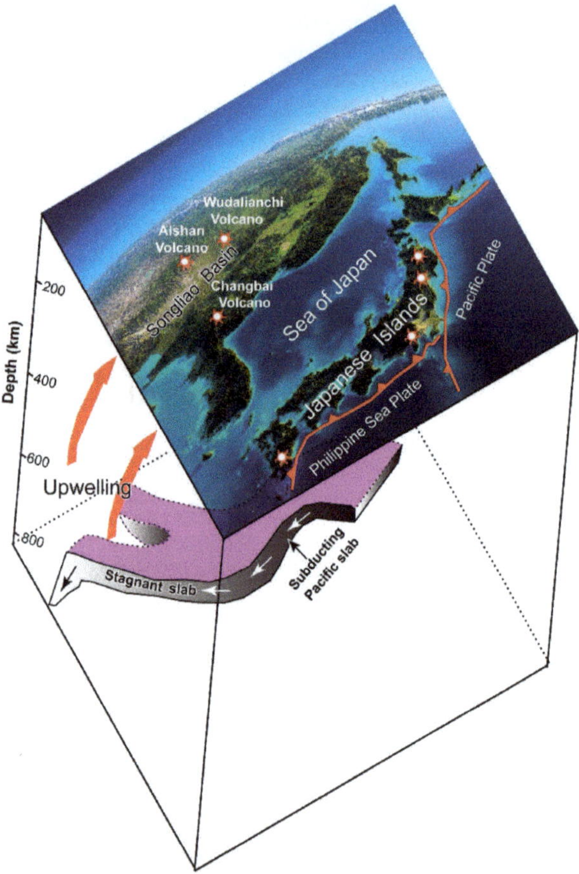

Figure 3. The three-dimensional (3D) geodynamic sketch map showing that the active volcanoes in Northeastern Asia are induced by the deep subduction of the western Pacific stagnant and subducting slabs.

The Wudalianchi volcano was successfully selected as one of the UNESCO (United Nations Educational, Scientific and Cultural Organization) Global Geoparks in 2003. The major reason is that the lava flow composition of the Wudalianchi volcanic belt is very special and is characterized by strong alkaline potassium-rich volcanic rock [38]. Its color ratio (40%–55%) is higher than the coarse porphyry and ring rock (35%–20%), while at the same time, it does not contain basic plagioclase, which basalt should have; thus, it is neither a rough rock nor an alkaline basalt.

Moreover, Laoheishan and Huoshaoshan are the latest volcanoes in the Wudalianchi volcanic group, whose most recent eruptions occurred in 1719–1721 AD and 1776 AD as recorded by historic documents. Thus, their lava flows formed by the eruptions are bare, well preserved, and the lava flows characteristics are clear. As the geomorphologic and lithologic features of the eruptions on Laoheishan and Huoshaoshan are well preserved, the Wudalianchi volcano has the reputation of "Volcano Natural Museum" [39,40], which is another important reason for its selection in the UNESCO Global Geoparks.

3. Methodology

Three scenes of relatively cloud-free ASTER data covering the Wudalinchi region were used in this study. The DEM data, which belongs to the ASTER Level-3A product, were derived from the ASTER Level-1A data, with a vertical accuracy of 20 m [10,12]. See Table 2 for details. The datasets used in this study is described in a flowchart (Figure 4).

Table 2. Details of ASTER data.

Obtained Date	Central Point	Product Level	Vertical Accuracy of Digital Elevation Model
7 March, 2002	48.87°N, 126.42°E	orthorectified Level-3A	20 m
6 September, 2002	48.73°N, 125.82°E	orthorectified Level-3A	20 m
6 April, 2004	48.60°N, 125.98°E	orthorectified Level-3A	20 m

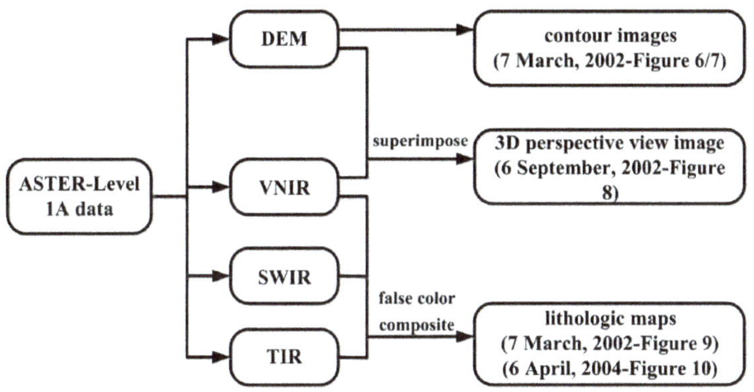

Figure 4. The flowchart of the datasets used in this study.

3.1. The Generation of Contour Image and 3D Perspective View Image

A contour image was derived automatically from ASTER Level-3A DEM data by using ER-Mapper (Earth Resource-Mapper) software. Three-dimensional (3D) perspective view image was generated through superimposing a three bands color composite image (the bands 2, 3, and 1 assigned in red, green, and blue, respectively) on the ASTER DEM data by using the same software.

3.2. Laboratory Measurement for the Emissivity of Lava Fows

The laboratory emissivity of typical lava rocks from the Wudalianchi region were measured by using portable FTIR (Fourier Transform Infrared Spectrometer) spectrometer (μ-FITR, Model 102) manufactured by Designs and Prototypes, Ltd., USA, in Nimoiya's Spectral Lab. of Geological Survey of Japan. The emissivity measurements of the samples of relatively low temperature (typically < 60 degree in Celsius) at the natural surfaces are difficult in achieving high S/N (Signal-Noise) ratio, thus, the measurements were generally made for the polished surfaces of the highly heated samples using the plate heating pot in the laboratory. This instrument provides spectral coverage from 2 to 16 μm with 6 cm^{-1} spectral resolution (which means band width) [41]. We collected three samples around Laoheishan and Huoshaoshan; the lithologic characters of these samples are listed in Table 3. As shown in Figure 5, the emissivity curve of sample WD-$_{2w}$ is displaying as flat one, which is the weathered surface of lava covered by some dry lichen-a symbiotic complex of algae and fungi [42]. As for the other samples, the high emissivity is located around ASTER band 10, and low emissivity of these lavas located around ASTER band 12 as shown in Figure 5. Moreover, the depth of low emissivity near 9.6 μm for these samples is quite different (such as WD-$_{3c}$ and WD-$_{3w}$), which indicates that there is a spectral difference with increase of weathered degree. So far, we can highlight these lava flows by using Lava Flow Index (LFI) of B10/B12 in order to distinguish whether the lava flows are fresh one or weathered one covered by dry lichen or low vegetation. This is different from using the Mafic Index (MI = B12/B13) to extract lithologic information for mafic-ultramafic rocks, as suggested by Ninomiya et al. [14].

Detailed description of samples WD$_1$, WD$_2$, and WD$_3$ see Table 3.

Table 3. Lithologic characters of samples collected from the Wudalianchi volcanoes.

Sample No.	Lithologic Characters	Location
WD-$_1$	fresh surface of black lava	South part of Huoshaoshan
WD-$_{2c}$	fresh cut surface of lava	North part of Laoheishan
WD-$_{2w}$	weathered surface of lava covered by some dry lichen	North part of Laoheishan
WD-$_{3c}$	fresh cut surface of lava	South part of Laoheishan
WD-$_{3w}$	weathered surface of lava	South part of Laoheishan

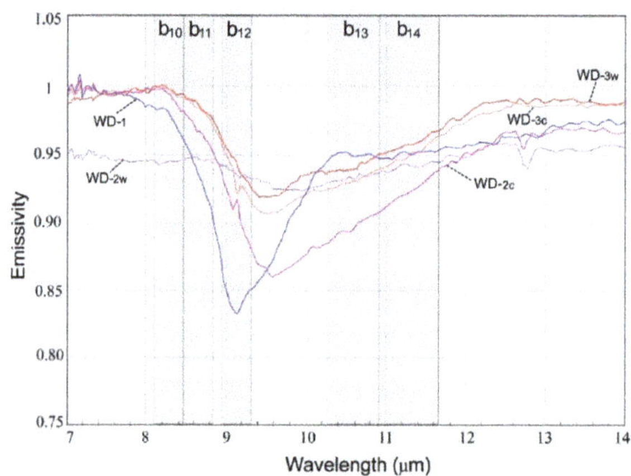

Figure 5. Emissivity of typical rock samples collected from Wudalianchi.

Furthermore, band ratios of 2/1 and 4/6 in the VNIR and SWIR regions are used to highlight hematite and altered Al(OH)$_3$ bearing minerals considering that weathered volcanic lavas may contain Hematite and Al(OH)$_3$. Moreover, the band ratios of B2/B1 for displaying as green could highlight the area covered by snow and ice, which is helpful to distinguish the lave flows from snow and ice.

4. Results

4.1. Topographic and Geomorphologic Features of Quaternary Volcanoes

The ASTER DEMs image shows that the study area is at an elevation range of 175–578 m above sea level as shown by elevation bar in Figure 6a. The 14 major conical volcanic craters are arranged as three sub-parallel zones, extending NE direction, which is similar to the direction of regional fault system [19] and thus, imply that they are associated with NE-striking normal fault system. The east volcanic zone consists of three shield volcanoes with six conical craters (Molabushan, East Longmenshan, West Longmenshan, Xiaogushan, East Jiaodebushan, West Jiaodebushan), the central zone composed of four shield volcanoes (Weishan, Laoheishan, Huoshaoshan, Wohushan) and two isolated cones (Bijiashan, Yaoquanshan), and the west zone consists of a large shield volcano (North Gelaqiushan, South Gelaqiushan) (Figure 7a). A shield volcano always has a low slope, generally less than 5°, while the diameter of the volcanic pedestal is large, generally more than 3 km. While the cone volcano is mainly composed of basalt and the volcanic cone has a relative height difference of 50 m to 750 m. The shape of cone volcano is just like a cone [43].

A contour image can highlight topographic features of these Quaternary volcanoes (Figures 6b and 7).

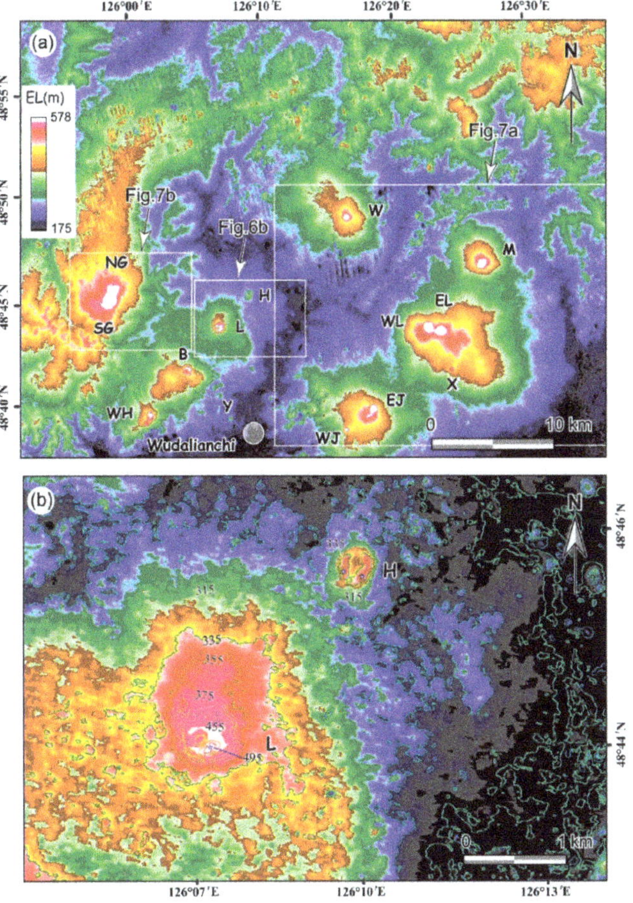

Figure 6. (a) ASTER DEM image, assigned a step color mode, showing the 14 cone-shaped volcanic craters in the Wudalianchi region. The DEM data obtained on 7 March, 2002. Letters refer to name of volcanic craters are same as Figure 1b. (b) Contour image derived from ASTER DEM data showing topographic features of the Laoheishan and Huoshaoshan volcanoes. Contour interval is 20 m, and summits of these craters are indicated by five-shaped stars.

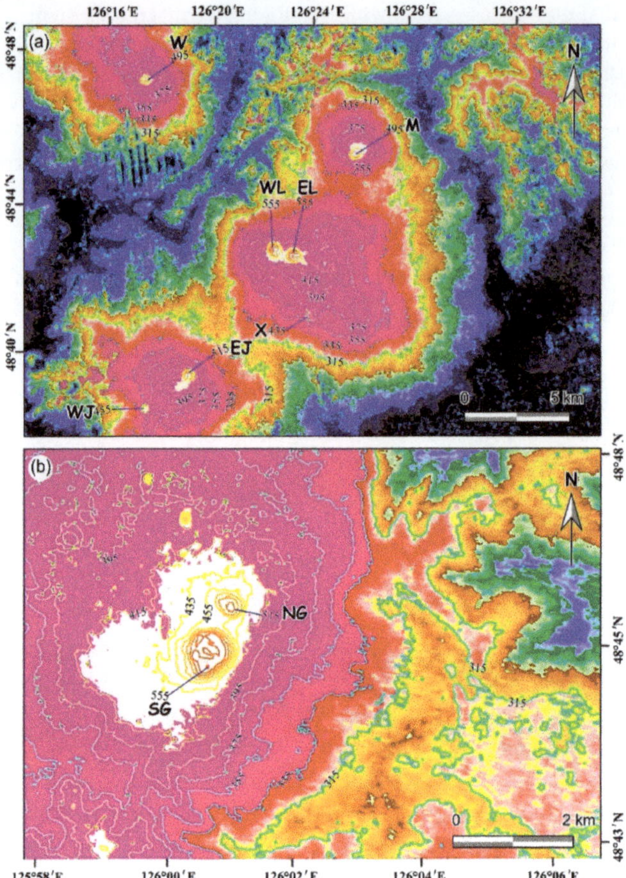

Figure 7. Enlarged contour image derived from ASTER data showing the topographic features of volcanic craters in the Wudalianchi volcanic zone. Letters refer to name of volcanic craters are same as Figure 1b, and contour interval is 20 m. (**a**) The east Wudalianchi volcanic zone. (**b**) The west Wudalianchi volcanic zone.

From Figures 6a and 7, Laoheishan volcano is located on southeast and it is about 505 m in height. Meanwhile, there is a deep crater with an elevation of 385 m to west of the summit. However, the Huoshaoshan volcano appears as a northward-facing horseshoe-shaped volcanic landform with two summits about 355 m in height (as indicated by five-shaped stars in Figure 6b). There are four striking shield volcanoes as shown in Figure 7a. The most remarkable shield volcano with a diameter of 2 km is located in middle of image, which consists of three craters (EL, WL, and X in Figure 7a). Two of them (WL and EL) appear as one pair of nearly eastwest extending glasses with a diameter of circa 400 m in the middle eastern part, and the crater of Xiaogushan (X in Figure 7a) in the south flank of EL crater appears as a northeastward-facing horseshoe shape with a diameter of 200 m. The highest elevations of these three carters are 565 m, 570 m, and 445 m (WL, EL, and X, respectively, in Figure 7a). In the north of this large shield volcanic landform, there is a ring shield volcano with a diameter of about 1000 m and summit height of 505 m (M in Figure 7a). The shield volcano in the southwest part of image consists of two craters with summits of 515 m and 465 m, respectively (EJ and WJ in Figure 7a). Another shield volcano has a diameter of 1250 m and a summit of 495 m, appearing in the top-left of image (W in Figure 7a). Figure 7b shows that an NNE striking shield landform with a pair of cones distributes along the western volcanic zone. The southern zone (SG) exhibits a westward-facing horseshoe shape and the northern zone (NG) exhibits a circular shape with a diameter of 500 m and 220 m, respectively. The summits are about 560 m and 525 m above sea level (SG and NG in Figure 7b) according to the ASTER DEMs contour image.

Three-dimensional (3D) perspective view image can provide an excellent view for geomorphologic features of geologic targets from different view directions [44]. Figure 8 represents an ASTER 3D perspective view image of the Laoheishan and Huoshaoshan volcanoes and adjacent region, which exhibits geomorphologic features of these historic volcanoes clearly (Figure 8).

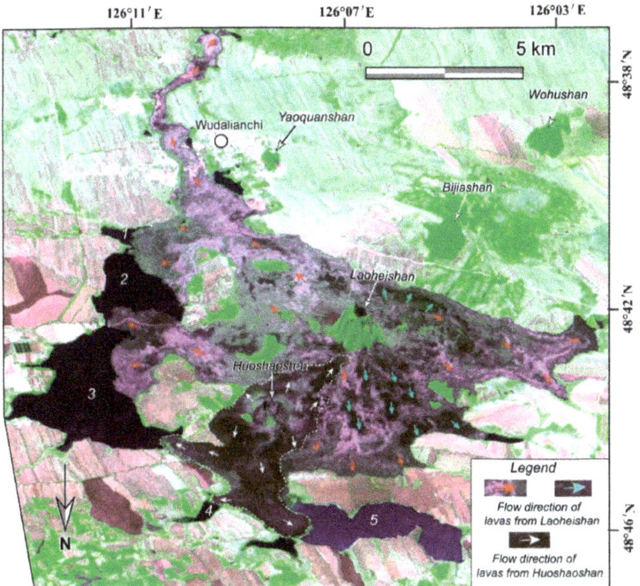

Figure 8. ASTER 3D perspective view image (taken on 6 September, 2002, with a three-time vertical exaggeration), showing lava flow and morphology of the Laoheishan and Huoshaoshan volcanoes. Three bands of ASTER VNIR data used to generate false color composite image (bands 2, 3, and 1 assigned in red, green, and blue, respectively). Green and dark blue color patterns represent vegetation and water body, and the five lakes are marked by numbers 1 to 5. Southward-looking view. The red and cyan arrow marks represent two different types of flow directions of lavas from Laoheishan crater and the white arrow marks represent the flow direction of lavas from Huoshaoshan crater.

Laoheishan crater covered with green vegetation exhibits a cone and locates in the central of the Three-dimensional (3D) perspective view image (Figure 8). About 2.5 km northeast of Laoheishan, Huoshaoshan crater shows a much more broken cone with a half-size of Laoheishan crater (Figure 8). Another geomorphologic feature is lava flows from these two historic volcanoes as shown in Figure 8. Based on different color tones and texture characteristics displayed on remote sensing images, different lavas of Laoheishan and Huoshaoshan can be distinguished. Lava flows from Laoheishan crater can be classified into two types: one is the reddish grey lava exhibiting a radial pattern and flowing as far as 15 km (indicated by red arrows in Figure 8), and the other is the darkish grey to black lava distributed around the crater (indicated by blue arrows in Figure 8). Black lava from the Huoshaoshan crater mainly flowed northward and southward (indicated by white arrows in Figure 8). Flow directions of lavas can be distinguished clearly according to the 3D perspective view image combined with the field observation, although over 280 years has passed. The 3D perspective view image also shows that several lakes distribute around the north and east part of these two volcanoes. The Wudalianchi is named after the Five Large Connected Pools, encircling Laoheishan and Huoshaoshan, as shown in Figure 8. The lakes were formed after the historic eruptions, when molten lava flows, pouring down from the crater, partially blocked the Baihe River.

4.2. Lithologic Mapping of Lava Products and Their Relative Age Dating

A false color composite image of B10/ B12, B2/B1, and B4/B6, assigned as R, G, and B, was derived to enhance subtle spectral change of products from lava flows in different stage of eruptions, as shown in Figure 9. Figure 9 shows that the volcanic lava flows have quite different color patterns (purplish red to bright red), which may represent lava flows formed in different eruptive stages as indicated by Kahle et al. [8] for Hawaiin lava flows by using airborne TIMS (Thermal Infrared Multispectral Scanner) images. In the Wudalianchi region, the lava rocks are strong alkaline potassium-rich volcanic rock with an average K (Potassium) content of 5.28% and average SiO_2 content of 50.46%. The mineral composition has not big difference. Therefore, we consider that the different eruptive stages are mainly responsible for these color patterns. During the field investigation, we had observed the difference in vegetation coverage, which can also affect these color patterns in the Wudalianchi area.

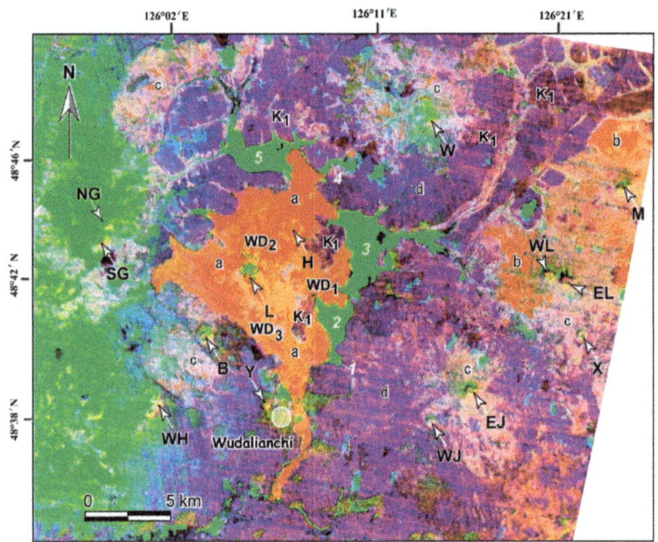

Figure 9. ASTER composite image (taken on 7 March, 2002), band ratios of 10/12, 2/1, and 4/6 displayed as red, green, and blue, respectively, showing the lithologic units in the Wudalianchi region. Letters refer to volcanic craters are same as Figure 1b. Letters "a", "b", "c", and "d" refer to relative dating for Quaternary lava flows. "WD$_1$", "WD$_2$", "WD$_3$" refer to the location of samples. "K$_1$" shows the Cretaceous sandstones and argillites. Bright green color patterns represent snow and ice. The five lakes of Wudalianchi are marked by numbers 1 to 5.

According to the results shown in Figure 9, the Wudalianchi Quaternary lava flows can be divided into at least four stages. Compared with the published maps of these flows in [27], the four stages are: (1) historic lava flows from Laoheishan and Huoshaoshan craters showing a bright red pattern (indicated by letter a in Figure 9), which formed by the most recent eruptions in 1719–1721 AD and 1776 AD as recorded by historic documents [17,28]; (2) lava flows distributed around the west flank of West Longmenshan (WL) and north flank of Molabushan showing a reddish pattern, which represent younger volcanoes (indicated by letter b in Figure 9) and belong to the Holocene [27]; (3) lava flows showing a pink pattern in Figure 9 (indicated by letter c), which may represent the product of third stage abruptions, which is the alkaline basalt belonging to the Late Pleistocene [27]; (4) the purplish blue pattern shows the products from the oldest eruption in the Wudalianchi region (indicated by letter d in Figure 9), which is the alkaline basalt belonging to the Early-middle Pleistocene [27]. In addition, the regions with a purplish red pattern show the Cretaceous sandstones and argillites as observed in the field (indicated by the letter K_1 in Figure 9). Our results are consistent with the regional geological map [27], except that they suggested the lava flows among Molabushan (M) belong to the Late Pleistocene, while our results demonstrated that they are at the same stage with the lava flows distributed around west flank of West Longmenshan (WL), which belong to the Holocene.

The enlarged ASTER image (Figure 10) displays more detailed lithologic features of the Laoheishan and Huoshaoshan volcanoes. Although the lava flows around the Laoheishan and Huoshaoshan volcanoes belong to the same stage, their color tones still have subtle differences.

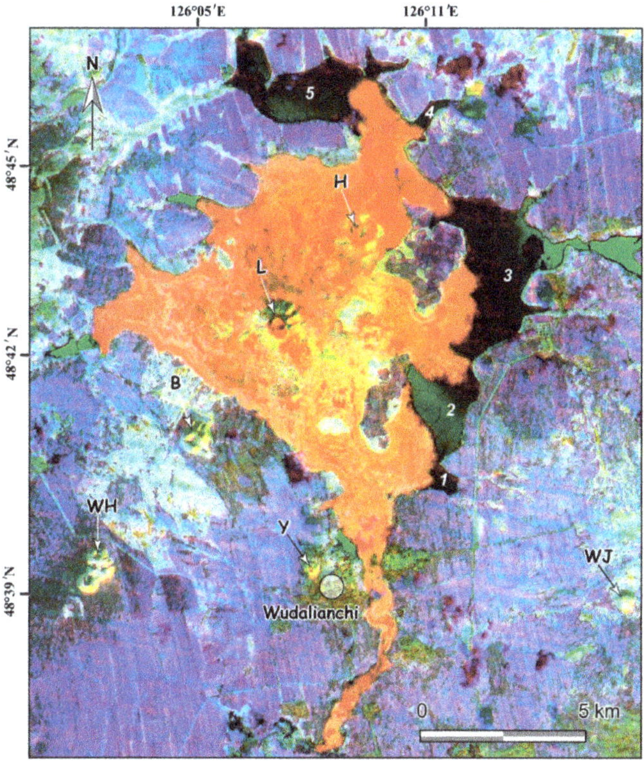

Figure 10. Enlarged ASTER composite image (taken on 6 April, 2004), band ratios of 10/12, 2/1, and 4/6 displayed as red, green, and blue, respectively, showing the detailed lithologic units of historical lava flows. Letters refer to volcanic craters are same as Figure 1b. The five lakes of Wudalianchi are marked by numbers 1 to 5.

5. Discussion

The quality of the available digital elevation models (DEMs) is crucial for the mapping topographic and geomorphologic features of Quaternary volcanoes. ASTER stereo imaging system can provide high quality DEMs data as shown in Figures 6 and 7. Contour images (Figures 6b and 7) in the Wudalianchi volcanic region demonstrate that ASTER DEMs data have a high vertical accuracy of 20 m without ground control points [12]. It can provide detailed topographic and geomorphologic features of the volcanic landforms with a low relief contrast (50–160 m) in the Wudalianchi region. The resolution of ASTER VNIR data up to 15 m is perfectly suited for large overviews of volcanoes. 3D perspective view image generated from ASTER VNIR and DEMs data provides excellent views for geomorphologic feature of the volcanoes (Figure 8).

According to the ASTER DEMs (Figures 2, 6 and 7), the most of volcanoes are distributed along the NE-trending faults. We inferred that these conjugate faults might provide the pathway for magma migration during the volcanic eruption [19].

As shown in Figure 9, lava flows from different eruption stages show different color patterns in ASTER multispectral ratio images. Interpretations of these images suggest that at least four stages of volcanic eruptions occurred during the Quaternary. This interpretation for relative dating has a good agreement with geologic mapping given by [27], except for the different judgments of what stage that the lava flows around Molabushan (M) should belong to. However, early studies suggested that there were three stages of lava flows: Early-middle Pleistocene (circa 1.33 ± 0.08 to 0.8 ± 0.02 Ma (million years), Late Pleistocene (circa 0.63–0.3 Ma), and historic periods [18,26]. Therefore, it is necessary to remeasure age for the lava flows in the Wudalianchi volcanic results revealed by this study.

Another point that needs to be addressed is the subtle differences of lava flows' color tones among the Laoheishan and Huoshaoshan volcanoes. Concerning the different color pattern in ASTER image of these lava flows represent the TIR spectral differences, we suggest that at least two different aspects may have been responsible for subtle spectral differences:

(1)   The effects of terrestrial weathering of lava flows. There is a spectral difference with increase of weathered degree as shown in Figure 5 and Table 3. Similar results also revealed that the weathering of Hawaiian basalts has caused their spectral changes [8,9].

(2)   The presence of vegetation on lava flows. Spectral measurement shows that the value of emissivity in band 12 increases comparing weathered surface covered by the lichen (WD-$_{2w}$) with cut surface of same sample (WD-$_{2c}$ in Figure 5). Field investigation shows that some lava flows are indeed covered by dry lichen or low brushes although vegetation is sparsely in the Wudalianchi volcanic region (Figure 11).

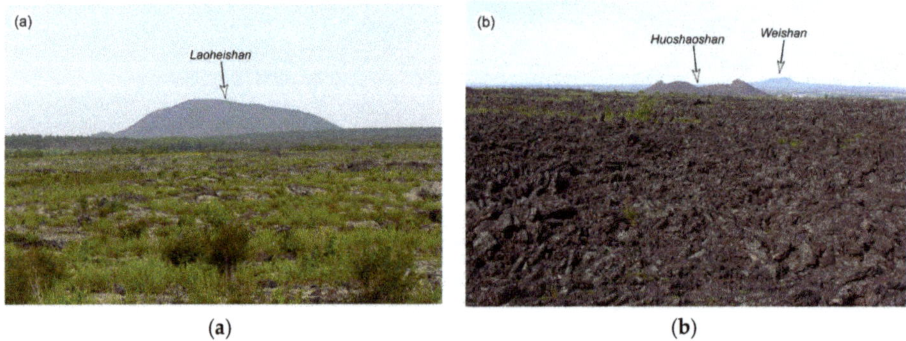

(a)　　　　　　　　　　　　　　　(b)

Figure 11. Field photographs (taken on August 2005). (**a**) The crater of the Laoheishan and lava flows from 1719–1721 AD eruptions. Northward-looking. (**b**) The lava flows of 1719–1721 AD eruptions from the Huoshaoshan volcano. The crater is the Huoshaoshan in the north part. Northeast-looking. Note lava flows covered by sparse vegetation.

6. Conclusions

ASTER DEMs data can provide detailed topographic features for describing volcanic landforms. The resolution of ASTER VNIR data up to 15 m is perfectly suited for large overviews of volcanoes. In this paper, through a series of processing of ASTER images and field investigation, it is implied that the Wudalianchi volcanoes are extending in a NE direction, which is likely influenced by the regional conjugate faults.

ASTER 3D perspective view images can provide excellent view for geomorphologic features of volcanoes, and thus, ASTER stereo imaging gives geoscientists a comprehensive tool for generating high quality topographic images of Quaternary volcanoes. Among the 14 volcanic craters in Wudalianchi region, the molten lava flows of Laoheishan and Huoshaoshan volcanoes partially blocked the Baihe River, which forms the Five Large Connected Pools, known as the Wudalianchi Lake.

Lithologic mapping indicates that at least four stages of volcanic eruptions are revealed in the Wudalianchi Quaternary volcano cluster. The lava flows from different stages of Quaternary volcanic eruptions are mapped successfully in the Wudalianchi volcanic cluster. These results demonstrate that ASTER multispectral data, particularly, the Lava Flow Index (LFI) (LFI = B10/B12) imagery, can be used to map subtle spectral variations caused by the surface weathering. Mapping of these lava flows on the basis of spectral properties may allow us to discriminate the relative age of the lava units in the sparsely vegetated region with arid and semi-arid climate conditions on the earth.

Although this study successfully distinguished four different formation stages of Wudalianchi volcanic lavas, it is still necessary to remeasure age and mineral component for the lava flows in the future research, to figure out what the major reason for these lava flows is with different color patterns in the ASTER image, which represent spectral differences.

In general, ASTER covers a wide spectral region with 14 bands from visible to thermal infrared with high spatial, spectral, and radiometric resolution. Therefore, ASTER can provide an effective approach for mapping the products from late Quaternary volcanoes.

Author Contributions: Conceptualization, B.F. and Y.N.; methodology, B.F. and Y.N.; software, B.F.; validation, B.F. and Y.N.; formal analysis, H.F. and B.F.; investigation, B.F. and Y.N.; resources, B.F., Y.N., H.F. and P.S.; data curation, B.F., Y.N., H.F. and P.S.; writing—original draft preparation, H.F. and B.F.; writing—review and editing, H.F. and B.F.; visualization, H.F. and B.F.; supervision, B.F.; project administration, B.F.; funding acquisition, B.F.

Funding: This work was supported by the Strategic Priority Research Program of Chinese Academy of Sciences (XDA 20070202) and the National Natural Science Foundation of China (No. 41761144071).

Conflicts of Interest: The authors declare no conflict of interest.

References

1. Ramsey, M.S.; Fink, J.H. Estimating silicic lava vesicularity with thermal remote sensing: A new technique for volcanic mapping and monitoring. *Bull. Volcanol.* **1999**, *66*, 32–39. [CrossRef]
2. Francis, P.; Rothery, D. Remote sensing of active volcanoes. *Annu. Rev. Earth Planet. Sci.* **2000**, *28*, 81–106. [CrossRef]
3. Abrams, M.; Glaze, L.; Sheridan, M. Monitoring Colima volcano, Mexico, using satellite data. *Bull. Volcanol.* **1991**, *53*, 571–574. [CrossRef]
4. Pieri, D.; Abrams, M. ASTER observations of thermal anomalies preceding the April 2003 eruption of Chikurachki volcano, Kurile Islands, Russia. *Remote Sens. Environ.* **2005**, *99*, 84–94. [CrossRef]
5. Urai, M.; Machida, S. Disclored seawater detection using ASTER reflectance products: A case study of Satsuma-Iwojima, Japan. *Remote Sens. Environ.* **2005**, *99*, 95–104. [CrossRef]
6. Urai, M.; Geshi, N.; Staudacher, T. Size and volume evaluation of the caldera collapse on Piton de la Fournaise volcano during the April 2007 eruption using ASTER stereo imagery. *Geophys. Res. Lett.* **2007**, *34*, L22318. [CrossRef]
7. Cater, A.J.; Ramsey, M.S.; Belousov, A.B. Detection of a new summit crater on Bezymianny Volcano lava dome: Satellite and field-based thermal data. *Bull. Volcanol.* **2007**, *69*, 811–815. [CrossRef]
8. Kahle, A.B.; Gllispie, A.R.; Abbott, E.A.; Abrams, M.J.; Walker, R.E.; Hoover, G.; Lockwood, J.P. Relative dating of Hawaiian lava flows using multispectral thermal infrared images: A new tool for geologic mapping of young volcanic terranes. *J. Geophys. Res.* **1988**, *93*, 15239–15251. [CrossRef]
9. Abrams, M.; Abbott, E.; Kahle, A. Combined use of visible, reflected infrared and thermal infrared images for mapping Hawaiian lava flows. *J. Geophys. Res.* **1991**, *98*, 475–484. [CrossRef]
10. Yamaguchi, Y.; Kahle, A.B.; Tsu, H.; Kawakami, T.; Pniel, M. Overview of Advanced Spaceborne Thermal Emission and Reflection Radiometer (ASTER). *IEEE Trans. Geosci. Remote Sens.* **1998**, *36*, 1062–1071. [CrossRef]

11. Abrams, M.; Yamaguchi, Y. Twenty years of ASTER contributions to lithologic mapping and mineral exploration. *Remote Sens.* **2019**, *11*, 1394. [CrossRef]

12. Fujisada, H.; Bailey, G.B.; Kelly, G.G.; Hara, S.; Abrams, M.J. ASTER DEM performance. *IEEE Trans. Geosci. Remote Sens.* **2005**, *43*, 2707–2714. [CrossRef]

13. Ninomiya, Y.; Fu, B. Extracting lithologic information from ASTER multispectral thermal infrared data in the northeastern Pamirs. *Xinjiang Geol.* **2003**, *21*, 22–30.

14. Ninomiya, Y.; Fu, B.; Cudahy, T.J. Detecting lithology with Advanced Spaceborne Thermal Emission and Refection Radiometer (ASTER) multispectral thermal infrared "radiance-at-sensor" data. *Remote Sens. Environ.* **2005**, *99*, 127–139. [CrossRef]

15. Rowan, L.C.; Mars, J.C. Lithologic mapping in the Mountain Pass, California area using Advanced Spaceborne Thermal Emission and Reflection Radiometer (ASTER) data. *Remote Sens. Environ.* **2003**, *84*, 350–366. [CrossRef]

16. Fu, B.; Zheng, G.; Ninomiya, Y.; Wang, C.; Sun, G. Mapping hydrocarbon-induced mineralogical alterations in the northern Tian Shan using ASTER multispectral data. *Terra Nova* **2007**, *19*, 225–231. [CrossRef]

17. Feng, M.; Whitford-Sttark, J.L. The 1719–1721 eruptions of potassium-rich lavas at Wudalianchi, China. *J. Volcanol. Geotherm. Res.* **1986**, *30*, 131–140. [CrossRef]

18. Wei, H.; Sparks, R.S.J.; Liu, R.; Fan, Q.; Wang, Y.; Hong, H.; Zhang, H.; Chen, H.; Jiang, C.; Dong, J.; et al. Three active volcanoes in China and their hazards. *J. Asian Earth Sci.* **2003**, *21*, 515–526. [CrossRef]

19. Wang, Y.; Chen, H. Tectonic controls on the Pleistocene-Holocene Wudalianchi volcanic field (northeastern China). *J. Asian Earth Sci.* **2005**, *24*, 419–431. [CrossRef]

20. Xu, S.; Zheng, G.; Nakai, S.; Wakita, H.; Wang, X.; Guo, Z. Hydrothermal He and CO_2 at Wudalianchi intra-plate volcano, NE China. *J. Asian Earth Sci.* **2013**, *62*, 526–530. [CrossRef]

21. Zhou, X.; Armstrong, R.L. Cenozoic volcanic rocks of eastern China-secular and geographic trends in chemistry and strontium isotopic composition. *Earth Planet. Sci Lett.* **1982**, *58*, 301–329. [CrossRef]

22. Liu, J.; Han, J.; Fyfe, W.S. Cenozoic episodic volcanism and continental rifting in northeast China and possible link to Japan Sea development as revealed from K-Ar geochronology. *Tectonophysics* **2001**, *339*, 385–401. [CrossRef]

23. Revenaugh, J.; Sipkin, S.A. Mantle discontinuity structure beneath China. *J. Geophys. Res.* **1994**, *99*, 21911–21927. [CrossRef]

24. Lei, J.; Zhao, D. P-wave tomography and origin of the Changbai intraplate volcano in Northeast Asia. *Tectonophysics* **2005**, *397*, 281–295. [CrossRef]

25. Fan, Q.; Liu, R.; Sui, J. Petrology and geochemistry of rift-type Wudalianchi K-rich volcanic rock zone. *Geol. Rev.* **1999**, *45*, 358–368. (In Chinese with English abstract)

26. Chen, H.; Ren, J.; Wu, X. Volcanic eruptive processes and characteristics of the current volcanoes in the Wudalianchi volcano clusters known from Manchurian-langanauge historical archives discovered at present. *Geol. Rev.* **1999**, *45*, 409–413. (In Chinese with English abstract)

27. Geological Museum of the Ministry of Geology. *Wudalianchi Volcano, China*; Shanghai Science and Technology Press: Shanghai, China, 1979; p. 3. (In Chinese with English introduction)

28. Ji, F.J.; Li, Q. The TL chronological evidence of the recent eruption in Wudalianchi volcanic cluster. *Seismol. Geol.* **1998**, *20*, 302–304. (In Chinese with English abstract)

29. Wu, X. Some Mongolian archives about eruption history in Wudalianchi volcanoes. *North. Cult. Relics* **1998**, *2*, 90–91. (In Chinese with English abstract)

30. Lu, Z. Modern volcanic structure and its formation mechanism of the Wudalianchi volcanic group in Heilongjiang Province. *J. Spectrosc. Miner.* **1994**, *1*, 5–21.

31. Mao, X.; Li, J.; Gao, W. A new understanding of the relationship between volcanic distribution and faults in the Wudalianchi volcanic group of Heilongjiang Province. *J. Geol. Geol.* **2010**, *16*, 226–235.

32. Tatsumi, Y.; Maruyama, S.; Nohda, S. Mechanism of backarc opening in the Japan sea: Role of asthenospheric injection. *Tectonophysics* **1990**, *181*, 299–306. [CrossRef]

33. Huang, J.; Zhao, D. High-resolution mantle tomography of China and surrounding regions. *J. Geophys. Res.* **2006**, *111*, B09305. [CrossRef]

34. Zhao, D.; Hasegawa, A.; Kanamori, H. Deep structure of Japan subduction zone as derived from local, regional, and teleseismic events. *J. Geophys. Res. Solid Earth* **1994**, *99*, 2313–22329. [CrossRef]

35. Zhao, D.; Xu, Y.; Wien, D. Depth extent of the lau back-arc spreading center and its relation to subduction processes. *Science* **1997**, *278*, 254–257. [CrossRef]

36. Zhao, D.; Mishra, O.P.; Sanda, R. Influence of fluids and magma on earthquakes: Seismological evidence. *Phys. Earth Planet. Inter.* **2002**, *132*, 249–267. [CrossRef]

37. Wei, W.; Hammond, J.O.S.; Zhao, D.; Xu, J.; Liu, Q.; Gu, Y. Seismic evidence for a mantle transition zone origin of the Wudalianchi and Halaha volcanoes in northeast China. *Geochem. Geophys. Geosyst.* **2019**, *20*, 398–416. [CrossRef]

38. Qiu, J.; Wu, Z.; Du, X. Study on the mantle inclusions in the Erkeshan-Wudalianchi-Kolofu potassium volcanic rocks in Heilongjiang Province. *Mod. Geol.* **1987**, *3*, 44–356.

39. Peng, N. Study on Dynamic Stability of Volcanic Magma Chamber in Wudalianchi New Period. Ph.D. Thesis, Capital Normal University, Beijing, China, 2004. (In Chinese with English abstract)

40. Qin, H. The Cracking Characteristics and Genesis of the Lava Flow in the Western Heilongjiang, Wudalianchi. Master's Thesis, Capital Normal University, Beijing, China, 2009. (In Chinese with English abstract)

41. Hook, S.J.; Kahle, A.B. The micro Fourier Transform Interferometer (μFTIR)—A new field spectrometer for acquisition of infrared data of natural surfaces. *Remote Sens. Environ.* **1996**, *56*, 172–181. [CrossRef]

42. Cornelissen, J.H.C.; Callaghan, T.V.; Alatalo, J.M.; Michelsen, A.; Graglia, E.; Hartley, A.E. Global change and arctic ecosystems: Is lichen decline a function of increases in vascular plant biomass? *J. Ecol.* **2001**, *89*, 984–994. [CrossRef]

43. Zou, Y.; Zhao, Y.; Fan, Y. Volcanic eruption methods and types of disasters in the Wudalianchi volcanic belt. *Seismol. Geol.* **2019**, *41*, 192–210. (In Chinese with English abstract)

44. Fu, B.; Ninomiya, Y.; Lei, X.; Toda, S.; Awata, Y. Mapping the active strike-slip fault triggered the 2003 Mw 6.6 Bam, Iran, earthquake with ASTER 3D images. *Remote Sens. Environ.* **2004**, *92*, 153–157. [CrossRef]

 *remote sensing*

Article

Lunar Calibration for ASTER VNIR and TIR with Observations of the Moon in 2003 and 2017

Toru Kouyama [1,*], Soushi Kato [1], Masakuni Kikuchi [2], Fumihiro Sakuma [2], Akira Miura [2], Tetsushi Tachikawa [2], Satoshi Tsuchida [3], Kenta Obata [4] and Ryosuke Nakamura [1]

[1] Artificial Intelligence Research Center, National Institute of Advanced Industrial Science and Technology, 2-4-7 Aomi, Koto-ku, Tokyo 135-0064, Japan; kato.soushi@aist.go.jp (S.K.); r.nakamura@aist.go.jp (R.N.)

[2] Research and Development Division, Japan Space Systems, 3-5-8 Shibakoen, Minato-ku, Tokyo 105-0011, Japan; Kikuchi-Masakuni@jspacesystems.or.jp (M.K.); fsakuma@jcom.home.ne.jp (F.S.); Miura-Akira@jspacesystems.or.jp (A.M.); Tachikawa-tetsushi@jspacesystems.or.jp (T.T.)

[3] Geological Survey of Japan, National Institute of Advanced Industrial Science and Technology, Central 7, Higashi 1-1-1, Tsukuba 305-8567, Japan; s.tuchidal@aist.go.jp

[4] Department of Information Science and Technology, Aichi Prefectural University, 1522-3 Ibaragabasama, Nagaute, Aichi 480-1198, Japan; obata@ist.aichi-pu.ac.jp

* Correspondence: t.kouyama@aist.go.jp; Tel.: +81-29-851-2506

Received: 21 October 2019; Accepted: 17 November 2019; Published: 19 November 2019

Abstract: The Advanced Spaceborne Thermal Emission and Reflection Radiometer (ASTER), which is a multiband pushbroom sensor suite onboard Terra, has successfully provided valuable multiband images for approximately 20 years since Terra's launch in 1999. Since the launch, sensitivity degradations in ASTER's visible and near infrared (VNIR) and thermal infrared (TIR) bands have been monitored and corrected with various calibration methods. However, a unignorable discrepancy between different calibration methods has been confirmed for the VNIR bands that should be assessed with another reliable calibration method. In April 2003 and August 2017, ASTER observed the Moon (and deepspace) for conducting a radiometric calibration (called as lunar calibration), which can measure the temporal variation in the sensor sensitivity of the VNIR bands enough accurately (better than 1%). From the lunar calibration, 3–6% sensitivity degradations were confirmed in the VNIR bands from 2003 to 2017. Since the measured degradations from the other methods showed different trends from the lunar calibration, the lunar calibration suggests a further improvement is needed for the VNIR calibration. Sensitivity degradations in the TIR bands were also confirmed by monitoring the variation in the number of saturated pixels, which were qualitatively consistent with the onboard and vicarious calibrations.

Keywords: ASTER; lunar calibration; radiometric calibration; VNIR; TIR

1. Introduction

The Advanced Spaceborne Thermal Emission and Reflection Radiometer (ASTER), which is a multiband-sensor suite composed of visible and near infrared (VNIR), shortwave infrared (SWIR), and thermal infrared (TIR) sensors onboard Terra [1], has successfully operated for 20 years since Terra's launch in 1999, and ASTER has provided numerous multiband images for those 20 years [2]. Similar to other spaceborne sensors, accurate radiometric calibration has been a center issue for the ASTER mission to provide reliable datasets. During the mission, characteristics of the radiometric performance of ASTER have been continuously monitored with several calibration methods, such as onboard calibration [3], vicarious calibration [4–6] and cross calibration [7,8], and sensitivity degradations in the VNIR, SWIR, and TIR have been confirmed. Since the measurement of the sensitivity degradations is used for correcting the observed brightness in ASTER products as radiometric correction coefficients

(RCC), the accurate measurement for the degradation is important to maintain the reliability of the ASTER products. The latest version of the RCC is version 4, which has been used since February 2014 for Level 1A processing, and the RCC was developed based on both onboard and vicarious calibration methods [9].

Due to the unresolved uncertainty in each calibration method, however, it has been confirmed there are large discrepancies among the results from the calibration methods for the VNIR bands [10]. For instance, the onboard calibration method measured a 10% larger sensitivity degradation for the shortest-wavelength band of the VNIR bands than that from the vicarious calibration method in 2015 [10]. Because the current RCC for the VNIR products uses a degradation trend based on both onboard and vicarious calibration methods, an additional calibration method for assessing the inconsistency between them has been expected to improve the RCC [10].

For the visible and near-infrared wavelength regions, lunar calibration has become a common calibration method and a standard calibration approach in an international satellite community [11], because it can accurately evaluate a temporal (i.e., relative) variation in a senor's sensitivity (on the order of 0.1%) [12–14]. Thus it is good for stability monitoring of a sensor [15], although the absolute accuracy of the lunar calibration method has been considered insufficient yet (5–10%) [16]. The better accuracy of the lunar calibration method for the relative degradation can be used for assessing the validity of onboard and vicarious calibration methods for the VNIR bands.

The lunar calibration method can be conducted through a comparison of the observed Moon brightness with the expected Moon brightness derived from lunar brightness models, such as a model developed based on radiance images acquired by the ground-based RObotic Lunar Observatory (ROLO) [17], and a model developed from observations by Spectral Profiler (SP) onboard SELENE, which was a Japanese lunar exploration orbiter. SP observed whole lunar surfaces with various illumination conditions without any significant degradation [18,19]. The ROLO models allows to simulate irradiance of the Moon observed from the Earth or from a low Earth orbit [17], and the SP model allows the simulate of any observation of the Moon with any observation geometry [20].

In addition to VNIR calibration, the Moon can be used to confirm the sensitivity degradation of the TIR bands by monitoring variation in the number of saturated pixels in a TIR image when the TIR sensor observes the Moon. This is because if sensitivity degradation of a TIR band occurs (i.e., the dynamic range of the TIR band is broadened) then the TIR band can capture areas of higher temperature without saturation. In 2003, because of the high temperature of the lunar surface around the sub-solar point (more than 120 °C [21]), the TIR images of the Moon had many saturated pixels. On the other hand, based on the degradation trend in the TIR bands [6], it is expected that the TIR sensor captured images of the Moon with a much smaller number of saturated pixels in 2017 if the TIR observes the Moon with a lunar surface temperature profile similar to that in 2003. Therefore, if we see a smaller number of saturated pixels in TIR images in 2017 as we expected, it supports the sensitivity degradation of the TIR bands, at least qualitatively.

In this study, we present a lunar calibration result for ASTER based on two lunar observations conducted in 2003 and 2017. For the VNIR bands, we will compare the observed lunar brightness with the expected lunar brightness from the lunar surface reflectance models (the SP and ROLO models). For the TIR bands, we will confirm the distribution of saturated pixels in 2003 and 2017. Each lunar calibration result will be used to assess validity of the results from other calibration approaches.

2. Lunar Observations and Data

Lunar calibration has a good ability for measuring relative variations in the sensor sensitivity of a sensor with a small uncertainty (the order of 0.1%) [13]. To measure the relative variation in a sensor sensitivity, at least two lunar observations are required. ASTER first observed the Moon in April 2003 with special pitch maneuver of Terra, which was required to observe the Moon with sensors that have narrow fields of view, especially for the ASTER VNIR, SWIR and TIR. Fourteen years later, in August 2017, ASTER successfully observed the Moon again. Therefore, we can apply the lunar calibration method for ASTER by using the two observations of the Moon in 2003 and 2017. Because the Moon is an irregular observation target for Terra, the lunar observation required careful discussion and well-prepared operations of the pitch maneuver in terms of reducing any operational risk. This is a part of reasons we needed the 14 years between the two lunar observations. In this study, we focused on the ASTER VNIR and TIR sensors that observed the Moon in both 2003 and 2017 (the SWIR sensor observed the Moon only in 2003 due to the rise in temperature of the detectors since May 2008, which has resulted in saturation and severe striping [22]).

Both the VNIR and the TIR sensors are multi band sensors; the VNIR sensor has three observation bands, and the TIR sensor has five bands. The specifications of both sensors are listed in Table 1. Because of the high spatial resolution of ASTER images, the Moon can be resolved with several hundred of pixels by the VNIR bands and ~100 pixels by the TIR bands. Note that to construct pairs of stereo images, the VNIR sensor is composed of two telescopes: a nadir-looking telescope and a backward-looking direction. The nadir-looking telescope has three bands (Band 1, Band2, and Band 3N), and the backward-looking telescope has one band (Band 3B) whose wavelength range is the same as that of Band 3N.

Table 1. Specifications of visible and near infrared (VNIR) and thermal infrared (TIR) bands of Advanced Spaceborne Thermal Emission and Reflection Radiometer (ASTER).

	VNIR	TIR
Wavelengths	Band 1: 520–600 nm Band 2 630–690 nm Band 3N/3B: 760–860 nm	Band 10: 8125–8475 nm Band 11: 8475–8825 nm Band 12: 8925–9275 nm Band 13: 10250–10950 nm Band 14: 10950–11650 nm
Ground sampling distance	15 m	90 m
Swath width	60 km	60 km
Bit depth	8 bits	12 bits
Spatial resolution for the Moon (from a 384,400 km distance)	8.2 km	25 km

In Table 2, the geometries of ASTER's observations of the Moon in 2003 and 2017 are summarized. Each parameter is required for using both the ROLO and SP models. Although the lunar surface condition is photometrically stable (e.g., changes that influence lunar irradiance at the 1% level are expected at intervals of the order of 10^8 years [23]), it has been known that different observation geometries may cause a calibration uncertainty of up to 1% due to the complicated surface features of the Moon [17]. To reduce this uncertainty, we choose the observation date of 2017 when ASTER could observe the Moon with an observation geometry similar to that from 2003. In both observations, deepspace was also in ASTER's field of view and was observed before and after capturing the Moon, which can be used for assessing the bias level of the observation.

Table 2. Observation geometries in the ASTER lunar observations on 14 April 2003 and 5 August 2017.

	14 April 2003	5 August 2017
Phase angle	−27.7°	−20.3°
Sub solar latitude	−0.9°	−0.3°
Sub solar longitude	22.1°	17.5°
Sub observer latitude	−6.8°	−4.2°
Sub observer longitude	−5.1°	−2.6°
Sub observer local time	10.2 h	10.7 h
Distance: Sun-Moon	1.005 AU	1.017 AU
Distance: Moon-Terra	359,021 km	394,856 km

2.1. Lunar Images Taken by the Visible and Near Infrared Bands

At the lunar observations, Terra changed its attitude with pitch maneuver from its nominal one to the attitude in which sensors onboard Terra could see the deepspace and the Moon. According to the pitch rotation, the VNIR and TIR scanned the Moon almost from north to south. To avoid the "undersampling" condition in which ASTER may miss observing some regions of the lunar surface, ASTER observed the Moon at a slower rotation speed for Terra than usual, so that the same region of the Moon was observed several times (called the "oversampling" condition). Figure 1 shows an example of lunar images obtained by the VNIR bands whose oversampling effect was not corrected. Due to the oversampling effect, the silhouette of the observed Moon was an ellipse, whereas the actual shape of the Moon is circle.

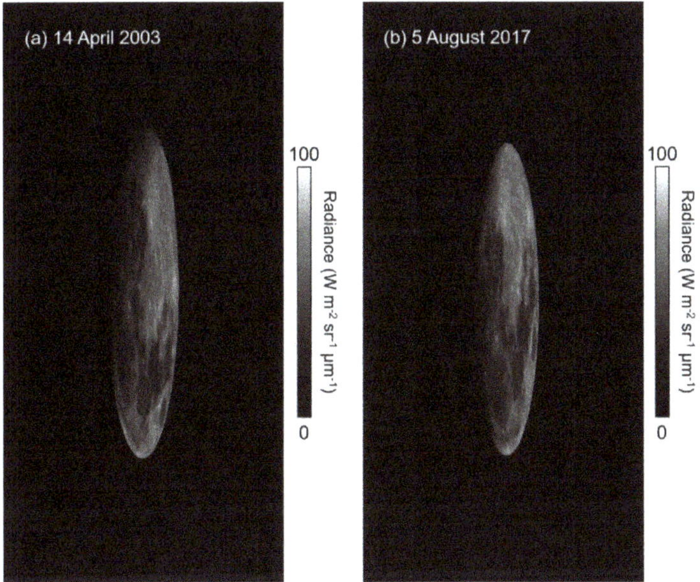

Figure 1. Images of the Moon obtained by ASTER VNIR Band 1 on 14 April 2003 and on 5 August 2017. Because the observation was conducted under oversampling conditions, the Moon's shape is elongated in the image frame.

The factor of the oversampling effect can be measured from information on the attitude control of a satellite in an observation of the Moon or can be measured with an image-processing method, such as fitting an ellipse to the elongated Moon (Appendix A). By adopting the ellipse-fitting method based on [24], we found that the oversampling factors were 4.57 in 2003 and 4.58 in 2017 whose uncertainties were less than 0.03 measured from residuals of the fitting, and we confirmed that these values were consistent with information on the attitude control provided by the Terra Flight Dynamics Team, 4.55 measured from the planned pitch maneuver rate of 0.122 degree s^{-1} [25] (Appendix A).

2.2. Lunar Images Taken by the Thermal Infrarred Bands

At the same time as the VNIR observations, the TIR sensor observed the Moon with Bands 10 to 14 in both 2003 and 2017. Figure 2 shows examples of the lunar images taken by the TIR sensor in 2003 and 2017 with Band 10. Since the TIR sensor is a whisk-broom sensor with 10 detectors for each band and the lunar observations were conducted under oversampling conditions same as the VNIR, the Moon was observed as a discrete shape, though the TIR sensor observed the whole lunar disk without any observation gaps (Appendix A). In 2003, due to the high temperature of the lunar surface around the sub-solar point (more than 120 °C), many pixels around the subsolar location were saturated. This saturation occurred because the specifications of the TIR bands were adjusted to monitor the temperature profile of the terrestrial surface. On the other hand, there were fewer saturated pixels in the images from 2017 and the saturation level was higher in 2017 than in 2003, which should reflect the sensitivity degradation over the 14 years. The validity of the saturation level is discussed in Section 4.

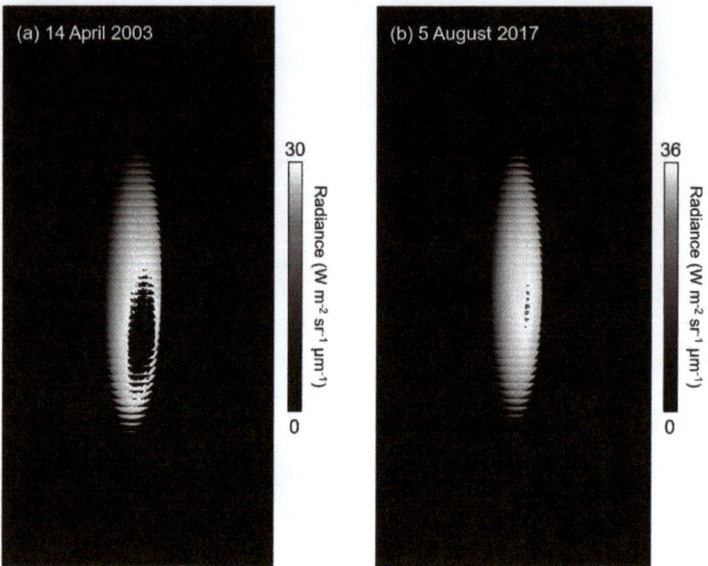

Figure 2. Images of the Moon taken by the TIR Band 10 on (**a**) 14 April 2003 and (**b**) 5 August 2017, whose oversampling effects were not corrected. Black pixels in the Moon disks indicate saturation due to the high temperature of the Moon that exceeds the observable range of the band.

3. Lunar Calibration for the Visible and Near Infrared Bands: Evaluating the Sensitivity Degradation

Because the brightness of the Moon depends on several geometric parameters (such as the distance between the Sun and the Moon, distance between the Moon and an observer, the phase angle between the Sun, the Moon, and an observer), it is difficult to investigate the VNIR's sensitivity variation simply by comparing the lunar brightness observed at different times. In the lunar calibration approach, to address the dependence of the lunar brightness on geometric parameters, a simulation of the lunar brightness that can reproduce the geometric dependence is required. Then, by comparing the ratio between the observed and the simulated lunar brightness (the former is affected by the sensor sensitivity degradation and the latter is not affected), we can investigate how much the sensor experiences a sensitivity degradation.

In addition to simulating the lunar brightness, several basic image-processing procedures are also required to achieve better accuracy in measuring the sensitivity variation. To convert the digital numbers in the image of the Moon, which are the original output from the VNIR sensor, to a physical quantity, i.e., the radiance (W m^{-2} sr^{-1} μm^{-1}), we used the latest version (version 4) of the RCC that was distributed by the ASTER Science Team and Japan Space Systems [22]. The RCC contains parameters for flat-fielding, removing bias, and correcting temporal variations in the sensor sensitivities of the VNIR bands, which are derived from ASTER's onboard calibration and a vicarious calibration. Note that in this study, the correction of the temporal variations in the sensitivities of the VNIR bands was not performed for investigating the variations in the sensitivity by lunar calibration.

In this section, we will first show the additional correction for the remaining bias that was found through analyzing the deepspace region, and then we will show the lunar calibration result based on the lunar reflectance models (the SP and ROLO models).

3.1. Bias Reduction by Deepspace Observations

Ideally, if no energy is input into a detector, no digital count is expected. However, due to dark currents, stray light, and other reasons, offset counts are usually observed, which may affect the uncertainty in the radiometric measurements of a target. Therefore, the temporal variations in the offset levels in the VNIR bands are monitored with an onboard calibration system and night-side observations [3].

Deepspace is an ideal target to validate offset monitoring because a detector receives no energy input from deepspace. Before and after scanning the Moon, ASTER also observed deepspace with several thousand lines. Figure 3 shows an example of averaged offsets in deepspace regions that were measured from a lunar and deepspace image taken by VNIR Band 1 in 2003. Small but nonzero offsets remained in all bands in both 2003 and 2017, even after the expected offsets based on the RCC were subtracted from the images. Even pixels and odd pixels have different amplification circuits, and the difference may affect profiles of the offset values for even and odd pixels [26]. Since the remaining offsets may affect the lunar calibration accuracy for which we assumed no remaining offset, we subtracted the offset at each pixel and successfully reduced the offsets to less than 10^{-3} W m^{-2} sr^{-1} µm^{-1} that can be ignored when measuring the lunar brightness.

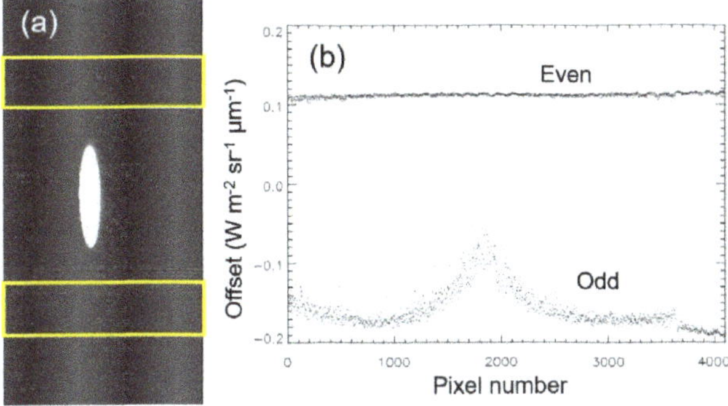

Figure 3. (**a**) An example of offset patterns observed in the VNIR images (Band 1, 2003). The brightness level was stretched to enhance the offset. (**b**) Measured offsets for each pixel obtained from averaging the apparent brightness in deepspace regions (indicated by the rectangular regions shown in (**a**)). "Even" and "Odd" represent offset values for even pixels and odd pixels, respectively.

3.2. Sensitivity Degradation from 2003 to 2017

Because the appearance of the Moon is different under different observation conditions (phase angles and libration conditions), it is difficult to evaluate the sensor sensitivity degradation by directly comparing the lunar brightness in different observations. Instead, in the lunar calibration method, the sensor sensitivity degradation is investigated by monitoring the variation in the ratio between the observed and expected brightness levels of the Moon. To evaluate the expected lunar brightness, we used two different lunar surface reflectance models, the SP model [19,20] and the ROLO model [17], to validate the consistency in the lunar calibration result.

The SP model is a map-base lunar surface reflectance model with a resolution of 0.5° × 0.5° that covers the whole lunar surface. Each grid in the SP model has hyper-spectral radiance factors that correspond to standardized lunar surface reflectance with a specified observation geometry (incident angle = 30°, emission angle = 0°, and phase angle = 30°). The wavelength coverage of the SP model is 512–1650 nm with 6–8 nm spectral intervals, which covers the wavelengths of the VNIR bands. In addition, the photometric dependence of the surface reflectance on incident, emission, and phase angles was also modeled in the SP model [19]. By utilizing the reflectance map and the modeled photometric dependence, we can simulate the lunar brightness as an image with any observation geometry [20]. On the other hand, the ROLO model provides an expected lunar irradiance that corresponds to integrated brightness over the whole lunar disk in an observed image. By modeling the dependence of disk-equivalent reflectance of the Moon on observation geometries as listed in Table 1, the ROLO model allows to simulate the lunar irradiance observed from the Earth or from a low-Earth orbit [17].

Figure 4a shows examples of the observed lunar images in 2003 and 2017 with VNIR Band 1, whose oversampling effects were corrected (Appendix B), and Figure 4b shows their SP simulations. Both the observed and simulated images were sufficiently similar (correlation coefficients > 0.99), which enables a pixel-by-pixel comparison between the observed and simulated lunar images [20].

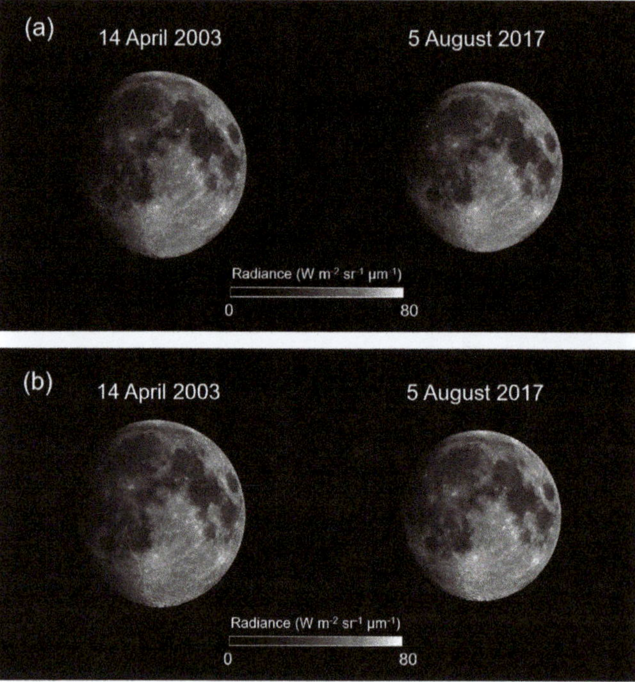

Figure 4. (a) Observed images of the Moon by VNIR Band 1 (520–600 nm) on 14 April 2003 and 5 August 2017 whose oversampling effects were corrected, whereas the effects from the sensor sensitivity degradation were not corrected. (b) Simulated images of the Moon for the two observations using the SP model.

Figure 5 shows the brightness ratios in 2003 and 2017 for Band 1 derived from the observed and simulated images of the Moon shown in Figure 4. As in Figure 5, the brightness ratio in 2017 was smaller than that in 2003 for the whole luanr disk, indicating that VNIR Band 1 experienced a sensitivity degradation during the 14 years. This tendency was the same in other bands. The amount of sensitivity degradation from 2003 to 2017 (i.e., the relative degradation) from the SP model can be measured by

$$r_{2003 \to 2017} = \overline{\left(\frac{I_{obs}}{I_{sim}}\right)}_{2017} \Bigg/ \overline{\left(\frac{I_{obs}}{I_{sim}}\right)}_{2003}, \tag{1}$$

where I_{obs} represents the observed brightness, and I_{sim} represents the simulated brightness at each observation. The overbars in (1) indicate averaging over the region we used (incident angle $< 60°$, and emission angle $< 45°$). Similarly, the degradation can be measured from the ROLO model by using the total irradiance of the Moon as

$$r_{2003 \to 2017} = \left(\frac{Irr_{obs}}{Irr_{ROLO}} \right)_{2017} \Big/ \left(\frac{Irr_{obs}}{Irr_{ROLO}} \right)_{2003},\tag{2}$$

where Irr_{obs} is the observed irradiance and Irr_{ROLO} is an expected irradiance of the Moon with the ROLO model (Appendix C).

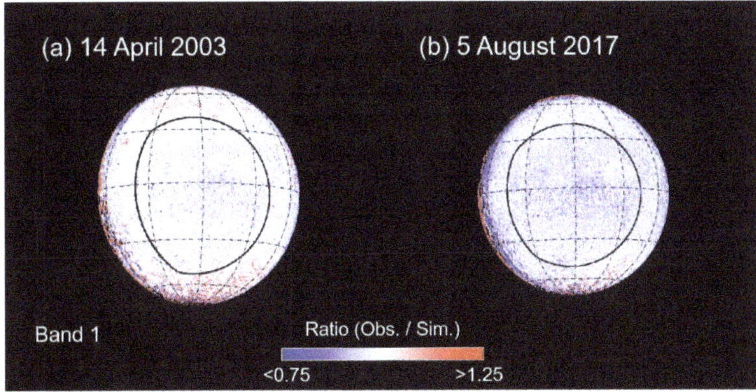

Figure 5. Brightness ratios for the observed and simulated brightness of the Moon for Band 1 in (**a**) 2003 and (**b**) 2017. The regions surrounded by solid black lines (incident angle $< 60°$, emission angle $< 45°$) were used for evaluating the sensitivity degradation.

Note that the reason why the lunar disk size in 2003 was larger than that in 2017 is that the distances to the Moon at the lunar observation in 2003 was 10% shorter than that in 2017 (see Table 2). Evaluations from both the ROLO and SP models are not affected by the distance variation, because the ROLO model includes a distance correction, whereas the SP models simulates a lunar image with a consideration of the distance.

Table 3 shows the sensor sensitivity degradations observed in VNIR Bands 1, 2, 3N, and 3B from 2003 to 2017 measured from the lunar calibration based on the SP and ROLO models. Note that following [20], only the limited regions (incident angle $< 60°$ and emission angle $< 45°$) in the lunar disks were used for lunar calibration with the SP model, where the SP model provides valid brightness values [19]. For comparing observations with simulations from the ROLO model, which simulates the irradiance of the Moon, we integrated the observed radiance of whole lunar-disk pixels in the VNIR images to measure the irradiance of the Moon (see Appendix C).

Table 3. Percentages of the sensor sensitivity degradation of the ASTER/VNIR bands from 2003 to 2017 measured from the lunar calibration based on Spectral Profiler (SP) and RObotic Lunar Observatory (ROLO) models. SP model (nominal) indicates the lunar calibration results using a limited region of the lunar disk according to the recommendation of [19], and SP model (whole Moon) indicates using the whole lunar disk.

Calibration Methods	ROLO Model	SP Model (Nominal)	SP Model (Whole Moon)
Band 1	3.0%	3.1%	3.6%
Band 2	5.4%	5.2%	5.6%
Band 3N	6.3%	5.8%	6.3%
Band 3B	3.0%	3.2%	3.9%

Considering that the error ranges in lunar calibration can be up to 1% from the discussion in [16], the lunar calibration results from SP and ROLO models were consistent with each other for all bands within the error range. Since the two models were developed independently, the consistency between the two models indicates the validity of the lunar calibration results. The lunar calibration results suggested that the VNIR bands experienced the sensitivity degradations of three to six percent from 2003 to 2017.

Although the SP model is expected not to be accurate at an observation condition with large incident and emission angles [19], it is worth to investigate how the result from the SP model changes when we use the whole lunar disk that includes both large incident and emission angle conditions to understand the performance of the SP model more. A planetary exploration mission, Hayabusa 2, conducted the lunar calibration with such a severe condition in which it had to use the whole lunar disk because of the small disk size of the Moon [27,28], although only the SP model was applicable because Hayabusa 2 observed the far-side of the Moon. We confirmed that the results from using the whole lunar disk provided basically lager degradation values, but the difference was only 1% from the results of the SP model (nominal case) and the ROLO model (Table 3). In addition, because an image registration technique is performed for simulation of the lunar image with the SP model [20], the accuracy of the lunar calibration with the SP model should not be sensitive to the accuracy of the oversampling factor. Indeed, even when we used a wrong oversampling factor (we tested with the factor of 4.2), we confirmed the measured degradation for the VNIR Band 1 from 2003 to 2017 was 3.1%, which was the same as the degradation when we used the correct oversampling factor. The robustness of the SP model against the accuracy of the oversampling factor should be worth to be investigated in future studies.

Finally, we confirmed the linearity of the VNIR bands by comparing the simulated and observed lunar brightness in 2003 and 2017 (Figure 6 for Band 1). In both years, the brightness distributions had clear linear relationship in all bands as in Figure 6, indicating no significant linearity variation happened in the VNIR bands.

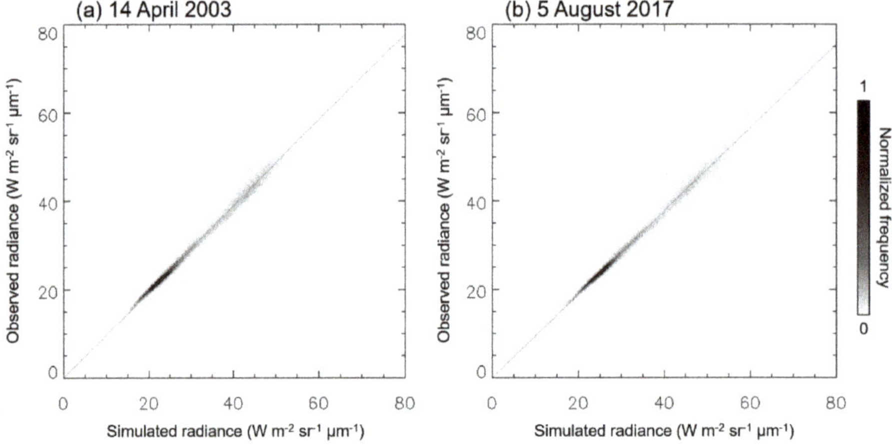

Figure 6. Comparison of simulated and observed radiance for Band 1 in (**a**) 2003 and (**b**) 2017. The simulated radiance was measured from the SP model. The gray scale represents the normalized frequency, and the gray line represents a line with a slope of the mean ratio between simulated and observed brightness in each plot.

4. Lunar Calibration for the Thermal Infrared Bands: Validation of the Measured Sensitivity Degradation

For the TIR bands, the amount of sensitivity degradation has been validated within 1 K uncertainty by onboard calibration and vicarious calibration, and both methods have shown consistent calibration results with each other since Terra's launch [6]. Based on the calibration, variation in the saturation levels can be expected in each band for 2003 and 2017. Considering that the highest temperature on the lunar surface can exceed 120 °C [21], there should be many saturated pixels in TIR images, especially in Band 10 in 2003 (Figure 7). In Figure 7, the saturation temperature was determined as the brightness temperature converted from the highest unsaturated digital counts of the TIR with the on-orbit calibration coefficients at each observation, thought the temperature could be an approximation value because 120 °C is out of the calibration range for the TIR [6].

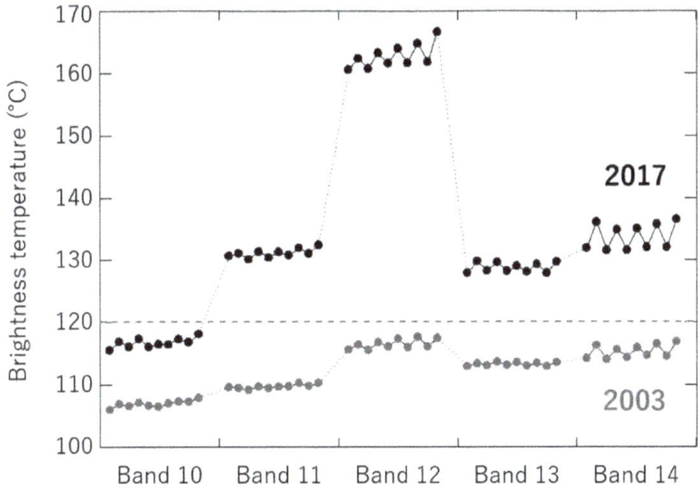

Figure 7. Expected saturation brightness temperature for each detector of each TIR band in 2003 (gray dots) and in 2017 (black dots). The higher saturated brightness temperatures in 2017 are due to the TIR sensor's sensitivity degradation that caused a wider dynamic range in the sensor. Note that the surface temperature of the Moon can exceed 120 °C (dashed line).

On the other hand, it is expected that the TIR sensor could observe the Moon surface with a smaller number of saturated pixels in 2017 even with Band 10. For this expectation, we assumed the temperature profile at the observation in 2017 was similar to that in 2003 at least around the sub-solar point, because the lunar surface has a good repeatability of the surface temperature in terms of the solar elevation [21] due to lack of atmosphere and water and the slow rotation speed of the Moon, and the observations in 2003 and 2017 were conducted with almost the same solar distance and the similar illumination and viewing conditions (Table 2).

Although it is difficult to determine the TIR's sensitivity degradation quantitatively with only the Moon due to the lack of a published surface brightness model in the TIR range, the degradation can be qualitatively confirmed by monitoring the variations in the saturation levels and the number of saturated pixels in the TIR images from 2003 to 2017.

Figure 8 shows sets of the TIR images of the Moon from 2003 and 2017 whose oversampling effects were corrected. In 2003, saturation occurred in all bands, whereas saturation occurred only in Band 10 in 2017, which has the lowest saturation temperature of the TIR bands. Although there were still saturated pixels in Band 10 in 2017, the number of saturated pixels was significantly reduced (from 3370 pixels in 2003 to 63 pixels in 2017). Saturation occurred only in Band 10 in 2017, which was consistent with the expectation of the TIR sensor's sensitivity degradation from the onboard and vicarious calibrations (Figure 6), assuming that the maximum temperature on the Moon is 120 °C. Quantitative evaluations will be possible when we conduct another lunar observation with the TIR sensor in the future for which we can compare TIR images that have no saturated pixels.

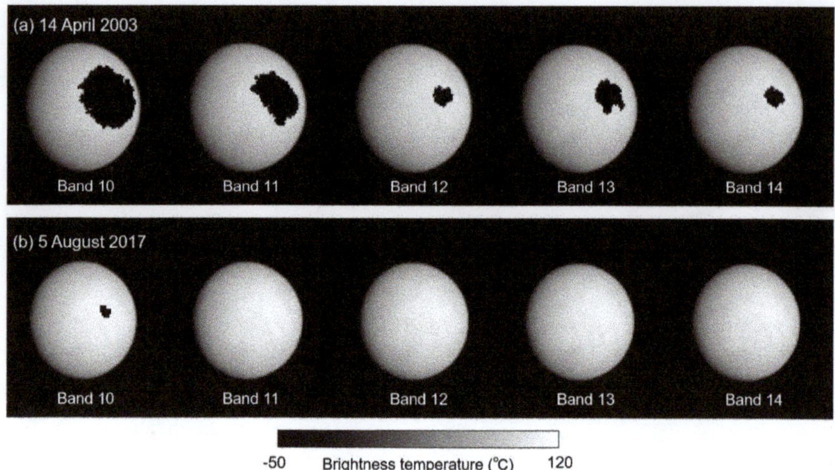

Figure 8. Observed images of the Moon in (**a**) 2003 and (**b**) 2017. The oversampling effects were corrected by averaging pixels whose field of views overlapped. The black pixels in the lunar disk represent regions where at least one pixel in the original image was saturated when averaging.

5. Discussion

Since lunar calibration can measure a temporal variation in a sensor sensitivity accurately (within 1% uncertainty), the validity of other calibration methods can be assessed through the comparison of their results with the result from the lunar calibration. Figure 9 shows a comparison of the relative sensitivity degradations from 2003 to 2017 measured by the lunar calibration with those from other calibration methods in Bands 1, 2, 3N, and 3B of the VNIR sensor. To estimate the relative degradation with onboard calibration, we used degradation values at the lunar observations measured by the onboard calibration system of ASTER, that is, 0.771, 0.844, and 0.895 for Bands 1, 2, and 3N on April 13 2003 and 0.694, 0.749, and 0.794 on August 11 2017, respectively. The onboard calibration for the VNIR bands is performed with onboard halogen lamps whose performances have been monitored with photodiodes [26]. For vicarious calibration, we used the degradation trends obtained from three selected observation sites (Railroad Valley, Alkari Lake, and Ivanpah Playa) as reported in [4,10].

From Figure 9, somewhat large discrepancies in the onboard calibration from other calibration methods can be confirmed. This finding indicates that the uncertainty in the onboard calibration increases with time. On the other hand, the lunar calibration results were more consistent with the results from the vicarious calibration. Although there were still a few percent discrepancies between them, especially in Band 3N, they were consistent with each other within the range of the uncertainties by considering the possible uncertainties of 3–5% in the vicarious calibration results [29]. The benefit from the lunar calibration is that we may determine the magnitude of the degradation within the smaller uncertainty, in other words, the vicarious calibration has a room for improvement in terms of the uncertainty. Obata et al. [10] proposed an approach for correcting vicarious calibrations by considering the surface reflectance spectra at the calibration sites with a band transition technique for inter-band comparisons (called "inter-band calibration"). The measured sensitivity degradations from the inter-band calibration were much more consistent with those from the lunar calibration than those from the vicarious calibration (Figure 8).

It should be noted that the current RCC for the VNIR bands is developed by mixing degradation trends from the onboard calibration and the vicarious calibration [9]. This feature is the reason that the magnitude of the degradation based on the RCC was different from that of others (Figure 9).

The discrepancy between the RCC and the lunar calibration results was beyond the error ranges of the lunar calibration, indicating that a further update is also required for the RCC.

Note that it is still unclear what things cause the large uncertainty in the onboard calibration. The possible candidates of the unresolved uncertainty are degradation of optical transparency of the calibration optics, sensitivity degradation of the monitoring photodiodes, and degradation of filters for the photodiodes [3] that have not been monitored, and thus their performances are highly uncertain.

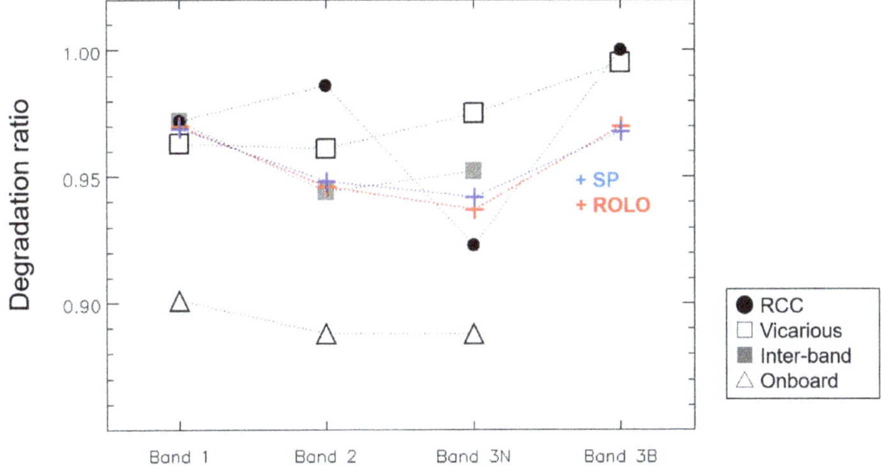

Figure 9. Comparison of the sensor sensitivity degradations in the VNIR bands from 2003 to 2017 measured from the onboard calibration (triangles), the vicarious calibration (white rectangles), and the lunar calibration (SP and ROLO models, crosses). The degradations measured from the inter-band calibration and the current RCC (version 4) are also shown (gray rectangles and black circles, respectively).

6. Conclusions

ASTER successfully observed the Moon in 2003 and 2017. Based on the two observations of the Moon, the temporal variations in the sensor sensitivities of ASTER's visible and near-infrared bands (Bands 1, 2, and 3N) were evaluated. The lunar calibration results indicate that sensor sensitivity degradations of several percent occurred in all the VNIR bands. The lunar calibration was basically consistent with the vicarious calibration, but there were still a few percent discrepancies between both results, which indicates further improvement is needed for ASTER calibration. For the TIR bands, the Moon can be used for validating the onboard and the vicarious calibration results by monitoring saturated pixels in images of the Moon. As expected of the onboard and the vicarious calibrations, there were fewer saturated pixels in 2017 than 2003, which supports the validity of the calibration for the TIR bands.

Although the lunar calibration for the VNIR bands can provide only one-point information about the sensitivity degradation from 2003 to 2017, the information can be used as a constraint when we fit a degradation trend measured from other calibration methods. In addition, other instruments onboard Terra, such as Moderate Resolution Imaging Spectroradiometer (MODIS), Multi-angle Imaging Spectroradiometer (MISR), Measurements Of Pollution In the Troposphere (MOPITT), observed the Moon at exactly the same time of ASTER's observations, indicating an ideal cross-calibration condition among these instruments via the Moon. Further analysis on the combination of all the calibration methods and collaborations with other calibration approaches, such as a calibration with the pseudo invariant calibration method [30] that can also provide high quality results for long-term stability trending, will provide a better RCC in the future and will enhance consistency among sensors.

Author Contributions: Conceptualization, T.K., M.K., and F.S.; methodology, T.K. and S.K.; validation, T.K., S.K. and F.S.; formal analysis, T.K., S.K. and F.S.; investigation, T.K., S.K., F.S. and K.O.; resources, T.T.; data curation, A.M., and T.T.; writing—original draft preparation, T.K.; writing—review and editing, T.K., S.T., F.S. and K.O.; visualization, T.K. and S.K.; supervision, M.K., S.T., and R.N.

Funding: This research was supported partly by JSPA/KAKENHI 16H02225, 19K14789.

Acknowledgments: The authors appreciate Kurtis Thome, Hirokazu Yamamoto, and Koki Iwao for their special support for the lunar calibration activity, and all members in ASTER Science team. The authors also appreciate Thomas Stone for confirmation of ROLO estimation, Xiaoxiong Xiong for commenting the cross calibration, and members in Terra Flight Dynamics Team for the ASTER attitude information. The authors thank Satoru Yamamoto who provided the onboard calibration trend. The authors also thank the three anonymous reviewers for providing valuable comments and suggestions on the draft of the paper.

Conflicts of Interest: The authors declare no conflict of interest. The funders had no role in the design of the study; in the collection, analyses, or interpretation of data; in the writing of the manuscript, or in the decision to publish the results.

Appendix A

In this section, we will show estimations of oversampling factors for the VNIR and the TIR bands based on attitude control information and image processing.

For the VNIR, which is a pushbroom sensor, if the pitch rotation rate is given from the attitude control information of Terra and the rate is constant during the lunar observation, the oversampling factor for the VNIR can be estimated as

$$f_{os}^{(VNIR)} = \frac{\theta_{iFOV}^{(VNIR)}}{r_p t_e^{(VNIR)}} \tag{A1}$$

where $\theta_{iFOV}^{(VNIR)}$ is the instantaneous field of view of the VNIR (i.e., angular size of one pixel), r_p is a pitch rotation rate, and $t_e^{(VNIR)}$ is a scan interval. From the specification of the VNIR [31], $\theta_{iFOV}^{(VNIR)}$ = 21.3 µrad, $t_e^{(VNIR)}$ = 2.199 ms, and from the attitude control information, r_p = 0.122 degree s^{-1} [25]. The estimated oversampling factor for the VNIR bands was 4.55. Note that the pitch rotation rate was common with the TIR bands.

For the TIR, which is a whiskbroom sensor, the oversampling factor can be estimated from the attitude control information as

$$f_{os}^{(TIR)} = \frac{\theta_{iFOV}^{(TIR)}}{r_p t_e^{(TIR)}} \times n_{detector} \tag{A2}$$

where $n_{detector}$ represents number of detectors of each TIR band aligned with an along-track direction. Because the TIR is a whiskbroom-type sensor, the TIR repeats cross-track scans by utilizing oscillation of an observation mirror, and t_e means an interval of cross-track scans. The configuration of the oversampling condition for the TIR bands is illustrated in Figure A1.

By substituting the TIR's specification [31], θ_{iFOV} = 127.8 µm, t_e = 131.94 ms, and $n_{detector}$ = 10, then we have $f_{os}^{(TIR)}$ = 4.55, which is same as the factor for VNIR. The effect from the orbital motion of Terra (~7 km s^{-1}) can be ignored in above calculation because the distance between the Moon and Terra at the observation was too large (more than 350,000 km).

On the other hand, since the VNIR has an enough resolution for capturing the Moon with a several hundred pixels, an ellipse fitting technique is applicable to estimate the oversampling factor. Figure A2a shows an example of our fitting results for the VNIR's lunar images with a robust ellipse fitting technique based on [24]. As illustrated in Figure A2b, by using the horizontal length of the ellipse, l_a, which represents an actual radius of the Moon in a VNIR image, and the vertical length, l_e, which is longer than l_a due to the oversampling condition, we can estimate the oversampling factor simply as

$$f_{os}^{(VNIR)} = \frac{l_e}{l_a}. \tag{A3}$$

The estimated $f_{os}^{(VNIR)}$ was 4.57 for 2003 images and 4.58 for 2017 images with an uncertainty of 0.03. The uncertainty was evaluated from a standard deviation of residuals between the fitted ellipse and the locations of the extracted limb points of the Moon. The standard deviation was up to 1.1 pixels whereas the size of the lunar disk was more than 400 pixels, thus we concluded that the uncertainty of the estimated oversampling factors from the image processing can be less than 0.5%.

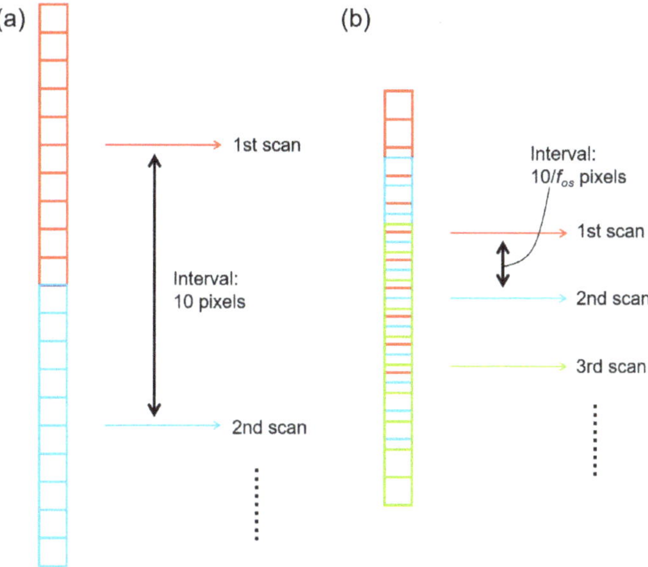

Figure A1. Schematic views of TIR's whiskbroom observations in (**a**) a nominal observation condition and (**b**) an oversampling condition. Note that each TIR band has 10 detectors and the actual TIR detectors are aligned with a stagger configuration.

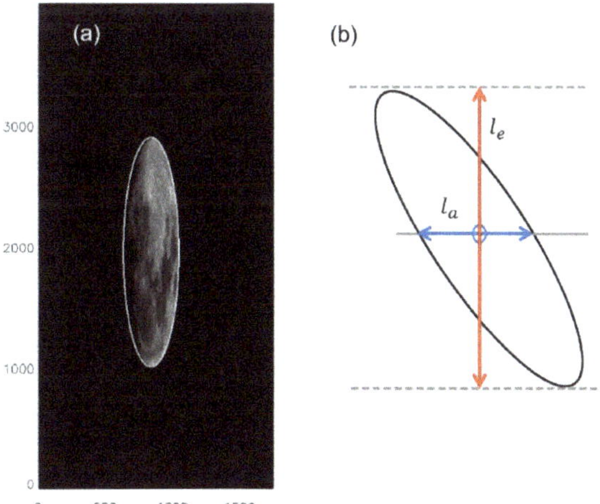

Figure A2. (**a**) An example of ellipse fitting for elongated lunar images (Band 1 for 2017). The standard deviation of residuals between the fitted ellipse and the locations of the extracted limb points in this example was 0.5 pixel whereas l_a and l_e as shown in (**b**) were 413 and 1893 pixels, respectively.

Appendix B

In this section, we will show our procedure of generating a corrected lunar image with a resampling technique for a lunar calibration with the SP model. To evaluate a brightness for a resampled pixel at (I, J) position in the corrected image, we calculated a weighted mean brightness as

$$I_c^{(I,J)} = \frac{\sum_j w_j I_r^{(I,j)}}{\sum_j w_j} \tag{A4}$$

where $I_c^{(I,J)}$ means the brightness for the resampled pixel, $I_r^{(I,j)}$ is an observed brightness at (I, j) position in a raw image, and w_j represents a weighting factor for j-th line in the raw image. w_j is determined from the projected position of a pixel in the corrected image as illustrated in Figure A3.

Because the oversampling factors for the VNIR and TIR bands are not integer, the values of the total weight $\sum_j w_j$ can be different at different J. We confirmed the values of the total weight were from 4.53 to 4.62 for the Band 1 lunar image in 2003. But since we used a weighted mean value for the resampling, the effect from the difference in the total weight was canceled. For the TIR bands, the correction of the oversampling effect can be done with the same manner.

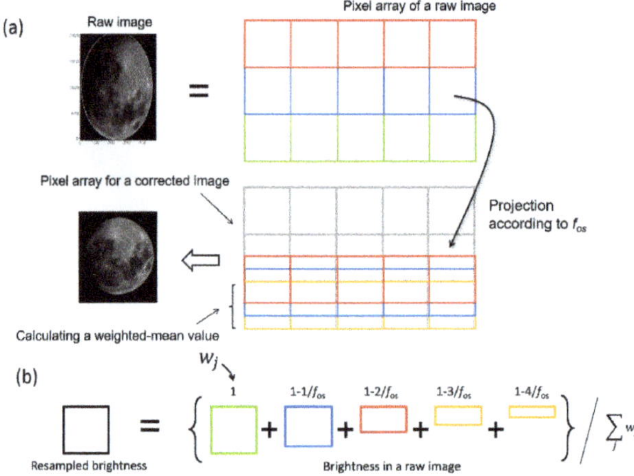

Figure A3. (**a**) Schematic view of correcting the oversampling effect for the VNIR bands in this study. (**b**) An example of calculating a weighted mean brightness for a pixel in a corrected image for the case 4 $< f_{os} < 5$. Note that the distributions of w_j are different for different lines in the corrected image due to the non-integer f_{os}.

Appendix C

We measured the observed lunar irradiance from the VNIR images by

$$Irr_{obs} = \frac{1}{f_{os}} \Omega_p \sum_{i,j} I_{i,j} \tag{A5}$$

where f_{os} is an oversampling factor, Ω_p is a solid angle of each pixel, and $I_{i,j}$ is the radiance of a pixel at (i, j). We integrated pixels not only in the lunar disk but also surrounding deepspace regions within approximately 200-pixel distance from the lunar disk limb. We confirmed the selection of the width for the deepspace did not so affect the measured irradiance (the order of 0.1%, we tested with 100, 200, and 300 pixels).

For computing the simulated irradiance from the ROLO model, we followed the calculation proposed in [17] as

$$I_k = \frac{A_k \Omega_M E_k}{\pi f_d} \tag{A6}$$

where I_k represents modeled irradiance for k-th ROLO band (k = 1–32, corresponding to 350–2383 nm), A_k is the disk-equivalent albedo for band k, Ω_M is the solid angle of the Moon observed from a standard distance (384,400 km), and E_k is the solar irradiance (derived from the model of Wehrli [32]) at 1 AU for the band k. f_d is a distance parameter for canceling the distance effect to the irradiance as

$$f_d = \left(\frac{D_{S-M}}{1 \text{ AU}}\right)^2 \left(\frac{D_{V-M}}{384,400 \text{ km}}\right)^2 \tag{A7}$$

Note that A_k is a function of geometry parameters [17], such as phase angle, sub-solar longitude, sub-observer latitude and longitude on the Moon that are listed in Table 3.

Above calculation is for ROLO bands (i.e., irradiance will be obtained at discrete wavelengths). To obtain a continuous irradiance spectrum, following to [17], we fitted a mixed spectrum of a returned Apollo soil sample [33] and a lunar breccia sample [34] to A_k. Then we performed (A6) and (A7) for the fitted reflectance spectrum. Finally, we obtained the modeled irradiance for a VNIR band as

$$Irr_{ROLO} = \frac{\int_\lambda I_{fit}(\lambda)\, S_{VNIR}(\lambda) \mathrm{d}\lambda}{\int_\lambda S_{VNIR}(\lambda)\mathrm{d}\lambda} \tag{A8}$$

where I_{fit} represents the fitted ROLO irradiance, S_{VNIR} is the spectral response function of the target band. Then the amount of sensitivity degradation from 2003 to 2017 (i.e., the relative degradation) can be measured by

$$r_{2003 \to 2017} = \left(\frac{Irr_{obs}}{Irr_{ROLO}}\right)_{2017} \bigg/ \left(\frac{Irr_{obs}}{Irr_{ROLO}}\right)_{2003}. \tag{A9}$$

References

1. Yamaguchi, Y.; Kahle, A.B.; Tsu, H.; Kawakami, T.; Pniel, M. Overview of advanced spaceborne thermal emission and reflection radiometer (ASTER). *IEEE Trans. Geosci. Remote Sens.* **1998**, *36*, 1062–1071. [CrossRef]
2. Abram, M.; Yamaguchi, Y. Twenty Years of ASTER Contributions to Lithologic Mapping and Mineral Exploration. *Remote Sens.* **2019**, *11*, 1394. [CrossRef]
3. Arai, K.; Ohgi, N.; Sakuma, F.; Kikuchi, M.; Tsuchida, S.; Inada, H. Onboard Trend Analysis of Onboard Calibration Data of Terra/ASTER/VNIR and One of the Suspected Causes of Sensitivity Degradation. *IJAS* **2011**, *2*, 71–83.
4. Tsuchida, S.; Yamamoto, H.; Kamei, A. Long-term vicarious calibration of ASTER VNIR bands. In Proceedings of the 52nd Conference of the Remote Sensing Society of Japan, Tokyo, Japan, 23 May 2012; pp. 85–86.
5. Thome, K.; Arai, K.; Tsuchida, S.; Biggar, S. Vicarious calibration of ASTER via the reflectance-based approach. *IEEE Trans. Geosci. Remote Sens.* **2008**, *46*, 3285–3295. [CrossRef]
6. Tonooka, H.; Palluconi, F.D.; Hook, S.J.; Matsunaga, T. Vicarious Calibration of ASTER Thermal Infrared Bands. *IEEE Trans. Geosci. Remote Sens.* **2005**, *43*, 2733–2746. [CrossRef]
7. Yamamoto, H.; Kamei, A.; Nakamura, R.; Tsuchida, S. Long-term cross-calibration of the Terra ASTER and MODIS over the CEOS calibration sites. In *Earth Observing Systems XVI*; International Society for Optics and Photonics: Bellingham, WA, USA, 2011; Volume 8153, p. 815318.
8. Obata, K.; Tsuchida, S.; Yamamoto, H.; Thome, K. Cross-Calibration between ASTER and MODIS Visible to Near-Infrared Bands for Improvement of ASTER Radiometric Calibration. *Sensors* **2017**, *17*, 1793. [CrossRef]
9. Tachikawa, T. ASTER Science Team Meeting. *Earth Obs.* **2014**, *26*, 18–21.
10. Obata, K.; Tsuchida, S.; Iwao, K. Inter-Band Radiometric Comparison and Calibration of ASTER Visible and Near-Infrared Bands. *Remote Sens.* **2015**, *7*, 15140–15160. [CrossRef]

11. Wagner, S.C.; Hewison, T.; Stone, T.; Lachérade, S.; Fougnie, B.; Xiong, X. A summary of the joint GSICS—CEOS/IVOS lunar calibration workshop: Moving towards intercalibration using the Moon as a transfer target. In *Sensors, Systems, and Next-Generation Satellites XIX*; International Society for Optics and Photonics: Bellingham, WA, USA, 2015; p. 96390Z.

12. Eplee, R.E.; Barnes, R.A.; Patt, F.S.; Meister, G.; McClain, C.R. SeaWiFS lunar calibration methodology after six years on orbit. In *Earth Observing Systems IX*; International Society for Optics and Photonics: Bellingham, WA, USA, 2004; pp. 1–13.

13. Eplee, R.E.; Sun, S.-Q.; Meister, G.; Patt, F.S.; Xiong, X.; McClain, C.R. Cross calibration of SeaWiFS and MODIS using on-orbit observations of the Moon. *Appl. Opt.* **2012**, *51*, 8702–8730. [CrossRef]

14. Kouyama, T.; Nakamura, R.; Kato, S.; Miyashita, N. One-year Lunar Calibration Result of Hodoyoshi-1, Moon as an Ideal Target for Small Satellite Radiometric Calibration. In Proceedings of the Small Satellite Conference, Logan, UT, USA, 4–9 August 2018. SSC18-III-04.

15. Choi, T.; Shao, X.; Cao, C.; Weng, F. Radiometric Stability Monitoring of the Suomi NPP Visible Infrared Imaging Radiometer Suite (VIIRS) Reflective Solar Bands Using the Moon. *Remote Sens.* **2016**, *8*, 15. [CrossRef]

16. Stone, T. Radiometric calibration stability and intercalibration of solar-band instruments in orbit using the Moon. In *Earth Observing Systems XIII*; International Society for Optics and Photonics: Bellingham, WA, USA, 2008; Volume 7081, p. 70810X.

17. Kieffer, H.H.; Stone, T.C. The spectral irradiance of the Moon. *Astron. J.* **2005**, *129*, 2887–2901. [CrossRef]

18. Yamamoto, S.; Matsunaga, T.; Ogawa, Y.; Nakamura, R.; Yokota, Y.; Ohtake, M.; Haruyama, J.; Morota, T.; Honda, C.; Hiroi, T.; et al. Preflight and In-Flight Calibration of the Spectral Profiler on Board SELENE (Kaguya). *IEEE Trans. Geosci. Remote Sens.* **2011**, *49*, 4660–4676. [CrossRef]

19. Yokota, Y.; Matsunaga, T.; Ohtake, M.; Haruyama, J.; Nakamura, R.; Yamamoto, S.; Ogawa, Y.; Morota, T.; Honda, C.; Saiki, K.; et al. Lunar photometric properties at wavelengths 0.5–1.6 μm acquired by SELENE Spectral Profiler and their dependency on local albedo and latitudinal zones. *Icarus* **2011**, *215*, 639–660. [CrossRef]

20. Kouyama, T.; Yokota, Y.; Ishihara, Y.; Nakamura, R.; Yamamoto, S.; Matsunaga, T. Development of an application scheme for the SELENE/SP lunar reflectance model for radiometric calibration of hyperspectral and multispectral Sensors. *Planet. Space Sci.* **2016**, *214*, 76–83. [CrossRef]

21. Williams, J.P.; Piage, D.A.; Greenhagen, B.T.; Sefton-Nash, E. The global surface temperatures of the Moon as measured by the Diviner Lunar Radiometer Experiment. *Icarus* **2017**, *283*, 300–325. [CrossRef]

22. ASTER Science Office ASTER SWIR Data Status Report. Available online: http://www.aster.jspacesystems.or.jp/en/about_aster/swir_en.pdf (accessed on 7 October 2019).

23. Kieffer, H.H. Photometric stability of the lunar surface. *Icarus* **1997**, *130*, 323–327. [CrossRef]

24. Ogohara, K.; Kouyama, T.; Yamamoto, H.; Sato, N.; Takagi, M.; Imamura, T. Automated cloud tracking system for the Akatsuki Venus Climate Orbiter data. *Icarus* **2012**, *217*, 661–668. [CrossRef]

25. Bruegge, C.J.; Abdou, W.; Diner, D.J.; Gaitley, B.J.; Helmlinger, M.; Kahn, R.A.; Martonchik, J.V. Validating the MISR radiometric scale for the ocean aerosol science communities. In Proceedings of the International Workshop on Radiometric and Geometric Calibration, Gulfport, MS, USA, 2 December 2004.

26. Sakuma, F.; Kikuchi, M.; Inada, H.; Akagi, S.; Ono, H. Onboard Calibration of the ASTER Instrument over Twelve Year. In Proceedings of the International Society for Optics and Photonics, Bellingham, WA, USA, 19 November 2012; Volume 8533, p. 853305.

27. Suzuki, H.; Yamada, M.; Kouyama, T.; Tatsumi, E.; Kameda, S.; Honda, R.; Sawada, H.; Ogawa, N.; Morota, T.; Honda, C.; et al. Initial inflight calibration for Hayabusa2 optical navigation camera (ONC) for science observations of asteroid Ryugu. *Icarus* **2018**, *300*, 341–359. [CrossRef]

28. Tatsumi, E.; Kouyama, T.; Suzuki, H.; Yamada, M.; Sakatani, N.; Kameda, S.; Yokota, Y.; Honda, R.; Morota, T.; Moroi, K.; et al. Updated inflight calibration of Hayabusa2's optical navigation camera (ONC) for scientific observations during the cruise phase. *Icarus* **2019**, *325*, 153–195. [CrossRef]

29. Biggar, S.; Slater, P.; Gellman, D. Uncertainties in the in-flight calibration of sensors with reference to measured ground sites in the 0.4–1.1 m range. *Remote Sens. Environ.* **1994**, *48*, 245–252. [CrossRef]

30. Helder, D.; Thome, K.J.; Mishra, N.; Chander, G.; Xiong, X.; Angal, A.; Choi, T. Absolute Radiometric Calibration of Landsat Using a Pseudo Invariant Calibration Site. *IEEE Trans. Geosci. Remote Sens.* **2013**, *51*, 1360–1369. [CrossRef]

31. ERSDAC. Algorithm Theoretical Basis Document for ASTER Level-1 Data Processing (Ver. 3.0). Available online: https://eospso.nasa.gov/sites/default/files/atbd/atbd-ast-01.pdf (accessed on 18 November 2019).

32. Wehrli, C. *Spectral Solar Irradiance Data, WMO ITD 149*; WMO: Geneva, Switzerland, 1986.
33. Pieters, C.M. The Moon as a Spectral Calibration Standard Enabled by Lunar Samples: The Clementine Example. In Proceedings of the New Views of the Moon II: Understanding the Moon through the Integration of Diverse Datasets, Flagstaff, AZ, USA, 22–24 September 1999; pp. 47–48.
34. Pieters, C.M.; Mustard, J.F. Exploration of crustal/mantal material for the Earth and Moon using reflectance spectroscopy. *Remote Sens. Environ.* **1998**, *24*, 151–178. [CrossRef]

Article

ASTER Cloud Coverage Assessment and Mission Operations Analysis Using Terra/MODIS Cloud Mask Products

Hideyuki Tonooka [1,*] and Tetsushi Tachikawa [2]

[1] Department of Computer and Information Sciences, Ibaraki University, Hitachi, Ibaraki 3168511, Japan
[2] Japan Space Systems, Minato-ku, Tokyo 1050011, Japan; Tachikawa-Tetsushi@jspacesystems.or.jp
* Correspondence: hideyuki.tonooka.dr@vc.ibaraki.ac.jp; Tel.: +81-294-38-5149

Received: 29 September 2019; Accepted: 23 November 2019; Published: 26 November 2019

Abstract: Since the Advanced Spaceborne Thermal Emission and Reflection Radiometer (ASTER) instrument cannot detect clouds accurately for snow-covered or nighttime images due to a lack of spectral bands, Terra/MODIS cloud mask (MOD35) products have been alternatively used in cloud assessment for all ASTER images. In this study, we evaluated ASTER cloud mask images generated from MOD35 products and used them to analyze the mission operations of ASTER. In the evaluation, ASTER cloud mask images from different MOD35 versions (Collections 5, 6, and 6.1) showed a large discrepancy in low- or high-latitude areas, and the rate of ASTER scenes with a high uncertain-pixel rate (≥30%) showed to be 2.2% in daytime and 12.0% in nighttime. In the visual evaluation with ASTER browse images, about 2% of cloud mask images showed some problems such as mislabeling and artifacts. In the mission operations analysis, the cloud avoidance function implemented in the ASTER observation scheduler showed a decrease in the mean cloud coverage (MCC) and an increase in the rate of clear scenes by 10% to 15% in each. Although 19-year-old time-series of MCC in five areas showed weather-related fluctuations such as the El Niño Southern Oscillation (ENSO), they indicated a small percent reduction in MCC by enhancement of the cloud avoidance function in April 2012. The global means of the number of clear ASTER scenes were 15.7 and 6.6 scenes in daytime and nighttime, respectively, and those of the success rate were 33.3% and 40.4% in daytime and nighttime, respectively. These results are expected to contribute not only to the ASTER Project but also to other optical sensor projects.

Keywords: Advanced Spaceborne Thermal Emission and Reflection Radiometer (ASTER); MODIS; MOD35; cloud mask; cloud coverage; uncertain flag; mission operations; observation scheduler; cloud avoidance; success rate

---

1. Introduction

The Advanced Spaceborne Thermal Emission and Reflection Radiometer (ASTER) is an optical sensor onboard NASA's Terra satellite launched on 18 December 1999 [1]. The ASTER instrument has 14 spectral bands divided by the following three subsystems: the visible and near-infrared (VNIR) subsystem (three bands), the shortwave infrared (SWIR) subsystem (six bands), and the thermal infrared (TIR) subsystem (five bands). The spatial resolution is 15 m, 30 m, and 90 m, for the VNIR, the SWIR, and the TIR subsystems, respectively, and the swath width of each subsystem is 60 km. Table 1 gives the spectral range and the ground resolution of each ASTER band [1]. ASTER products have been widely used in various fields such as surface mineralogical mapping, long-term global monitoring of glaciers, volcanoes, and coral reefs, as well as regional surface heat balance analysis [2].

In the generation of such products, cloud coverage (CC) assessment for each scene is a fundamental procedure. An assessment result is used not only for higher product generation but also for purposes such as image search and mission operations analysis. In the ASTER level-1 processing, the ASTER cloud cover assessment algorithm (ACCAA) has been implemented [3,4]. The original ACCAA which is a modified version of the automatic cloud cover assessment (ACCA) of Landsat-5 TM [5] uses three filters (i.e., threshold tests) with ASTER bands 2 (VNIR), 4 (SWIR), and 11 (TIR). Since 28 October 2000, an improved version of ACCAA has been used for all ASTER scenes observed in the full mode in which all bands are on [6]. This improved version which was developed based on the ACCA of Landsat-7 ETM+ [5] has eight filters with bands 1 to 3 (VNIR), band 4 (SWIR) and band 13 (TIR) and two passes, where the pass one goal is to develop a reliable cloud signature for use in pass two where the remaining clouds are identified [5]. The improved ACCAA is more accurate than the original ACCAA, but it performs less well in some combinations of surface type and sun elevation angle such as desert observation under a low sun elevation [6]. In addition, the original and the improved ACCAAs are not accurate enough for nighttime scenes due to thresholding with single TIR band [7], and also became unreliable particularly in the cryosphere after April 2008, because SWIR imaging has been saturated due to anomalously high SWIR detector temperature [8]. Although Hulley and Hook have proposed a new methodology for ASTER cloud detection and classification [9], on the one hand, it is difficult to use in operation due to a lack of SWIR bands. On the other hand, the Moderate Resolution Imaging Spectroradiometer (MODIS) onboard the Terra satellite provides cloud mask (MOD35) products [10,11] by simultaneous observations with ASTER using effective bands for cloud detection, although the spatial resolution is lower than ASTER. Thus, the ASTER Project has operated the ASTER CC reassessment system using MOD35 products since June 2009 [6,12]. In this system, ASTER cloud mask images are generated from only collocated MOD35 products and used for recalculation of ASTER CC values. ASTER cloud mask images generated are provided to the public through the Internet [13].

Thus, in this study, we evaluate the reliability of MOD35-based ASTER cloud mask images, and, then, perform ASTER mission operations analysis using those cloud mask images. As for the accuracy of MOD35 products, Ackerman et al. performed a comprehensive study for Collection 5 (C5) MOD35 products using lidar data [14], and Wilson et al. reported that C5 MOD35 products have a bias induced by land cover types [15]. Wang et al. evaluated C6 Aqua/MODIS cloud mask (MYD35) products using Cloud-Aerosol Lidar and Infrared Pathfinder Satellite Observations (CALIPSO) products [16], and Moeller et al. described improvements of MOD35 products in updating from C6 to C6.1 [17]. These studies, however, do not focus on usage of MOD35 products in cloud assessment for a type of high-resolution sensor such as ASTER. For the evaluation of the reliability of MOD35-based ASTER cloud mask images, we, therefore, report results for a comparison among C5, C6, and C6.1 MOD35 products, an evaluation of uncertain pixels in MOD35 products, and a visual evaluation with ASTER browse images. Although the visual evaluation of MOD35-based ASTER cloud masks was reported by Tonooka et al. [6], we demonstrate an updated result using C6.1 MOD35 products as of August 2019.

With respect to ASTER mission operations analysis using cloud mask images, we report results of an effectiveness evaluation of the cloud avoidance function implemented in the ASTER observation scheduler [18]. The cloud avoidance function of the ASTER scheduler has not been evaluated in the past. We also report results from the time-series analysis of the mean cloud coverage, and from global mapping of the number of clear scenes and the success rate, where, in this study, the success rate is defined as the rate of clear scenes with a CC of 20% or less. Although King et al. reported the global maps of the seasonal mean cloud fraction using MODIS C5.1 cloud products from 2000 to 2011 [19], our time-series analysis and global mapping with MOD35 products differs from their study in that we use only MOD35 products accompanied by ASTER observations mainly over land areas. Since Tonooka et al. have already performed global mapping of a number of clear scenes and the success rate for ASTER using C5 MOD35 products as of 2010 [6], we demonstrate updated results using C6.1 MOD35 products as of August 2019.

Table 1. Spectral range and ground resolution of each Advanced Spaceborne Thermal Emission and Reflection Radiometer (ASTER) spectral band [1].

Subsystem	Spectral Range [μm]	Ground Resolution
VNIR	Band 1: 0.52–0.60 Band 2: 0.63–0.69 Band 3N, 3B: 0.78–0.86	15 m
SWIR	Band 4: 1.600–1.700 Band 5: 2.145–2.185 Band 6: 2.185–2.225 Band 7: 2.235–2.285 Band 8: 2.295–2.365 Band 9: 2.360–2.430	30 m
TIR	Band 10: 8.125–8.475 Band 11: 8.475–8.825 Band 12: 8.925–9.275 Band 13: 10.25–10.95 Band 14: 10.95–11.65	90 m

2. Materials and Methods

2.1. ASTER Cloud Coverage Reassessment Using MOD35 Products

2.1.1. Overview of MOD35 Products

The MOD35 product is a level-2 product with a spatial resolution of 1 km or 250 m [10,11]. In the ASTER Project, the MOD35 product with a spatial resolution of 1 km is used for CC reassessment, because a spatial resolution of 250 m is not available in nighttime [6]. In the MOD35 algorithm, each pixel is classified to a particular domain according to surface type and solar illumination, and then tested by a series of threshold tests prepared for each domain. According to the confidence flag calculated through these tests, one of four levels (0, cloudy; 1, uncertain; 2, probably clear; and 3, clear) is assigned to each pixel as a cloud mask value. The details on this algorithm can be found in [10,11].

The MOD35 algorithm was significantly improved in C4 to C5 updating, particularly for polar regions and oceans in nighttime [20]. Ackerman et al. validated C5 MOD35 products through lidar observations from ground, aircraft, and spaceborne platforms, concluding that the MODIS algorithm agreed with the lidar about 85% of the time, the optical depth limitation of the MODIS cloud mask was approximately 0.4, and approximately 90% of the mislabeled scenes had optical depths less than 0.4 [14]. Wilson et al. pointed out that C5 MOD35 cloud mask had a systematic land cover bias, and this bias partly remained also in C6 MOD35 cloud mask [15].

In updating from C5 to C6, various improvements such as use of normalized difference vegetation index (NDVI) background maps and addition of a new night ocean test were applied [11,21]. Wang et al. evaluated C6 MYD35 products using Cloudsat CALIPSO products, showing that the total agreement (the sum of the clear- and cloud-hit rates) between them was 77.8%, and MODIS mislabeled cloud as clear for 9.1% and clear as cloudy for 1.8% [16].

Updating of MOD35 from C6 to C6.1 was performed due to reprocessing of level-1B (L1B) radiance products. C6 L1B products of Terra MODIS had several calibration issues including an electrical crosstalk contamination in TIR bands [22], and these issues have been solved in C6.1 L1B products [23].

2.1.2. ASTER Cloud Coverage Reassessment System

The ASTER CC reassessment system generates a cloud mask image of each ASTER scene from the MOD35 product simultaneously observed with ASTER, and calculates the CC of that ASTER scene [6]. The system has been operated by Ibaraki University since June 2009, now providing a reassessment result in two days after each ASTER observation. The system has the following procedures:

1. Obtain daily ASTER catalogue information from Japan Space Systems (JSS);
2. Obtain MOD35 products from NASA Goddard Space Flight Center (GSFC);
3. Generate cloud mask images for all ASTER scenes in the catalogue;
4. Send the reassessment results (CC etc.) to JSS, Land Processes Distributed Active Archive Center (LP DAAC), and Institute of Advanced Industrial Science and Technology (AIST);
5. Release ASTER cloud mask images to public through Internet [13].

The cloud mask for each scene has the same size with an ASTER/TIR level-1A image (700 × 700 pixels) by assignment of MOD35 results to ASTER/TIR pixels with the nearest neighbor interpolation. The CC of each ASTER scene is calculated from each cloud mask by giving 100% to cloud pixels, 50% to uncertain pixels, and 0% to probably clear or clear pixels, and averaging them over the scene. Since the rate of uncertain pixels in each scene can be used as a reliability index, this value is added as an ancillary information to the delivered result.

Although the system started with C5 MOD35 products, C6.1 MOD35 products have being used since January 2017. Reprocessing with C6.1 MOD35 products was completed for all ASTER archives.

As of 10 August 2019, the total number of ASTER scenes accompanied with MOD35-based cloud masks is 3,618,286, 99.4% of all ASTER archives (equals 3,640,027). The rate of ASTER scenes without MOD35-based cloud masks is 0.71% in daytime observations, and 0.23% in nighttime observations. The reason why daytime observations have more losses is because Terra MODIS had more observation losses in the period of 2000 to 2001 in which ASTER did not have many observations in nighttime. Main causes of MODIS observation losses which affected ASTER cloud reassessment are events such as Solid State Recorder (SSR) anomaly, Science Formatting Equipment (SFE) anomaly, and Terra's maneuver [24].

2.2. Evaluation of MOD35-Based ASTER Cloud Masks

2.2.1. Comparison of ASTER Cloud Coverage Among MOD35 Versions

In the ASTER CC reassessment system, three versions of MOD35 products, C5, C6, and C6.1, have been used since June 2009. Since the CC values based on C5 or C6 MOD35 products were distributed to ASTER users before January 2018, we randomly selected 17,845 daytime scenes and 8521 nighttime scenes from ASTER scenes observed in 2016 (13.3% and 14.2% of daytime and nighttime scenes in 2016, respectively), and compared the CC values among C5, C6, and C6.1. The results are described in Section 3.1.1.

2.2.2. Rate of High-Uncertain Scenes

Since a MOD35 product with many pixels labeled as "uncertain" will be less reliable, the rate of uncertain pixels in an ASTER cloud mask image can be used as an index of reliability for that cloud mask image. We, therefore, mapped the rate of high-uncertain scenes (referred to as the high-uncertain rate) over 0.1 × 0.1 degree grids for daytime and nighttime using all ASTER archives as of 10 August 2019, where a high-uncertain scene is defined as a scene that the rate of uncertain pixels in that scene is 30% or greater. The results can be found in Section 3.1.2.

2.2.3. Visual Evaluation of ASTER Cloud Mask Images Using Browse Images

C5-based ASTER cloud mask images have been visually evaluated with ASTER browse images by human analysts [6]. In this study, we evaluated C6.1-based ASTER cloud mask images in the same way, although this approach is affected by ambiguity in human judgment.

First, we selected daytime and nighttime scenes randomly from all ASTER archives, and acquired the VNIR, SWIR, and TIR browse images for each daytime scene, and the TIR browse image for each nighttime scene. Then, we visually evaluated consistency between the C6.1-based ASTER cloud mask image and the browse image(s) for each of evaluable scenes in the selected scenes. The total number of evaluated scenes is 24,706 for daytime, and 8006 for nighttime, corresponding to about 0.9% and 1.0% of all daytime and nighttime scenes, respectively. The results are described in Section 3.1.3.

2.3. ASTER Mission Operations Analysis Using MOD35-Based ASTER Cloud Masks

2.3.1. Validation of the Cloud Avoidance Function in the ASTER Observation Scheduler

The ASTER instrument has the pointing function because the swath width of 60 km is narrower than the orbit-to-orbit distance at the equator (172 km) [1]. Using this function effectively, on the one hand, ASTER needs to respond to various observation requests such as global mapping, scientific studies, disaster monitoring, and calibration. On the other hand, the ASTER instrument has a limited duty cycle of 8%, corresponding to the two-orbit maximum data acquisition time of 16 min, mainly due to limitations in the data volume allocated in Terra's SSR, and the communications link with the Tracking and Data Relay Satellite System (TDRSS) and ground stations [1,18]. In addition, ASTER has several instrumental limitations such as the operating time and the total number of pointing changes. Thus, the ASTER Project has developed and used the ASTER observation scheduler for responding to many observation requests optimally under various limitations throughout the mission period [18].

Basically, an observation schedule is generated on a day-by-day basis using the ASTER observation scheduler. This is called a one-day schedule (ODS). Each ODS is generated as of two or more days before each observation day. This is called a normal ODS (referred to as N-ODS in this paper). Each ODS is regenerated as of 16 hours before each observation time, and each N-ODS is replaced by that. This is called a late-change ODS (referred to as LC-ODS in this paper). In generation of N-ODS and LC-ODS, the priority function for each observation request is first calculated by multiplying eleven subfunctions from f1 to f11. Next, the score of each observation request is calculated by multiplying the priority function of that request with the overlapping area of the requested and the overpassed areas. The derived score is then compared among conflicting requests, and the observation request with the maximum score is finally selected and included in ODS. In this procedure, various instrumental limitations are also considered.

In the eleven subfunctions, f4 is the subfunction prepared for avoiding clouds and increasing cloud-free scenes. This subfunction is calculated by

$$f4 = b + a \times \arctan\{\kappa \times (1 - C_{max}) \times (C_{max} - C_{pred})\}, \tag{1}$$

where C_{max} is the maximum CC accepted by each observation request; C_{pred} is the predicted CC for each observation request; and a, b, and κ are fixed coefficients. Although the values of a and b have not been changed since the launch (a = 1 and b = 1.5708), the value of κ was changed from 50 to 10,000,000 in April 2012 for enhancing the effect of the cloud avoidance function. Using the updated value of κ, the value of f4 is about $10^{-5.4}$ for $C_{max} < C_{pred}$, and about $10^{+0.6}$ for $C_{max} > C_{pred}$. A notable point here is that f4 is calculated for LC-ODS, but not for N-ODS, because the cloud prediction is generally less accurate as of two days before each observation day. Thus, a cloud prediction result is used in only LC-ODS, where the cloud prediction dataset used for LC-ODS is the total cloud coverage (TCC) which is the "TCDC: entire atmosphere" layer extracted from the "T1534 Semi-Lagrangian grid" data produced by the Global Forecast System (GFS) of National Centers for Environmental Prediction (NCEP) [25]. TCC provides a 6 h averaged CC with the horizontal resolution of 3072 E-W grids and 1536 N-S grids at 6 h intervals until 72 hours after the present.

As an example, Figure 1 shows observation areas scheduled on 7 April 2018, where the background is the daily cloud fraction image from the MODIS MOD08 product on the same day. "Only N-ODS" and "only LC-ODS" show scheduled areas included in only N-ODS and in only LC-ODS, respectively, and "Common" shows scheduled areas included in both N-ODS and LC-ODS. The common scenes of N-ODS and LC-ODS account for about a half of the total scenes of LC-ODS on average. Figure 1 indicates that the cloud avoidance function worked well as expected, but it should be more quantitatively evaluated by a comparison of CC between actually observed ASTER scenes included in only LC-ODS and not-observed (fictitious) ASTER scenes included in only N-ODS on a day-by-day basis. Although fictitious ASTER scenes do not exist, the CCs of them can be calculated by a combination of N-ODS data and MOD35 products. Thus, we evaluated the effectiveness of the cloud avoidance function in the ASTER observation scheduler by this approach using the ASTER cloud reassessment system.

First, we investigated which areas were frequently cancelled from N-ODS by the cloud avoidance function. In this analysis, we defined the cancellation rate as

$$\text{cancellation rate} = N_{\text{only-N}} / \{ N_{\text{only-N}} + N_{\text{com}} \}, \tag{2}$$

where $N_{\text{only-N}}$ is the number of scenes included in "only N-ODS", and N_{com} is that included in both N-ODS and LC-ODS, on a day-by-day basis. Next, we compared the mean cloud coverage (MCC) of N-ODS and that of LC-ODS for the three years (1 May 2016 to 30 April 2019). The results obtained by these analyses are described in Section 3.2.1.

2.3.2. Time-Series Analysis of the Mean Cloud Coverage

The MCC value is affected by climate classification and also by factors such as seasonal change, abnormal weather, long-term climate change, and MOD35 product error. Moreover, is reduced if the cloud avoidance function works effectively, indicating that the MCC value is affected also by scheduling ways. We, therefore, performed time-series analysis of MCC over 10×10 degree grids at three-month intervals in the period from April 2000 to June 2019.

First, we selected grid cells which had been repeatedly observed by the ASTER instrument in all seasons throughout 19 years since the launch, because such grid cells are suitable for time-series analysis at three-month intervals. Then, we performed seasonal decomposition with moving averages [26] for the MCC data at three-month intervals in the selected grid cells, and divided the time-series data of each cell into three components, the trend, the seasonal change, and the residual.

Next, we investigated whether the three-monthly MCC values were affected by updating of the coefficient κ in Equation (1), in April 2012. In order to reduce the effects of the El Niño Southern Oscillation (ENSO) [27], we derived the three-monthly Niño 3.4 index from April 2000 to June 2019, where the monthly Niño 3.4 index is the monthly mean of the deviation of sea surface temperature (SST) from the average in a single fixed 30-year base period in the Niño 3.4 region (5N to 5S in latitude, 120 to 170W in longitude) [28]. We, then, selected only the three-monthly MCC values with a three-monthly Niño 3.4 index within ±0.5 (i.e., less effect of ENSO), and investigated the effect of the updating of κ to the three-monthly MCC values.

The results obtained by these analyses are described in Section 3.2.2.

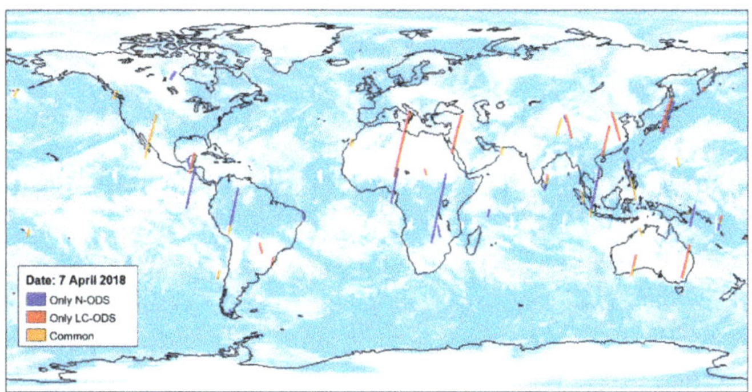

Figure 1. Scheduled areas included in only normal one-day schedule "(N-ODS)", only late change one-day schedule "(LC-ODS)", and both N-ODS and LC-ODS ("common") on 7 April 2018. The background is the daily cloud fraction image from the MODIS MOD08 product on the same day (blue, cloud and white, clear).

2.3.3. Mapping of the Number of Clear Scenes and the Success Rate

Tonooka et al. generated the global maps of the number of clear scenes and the success rate in daytime and nighttime as of September 2010, using ASTER cloud mask images based on the C5 MOD35 product [6]. In this study, we generated those maps as of August 2019 using C6.1 MOD35 products. The results can be found in Section 3.2.3.

3. Results and Discussion

3.1. Evaluation of MOD35-Based ASTER Cloud Masks

3.1.1. Comparison of ASTER Cloud Coverage Among MOD35 Versions

Figure 2 shows the difference of C6-based CC and C5-based CC (C6−C5), and that of C6.1-based CC and C6-based CC (C6.1−C6) in daytime and nighttime, as a function of latitude. Table 2 gives the mean of the difference (bias), and the rates of ASTER scenes with the CC difference of 10% or greater ($R_{10\%}$), and with that of 50% or greater ($R_{50\%}$), for each of C6−C5, and C6.1−C6 in daytime and nighttime. In updating from C5 to C6, $R_{10\%}$ and $R_{50\%}$ are 18.5% and 6.5% in daytime, respectively, and 16.3% and 2.8% in nighttime, respectively. The most likely cause that nighttime shows smaller rates than daytime is that snow-covered and vegetated areas giving a large CC-difference in daytime have not been prioritized in ASTER nighttime scheduling. Figure 2 indicates that C5 has a larger CC than C6 in low latitude areas, whereas C6 has a larger CC than C5 in Antarctica.

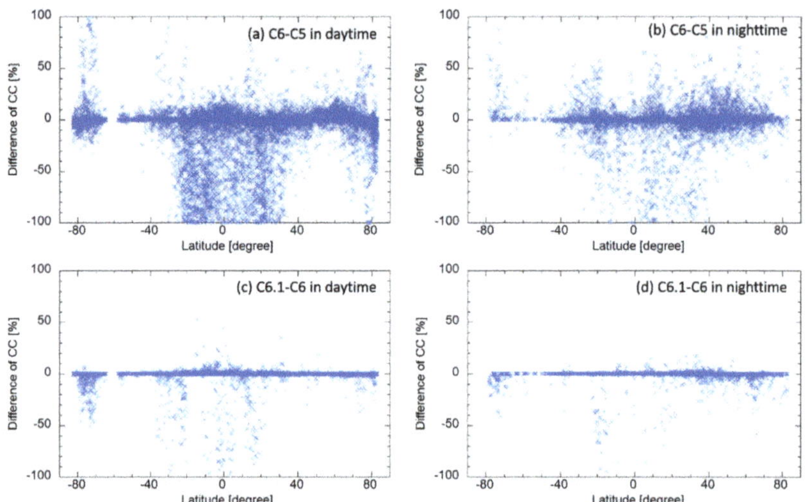

Figure 2. Difference of ASTER CC between different MOD35 versions as a function of latitude. (**a**) C6 minus C5 in daytime, (**b**) C6 minus C5 in nighttime, (**c**) C6.1 minus C6 in daytime, and (**d**) C6.1 minus C6 in nighttime. A total of 17,845 daytime scenes and 8521 nighttime scenes observed in 2016 were used.

In comparison to C6−C5, the difference of CC between C6.1 and C6 is smaller, because updating to C6.1 is a change in input L1B radiances; the bias is −0.6%, and $R_{10\%}$ and $R_{50\%}$ are about 2% and 0.4%, respectively, in both daytime and nighttime. Since the crosstalk issue in C6 L1B products gave an impact to the test for detecting ice clouds over water surfaces in the MOD35 algorithm, the improvements by C6.1 updating were seen mainly over oceans [17]. Thus, another reason why no large differences between C6 and C6.1 are seen in Figure 2c,d is because the rate of ASTER ocean scenes is low. Figure 2 also shows that C6 gives a larger CC than C6.1 in most inconsistent cases seen in low- or high-latitude areas, where it should be noted that such inconsistency was caused from a time-dependent error on C6 L1B radiances. Thus, C6.1-based cloud mask images should be used for ASTER scenes.

Table 2. Mean of the CC difference (bias), $R_{10\%}$, and $R_{50\%}$ in daytime and nighttime for C6 minus C5 and C6.1 minus C6.

Parameter	C6 Minus C5		C6.1 Minus C6	
	Daytime	Nighttime	Daytime	Nighttime
bias	−5.7%	−1.4%	−0.6%	−0.6%
$R_{10\%}$	18.5%	16.3%	1.9%	2.1%
$R_{50\%}$	6.5%	2.8%	0.4%	0.3%

3.1.2. Rate of High-Uncertain Scenes

Figure 3 displays the C6.1-based high-uncertain rate maps generated for daytime and nighttime. Globally, the high-uncertain rate is 2.2% in daytime, 12.0% in nighttime, and 4.4% in total; the high-uncertain rate in nighttime is five-times larger than that in daytime. On the one hand, in daytime, the high-uncertain rate is larger particularly in low latitude areas such as Venezuela to the Amazon River estuary area, the Yucatan Peninsula, the Gulf of Guinea, and the Congo areas, Indonesia, and the Mekong River estuary area, or in snow-covered areas such as the south part of Greenland and the inland of Antarctica. In addition, the high-uncertain rate is large in a part of land areas such as the north part of Mozambique, the Yellow River basin, and the Ganges River basin, and in ocean areas such as the offshores of San Francisco and Newfoundland, and the Sea of Okhotsk. On the other hand, the high-uncertain rate in nighttime is large particularly in mountainous areas such as Tibet and the north of the Andes basin (Colombia to Peru), and in glacier lake areas such as a part of Canada, the south part of Chile, the south part of Scandinavia, and Scotland. Such regionality indicates that there is a cloud fraction bias or higher uncertainty associated with certain land cover types, as pointed out by Wilson et al. [15].

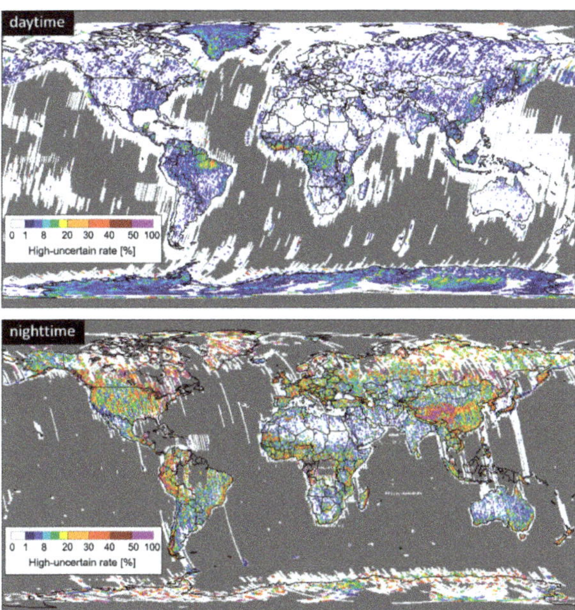

Figure 3. C6.1-based high-uncertain rate maps over 0.1 × 0.1 degree grids for daytime and nighttime using all ASTER archives as of 10 August 2019.

On the basis of the above results, it should be highlighted that the reliability of MOD35-based cloud masks is strongly dependent upon location and day and night difference.

3.1.3. Visual Evaluation of ASTER Cloud Mask Images Using Browse Images

The visual evaluation of ASTER cloud mask images with browse images was performed for 24,706 daytime and 8006 nighttime scenes. As a result, 98.3% of daytime scenes (24,288) and 96.6% of nighttime scenes (7737) showed high consistency between cloud mask and browse images, leaving 1.7% of daytime and 3.4% of nighttime scenes with differences. As for 687 scenes with a problem, 31% of them were due to mislabeling in snow-covered areas (case 1), 19% of them were due to mislabeling of a water surface as a cloud in a coastal area (case 2), 15% of them were due to artifacts probably associated with ancillary data used in the MOD35 algorithm (case 3), and 35% of them were due to a large discrepancy in CC over surfaces such as ocean and desert (case 4). Figure 4 demonstrates examples of cases 1 to 4.

Thus, it should be noted that the above type of problems can be seen in a limited number of MOD35-based ASTER cloud masks, particularly in snow-covered or coastal areas.

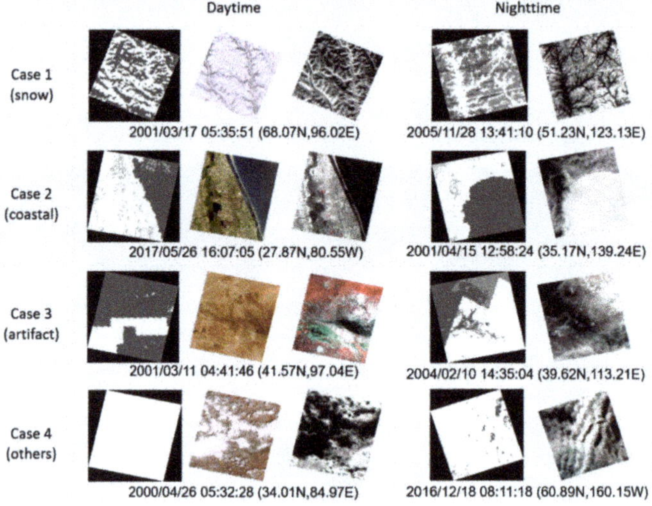

Figure 4. Examples of ASTER cloud mask images with a problem in visual evaluation for daytime and nighttime for case 1 (snow/ice area), case 2 (coastal waters), case 3 (artifact), and case 4 (others including ocean and desert). Images in each cell are the cloud mask image (**left**), the visible and near-infrared (VNIR) browse image (**middle**, only daytime; BGR = bands 1, 2, and 3N), and the thermal infrared (TIR) browse image (**right**; BGR = bands 10, 12, and 14). White, dark gray, and black on each cloud mask image indicate clear, cloud, and background, respectively.

3.2. ASTER Mission Operations Analysis Using MOD35-Based ASTER Cloud Masks

3.2.1. Validation of the Cloud Avoidance Function in the ASTER Observation Scheduler

Figure 5 displays the cancellation rate of N-ODS for three years from 1 May 2016 to 30 April 2019 over 10 × 10 degree grids. As shown, observation cancels caused by the cloud avoidance function occur frequently in most of global land areas. In the west coast of USA, the cancellation rate is relatively low because this area is often observed by calibration-related requests without the cloud avoidance. As for Antarctica and Greenland, the cancellation rate in daytime is significantly low because these areas are frequently observed by requests without the cloud avoidance from the Global Land Ice Measurements from Space (GLIMS) Project [29].

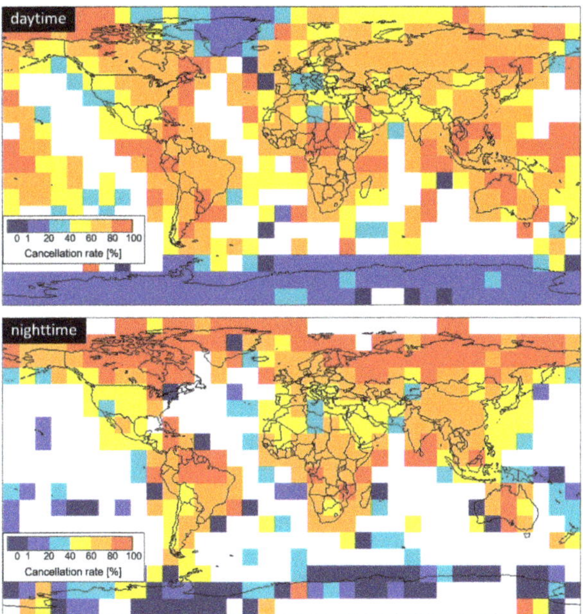

Figure 5. Cancellation rate of N-ODS for three years from 1 May 2016 to 30 April 2019 over 10 × 10 degree grids for each of daytime and nighttime.

Figure 6 gives the daily MCC values of "only N-ODS" and "only LC-ODS" as a function of date in daytime and nighttime. As shown, the TCC-based cloud avoidance function reduced the MCC significantly throughout the period, while it scatters more largely in nighttime. Table 3 gives the root mean square (RMS) differences of MCC and the clear-scene rate between N-ODS and LC-ODS for daytime and nighttime in the three years, where the clear-scene rate is defined as the rate of scenes with a CC of 20% or less. The table shows two cases that common scenes are included or not. Since the common scenes of N-ODS and LC-ODS account for about a half of the total scenes of LC-ODS in average as mentioned, the RMS differences with common scenes reduce by half, but the results show that the cloud avoidance function decreased MCC by 10% to 15% and increased the clear-scene rate by 10% to 15% even if common scenes are considered. The likely reason why the cloud avoidance function showed a better result in nighttime is probably because the subfunction f4 would affect ODS more strongly in nighttime due to fewer conflicts among fewer observation requests than in daytime.

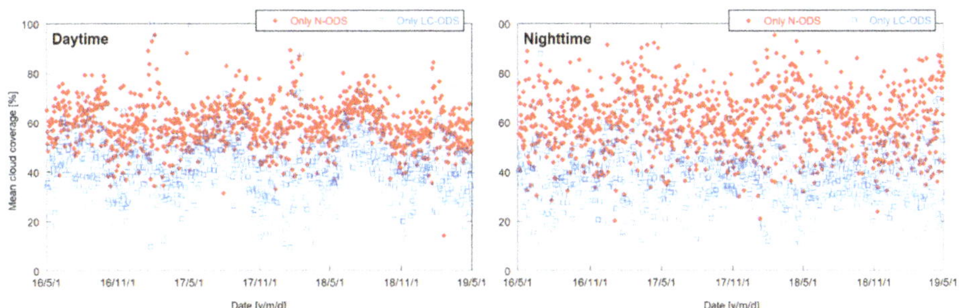

Figure 6. Daily mean cloud coverage (MCC) values of "only N-ODS" and "only LC-ODS" as a function of date in the three years from 1 May 2016 to 30 April 2019.

Table 3. RMS differences of MCC and the clear-scene rate between N-ODS and LC-ODS for daytime and nighttime in the three years. Two cases that common scenes are included or not are shown.

Day/Night	RMS Difference of MCC		RMS Difference of Clear-scene Rate	
	no Common	with Common	no Common	with Common
daytime	20.1%	11.4%	21.5%	12.3%
nighttime	25.3%	15.4%	25.8%	15.8%

Figure 7 displays the difference of MCC between "only LC-ODS" and "only N-ODS" over 10×10 degree grids for daytime and nighttime in the three years. If a grid cell has a negative difference, it means that LC-ODS has a smaller MCC than N-ODS, indicating the success of the cloud avoidance for that grid cell. As demonstrated by Figure 7, many grid cells are bluish, and reddish grid cells are limited in a small part of high latitude areas. This indicates that the cloud avoidance function performs well in most areas.

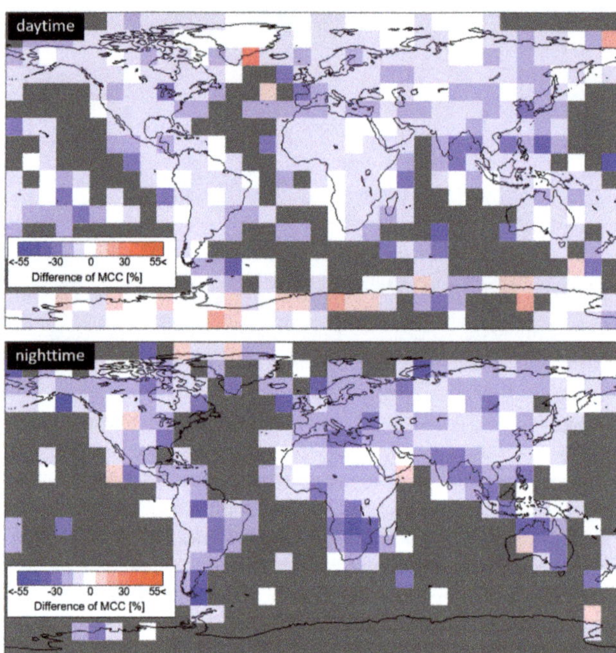

Figure 7. Difference of MCC between "only LC-ODS" and "only N-ODS" over 10×10 degree grids for daytime and nighttime in the three years from 1 May 2016 to 30 April 2019. In bluish cells, "only LC-ODS" has a smaller MCC than "only N-ODS", indicating the success of the cloud avoidance function.

3.2.2. Time-Series Analysis of the Mean Cloud Coverage

Figure 8 shows the time-series satisfaction rate for global grid cells in daytime and nighttime, where the time-series satisfaction rate is defined as the rate of three-month terms including 10 scenes or more in a grid cell. The red grid cells are suitable for time-series analysis because they have been repeatedly observed in all seasons throughout 19 years, while the light purple or white grid cells are not suitable because observations are limited to a specific season or a part of the mission period. According to this result, we selected the following five grid cells, as shown in the figure: (A) USA, (B) Argentina, (C) Mozambique, (D) Indonesia, and (E) Japan.

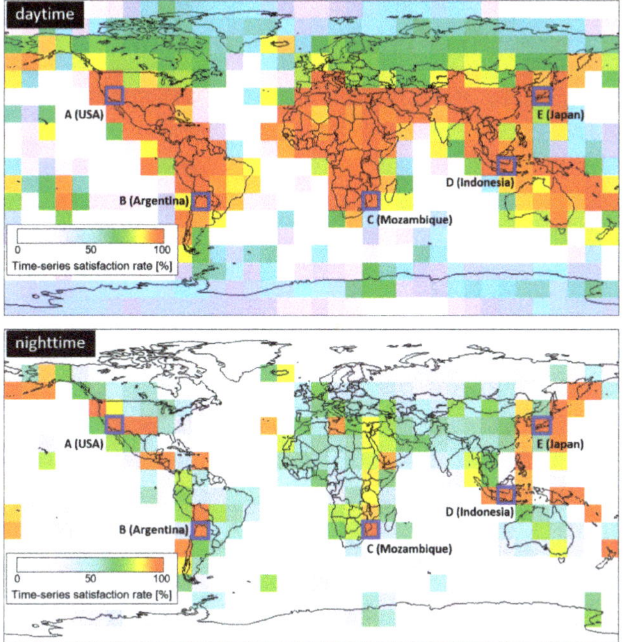

Figure 8. Time-series satisfaction rate for global grid cells in daytime and nighttime. The five grid cells used for the time-series analysis are also shown.

Next, we performed seasonal decomposition with moving averages [26] for the MCC data at three-month intervals in the grid cells A to E, and divided the time-series data of each cell into the three components. As an example, Figure 9 displays the obtained time-series data at the grid cell E in daytime as follows: (a) the original MCC data, (b) the trend, (c) the seasonal change, and (d) the residual.

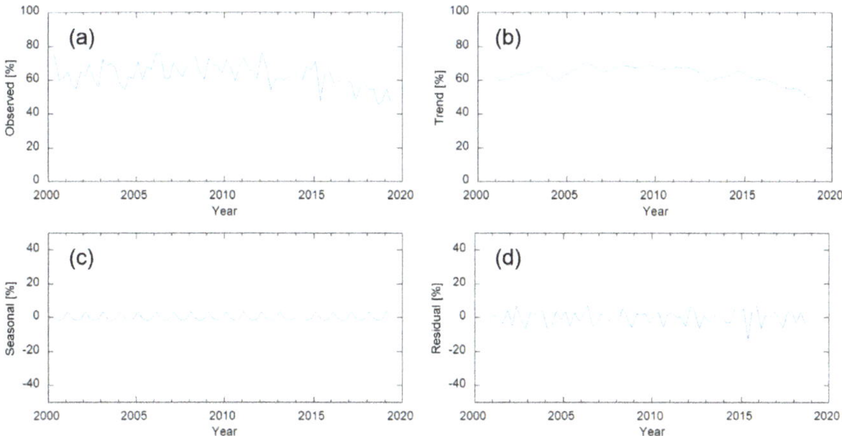

Figure 9. Time-series data at the grid cell E in daytime as an example: (**a**) the original MCC data, (**b**) the trend, (**c**) the seasonal change, and (**d**) the residual.

Figure 10 gives the trend plots of the three-monthly MCC values for the grid cells A to E in daytime and nighttime, also showing the trend plot of the monthly Niño 3.4 index. As shown, it appears that the fluctuations seen in the MCC trends can be partly explained by ENSO. For example, El Niño is often associated with warm and dry conditions around Indonesia [27]. The grid cell D, therefore, seems to have smaller MCC values in Niño events (positive in the monthly Niño 3.4 index), and larger MCC values in La Niña events (negative in the index).

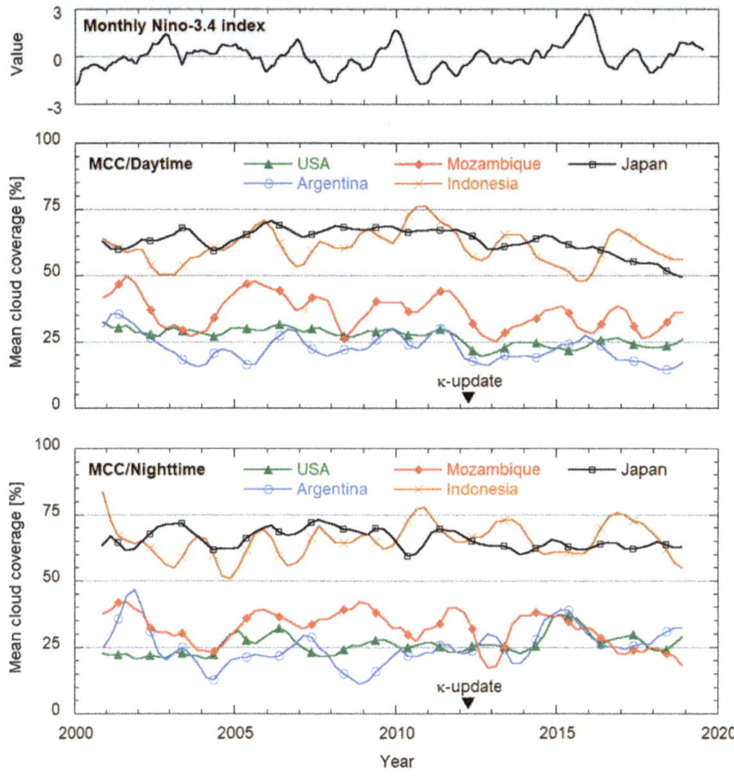

Figure 10. Trend plots of the three-monthly MCC values for the grid cells A to E in daytime and nighttime, also showing the trend plot of the monthly Niño 3.4 index.

Table 4 shows the means of the three-monthly MCC values for the five grid cells in daytime and nighttime over the period before the updating of κ (Period 1) and over the period after the updating (Period 2). The table also shows the difference between Periods 1 and 2 for each grid cell in daytime and nighttime. As demonstrated, the mean of the three-monthly MCC value in daytime is 0.5% to 7.9% smaller in Period 2 than in Period 1 for all grid cells in daytime and two grid cells (C and E) in nighttime, probably due to enhancement of the cloud avoidance function by updating of κ. The likely reason why the grid cells A, B, and D gave a positive difference in nighttime is due to weather fluctuations, although it is also noted that the high-uncertain rate is high around the grid cell D in nighttime (see Figure 3), and that the cancellation rate of N-ODS is low at the grid cell A (see Figure 5).

Table 4. The means of the three-monthly MCC values for the grid cells A to E in daytime and nighttime over the period before the updating of κ in Equation (1) (Period 1) and over the period after the updating (Period 2). The difference between Periods 1 and 2 in each case is also shown.

Grid Cell	Daytime			Nighttime		
	Period 1	Period 2	diff.	Period 1	Period 2	diff.
(A) USA	29.1	23.3	−5.8	24.2	26.0	1.8
(B) Argentina	24.0	18.3	−5.7	25.6	26.3	0.7
(C) Mozambique	39.0	31.2	−7.9	33.7	27.4	−6.3
(D) Indonesia	60.7	60.3	−0.5	64.0	67.8	3.8
(E) Japan	64.9	59.4	−5.4	66.7	62.8	−3.9

3.2.3. Mapping of the Number of Clear Scenes and the Success Rate

Figure 11 displays the daytime and the nighttime maps of the number of clear scenes over 10×10 degree grids in 19.5 years from March 2000 to August 2019, where a clear scene is defined as a scene with a CC of 20% or less. The overall mean values are 15.7 scenes in daytime, and 6.6 scenes in nighttime. In daytime, a large number of clear scenes can be seen in more frequently observed areas such as the west part of USA and Japan, in arid areas such as the Middle East, Sahara, the south of Africa, and the west coast of Central and South America, and in the polar areas with a narrower orbit spacing. Although the same tendency can be seen also in nighttime, the number of clear scenes is smaller in snow-covered or dense vegetated areas, because these areas have not been prioritized in nighttime scheduling.

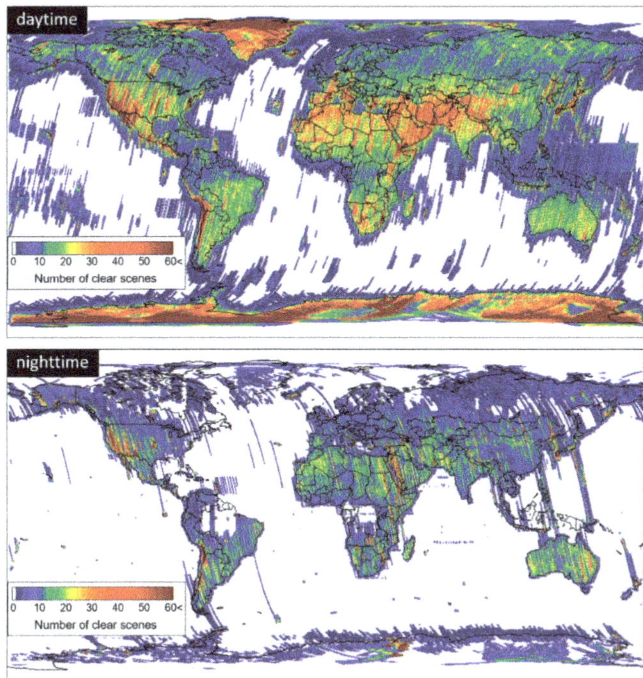

Figure 11. Daytime and nighttime maps of the number of clear scenes with a CC of 20% or less over 10×10 degree grids in 19.5 years from March 2000 to August 2019.

Figure 12 shows the daytime and the nighttime maps of the success rate over the same grids in the same period. The mean success rates in daytime and nighttime are 33.3% and 40.4%, respectively.

The spatial distribution of the success rate is similar between daytime and nighttime, the rate is higher in arid areas, and lower in humid areas. The success rate in arid areas such as Sahara, Middle East, Australia, and Namibia is somewhat smaller in nighttime than in daytime, while that in high latitude areas is somewhat larger in nighttime.

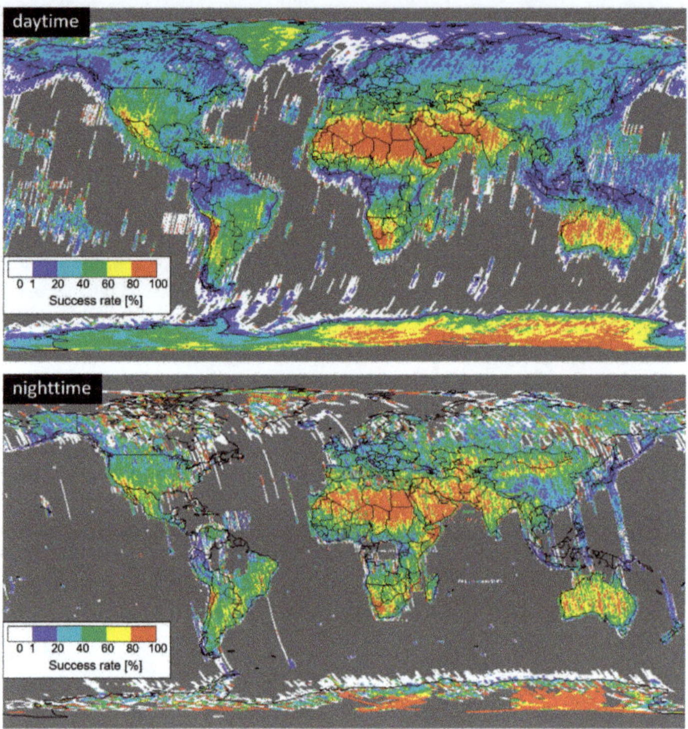

Figure 12. Daytime and nighttime maps of the success rate over 10 × 10 degree grids in 19.5 years from March 2000 to August 2019.

4. Summary and Conclusions

Since the ASTER instrument cannot detect clouds accurately for snow-covered or nighttime images due to lack of spectral bands, the MOD35 product has been alternatively used in cloud assessment for all ASTER images. The most advantageous point of this approach is that the MODIS instrument onboard the same platform (Terra) always provides coincident and collocated observations using effective bands for cloud detection, although spatial resolution differences between ASTER and MODIS should be noted. In this study, we evaluated the reliability of MOD35-based ASTER cloud mask images, and then performed ASTER mission operations analysis using these cloud mask images.

For the evaluation of ASTER cloud mask images, we first compared them among different MOD35 versions. In updating from C5 to C6, the rates of ASTER scenes with the CC difference of 10% or greater ($R_{10\%}$) and with that of 50% or greater ($R_{50\%}$) were 18.5% and 6.5% in daytime, respectively, and 16.3% and 2.8% in nighttime, respectively. The likely cause of discrepancy between daytime and nighttime is a difference of observation areas. In updating from C6 to C6.1, $R_{10\%}$ and $R_{50\%}$ were about 2% and 0.4%, respectively, in both daytime and nighttime. Since C6 MOD35 products have been affected by an error on input L1B radiances, the ASTER Project has been generating cloud masks using C6.1 MOD35 products for all ASTER scenes.

Next, we analyzed the rate of uncertain pixels in each MOD35-based ASTER cloud mask image, because the uncertain flag in the MOD35 product can be used as a reliability index of cloud masks. The high-uncertain rate (the rate of scenes with the rate of uncertain pixels ≥30%) was 2.2% in daytime, and 12.0% in nighttime, indicating that the reliability of cloud masks is lower in nighttime. The results also indicate that the high-uncertain rate depends on location; daytime scenes have a large value in low-latitude areas, snow-covered areas, or some localized areas such as the north of Mozambique and the basin of Yellow River, and nighttime scenes have a large value in mountainous or glacier-lake areas. In addition, we evaluated MOD35-based ASTER cloud mask images visually using ASTER browse images for 24,706 daytime scenes and 8006 nighttime scenes. As a result, about 2% of the evaluated scenes showed some problems such as mislabeling in snow-covered or coastal areas, artifacts probably associated with ancillary data used in the MOD35 algorithm and mismatching in the cloud coverage.

With respect to ASTER mission operations analysis using MOD35-based ASTER cloud mask images, we first investigated the effectiveness of the TCC-based cloud avoidance function implemented in the ASTER observation scheduler. The analysis using N-ODS and LC-ODS data in the period from May 2016 to April 2019 showed that the cloud avoidance function canceled observations frequently in most of global land areas except for some areas frequently observed without the cloud avoidance function, such as the west area of North America, Antarctica and Greenland. Such canceled observation reduced MCC effectively in most areas, even if uncanceled scenes were considered, the cloud avoidance function decreased MCC by 10% to 15% and increased the rate of clear scenes by 10% to 15%. The time-series analysis of the three-monthly MCC values based on actual observations since the launch was also performed for the five grid cells (USA, Argentina, Mozambique, Indonesia, and Japan) in daytime and nighttime. Although the trend plots of MCC seem to have been partly affected by weather fluctuations, such as the ENSO, those plots showed a reduction of 0.5% to 7.9% in MCC between before and after April 2012 in which the coefficient κ in the cloud avoidance function was updated, except for three grid cells in nighttime. Finally, we generated the global maps of the number of clear scenes and the success rate using MOD35-based ASTER cloud mask images. The maps showed that the mean of the number of clear scenes was 15.7 scenes in daytime, and 6.6 scenes in nighttime, the mean success rate was 33.3% in daytime, and 40.4% in nighttime, and the success rate in arid areas was somewhat lower in nighttime than in daytime and that high-latitude areas had the opposite tendency.

Almost twenty years have passed since the ASTER instrument was launched in December 1999, and there are not many optical sensors with such a long-operation life. The results reported in this study are expected to contribute not only to the ASTER Project but also to other optical sensor projects.

Author Contributions: Conceptualization, H.T. and T.T.; methodology, H.T. and T.T.; software, H.T.; validation, H.T. and T.T.; formal analysis, H.T.; investigation, H.T. and T.T.; resources, H.T. and T.T.; data curation, H.T.; writing—original draft preparation, H.T.; writing—review and editing, H.T. and T.T.; visualization, H.T.; funding acquisition, T.T.

Funding: This research was funded by NASA Jet Propulsion Laboratory (USA).

Acknowledgments: The authors would like to thank the ASTER Science Team, particularly Michael Abrams (Jet Propulsion Laboratory, USA) and Yasushi Yamaguchi (Nagoya University, Japan) for giving comments on this study. Additional thanks to Yasushi Horiguchi and Akira Miura (Japan Space Systems) for providing useful information and data on the cloud avoidance function, and also to Yuki Koike, Mihiro Tsuda, Kazuki Ayuzawa, Yuki Kawamura, and Atsushi Tokida (Ibaraki University, Japan) for their supports with the visual evaluation of cloud mask images. The authors would also like to thank the editor and three anonymous reviewers for their comments and suggestions that greatly improved this manuscript.

Conflicts of Interest: The authors declare no conflict of interest.

References

1. Yamaguchi, Y.; Kahle, A.B.; Tsu, H.; Kawakami, T.; Pniel, M. Overview of the advanced spaceborne thermal emission and reflectance radiometer (ASTER). *IEEE Trans. Geosci. Remote Sens.* **1998**, *36*, 1062–1071. [CrossRef]

2. Yamaguchi, Y.; Abrams, M.J.; Kato, M.; Watanabe, H.; Tsu, H. ASTER science outcome and operation status. In Proceedings of the 2005 IEEE International Geoscience and Remote Sensing Symposium, Seoul, Korea, 29 July 2005; pp. 5578–5579. [CrossRef]

3. Level 1 Data Working Group, ASTER Science Team. *ASTER Algorithm Theoretical Basis Document for ASTER Level-1 Data Processing*; Ver. 3.0; ERSDAC LEL/8-9; Earth Remote Sensing Data Analysis Center: Tokyo, Japan, 1996. Available online: http://eospso.gsfc.nasa.gov/sites/default/files/atbd/atbd-ast-01.pdf (accessed on 15 September 2019).

4. Fujisada, H. ASTER level 1 data processing algorithm. *IEEE Trans. Geosci. Remote Sens.* **1998**, *36*, 1101–1112. [CrossRef]

5. Irish, R. Landsat 7 automatic cloud cover assessment. In *Algorithms for Multispectral, Hyperspectral, and Ultraspectral Imagery VI*; International Society for Optics and Photonics: Bellingham, WA, USA, 2000; Volume 4049, pp. 348–355. [CrossRef]

6. Tonooka, H.; Omagari, K.; Yamamoto, H.; Tachikawa, T.; Fujita, M.; Paitaer, Z. ASTER cloud coverage reassessment using MODIS cloud mask products. In *Earth Observing Missions and Sensors: Development, Implementation, and Characterization*; International Society for Optics and Photonics: Bellingham, WA, USA, 2010; Volume 7862. [CrossRef]

7. Tonooka, H. ASTER nighttime cloud mask database using MODIS cloud mask (MOD35) products. In *Remote Sensing of Clouds and the Atmosphere XIII*; International Society for Optics and Photonics: Bellingham, WA, USA, 2008; Volume 7107. [CrossRef]

8. Meyer, D.; Siemonsma, D.; Brooks, B.; Johnson, L. *Advanced Spaceborne Thermal Emission and Reflection Radiometer Level 1 Precision Terrain Corrected Registered At-Sensor Radiance (AST_L1T) Product, Algorithm Theoretical Basis Document*; Open-File Report 2015-1171; U.S. Geological Survey: Reston, VA, USA, 2015. [CrossRef]

9. Hulley, G.C.; Hook, S.J. A new methodology for cloud detection and classification with ASTER data. *Geophys. Res. Lett.* **2008**, *35*, L16812. [CrossRef]

10. Ackerman, S.A.; Strabala, K.I.; Menzel, W.P.; Frey, R.A.; Moeller, C.C.; Gumley, L.E. Discriminating clear sky from clouds with MODIS. *J. Geophys. Res.* **1998**, *103*, 32141–32157. [CrossRef]

11. Ackerman, S.; Frey, R.; Strabala, K.; Liu, Y.; Gumley, L.; Baum, B.; Menzel, P. *Discriminating Clear-Sky from Cloud with MODIS Algorithm Theoretical Basis Document (MOD35)*; Version 6.1; University of Wisconsin–Madison: Madison, WI, USA, 2010. Available online: https://modis-atmos.gsfc.nasa.gov/sites/default/files/ModAtmo/MOD35_ATBD_Collection6_0.pdf (accessed on 15 September 2019).

12. LP DAAC. ASTER Cloud Cover Accuracy Improved. Available online: https://lpdaac.usgs.gov/news/aster-cloud-cover-accuracy-improved/ (accessed on 15 September 2019).

13. Tonooka, H. ASTER Cloud Mask Database. Available online: http://tonolab.cis.ibaraki.ac.jp/ASTER/cloud/ (accessed on 15 September 2019).

14. Ackerman, S.A.; Holz, R.E.; Frey, R.; Eloranta, E.W.; Maddux, B.C.; McGill, M. Cloud detection with MODIS. Part II: Validation. *J. Atmos. Ocean. Technol.* **2008**, *25*, 1073–1086. [CrossRef]

15. Wilson, A.M.; Parmentier, B.; Jetz, W. Systematic land cover bias in Collection 5 MODIS cloud mask and derived products: A global overview. *Remote Sens. Environ.* **2014**, *141*, 149–154. [CrossRef]

16. Wang, T.; Fetzer, E.J.; Wong, S.; Kahn, B.H.; Yue, Q. Validation of MODIS cloud mask and multilayer flag using CloudSat-CALIPSO cloud profiles and a cross-reference of their cloud classifications. *J. Geophys. Res. Atmos.* **2016**, *121*, 11620–11635. [CrossRef]

17. Moeller, C.; Frey, R. Terra MODIS Collection 6.1 Calibration and Cloud Product Changes, Version 1.0 (27 June 2017). Available online: https://modis-atmos.gsfc.nasa.gov/sites/default/files/ModAtmo/C6.1_Calibration_and_Cloud_Product_Changes_UW_frey_CCM_1.pdf (accessed on 15 September 2019).

18. Muraoka, H.; Cohen, R.H.; Ohno, T.; Doi, N. ASTER observation scheduling algorithm. In Proceedings of the International Symposium Space Mission Operations and Ground Data Systems, Tokyo, Japan, 1–5 June 1998.

19. King, M.D.; Platnick, S.; Menzel, W.P.; Ackerman, S.A.; Hubanks, P.A. Spatial and temporal distribution of clouds observed by MODIS onboard the Terra and Aqua satellites. *IEEE Trans. Geosci. Remote Sens.* **2013**, *51*, 3826–3852. [CrossRef]

20. Frey, R.A.; Acherman, S.A.; Liu, Y.; Strabala, K.I.; Zhang, H.; Key, J.R.; Wang, X. Cloud detection with MODIS. Part I: Improvements in the MODIS cloud mask for Collection 5. *J. Atmos. Ocean. Technol.* **2008**, *25*, 1057–1072. [CrossRef]

21. Frey, R. Collection 6 Updates to the MODIS Cloud Mask (MOD35). 2010. Available online: https://modis.gsfc.nasa.gov/sci_team/meetings/201001/presentations/atmos/frey.pdf (accessed on 15 September 2019).
22. Wilson, T.; Wu, A.; Shrestha, A.; Geng, X.; Wang, Z.; Moeller, C.; Frey, R.; Xiong, X. Development and implementation of an electronic crosstalk correction for bands 27–30 in Terra MODIS collection 6. *Remote Sens.* **2017**, *9*, 569. [CrossRef]
23. Level-1B (L1B) Calibration, Collection 6.0 and Collection 6.1 Changes, Terra and Aqua MODIS. Available online: https://atmosphere-imager.gsfc.nasa.gov/sites/default/files/ModAtmo/C061_L1B_Combined_v10.pdf (accessed on 15 September 2019).
24. MODIS/Terra Data Outages. Available online: https://modaps.modaps.eosdis.nasa.gov/services/production/outages_terra.html (accessed on 15 September 2019).
25. NOAA/National Weather Service, National Centers for Environmental Prediction, Environmental Modeling Center. Global Forecast System (GFS). Available online: https://www.emc.ncep.noaa.gov/GFS/ (accessed on 15 September 2019).
26. Seabold, S.; Perktold, J. Statsmodels: Econometric and statistical modeling with python. In Proceedings of the 9th Python in Science Conference (SciPy 2010), Austin, TX, USA, 28–30 June 2010.
27. World Meteorological Organization. El Niño/Southern Oscillation. WMO-No.1145. 2014. Available online: http://www.wmo.int/pages/prog/wcp/wcasp/documents/JN142122_WMO1145_EN_web.pdf (accessed on 15 September 2019).
28. NOAA/National Weather Service, Climate Prediction Center. Historical El Nino/La Nina episodes (1950–present). Available online: https://origin.cpc.ncep.noaa.gov/products/analysis_monitoring/ensostuff/ONI_v5.php (accessed on 15 September 2019).
29. Raup, B.H.; Racoviteanu, A.; Khalsa, S.J.S.; Helm, C.; Armstrong, R.; Arnaud, Y. The GLIMS geospatial glacier database: A new tool for studying glacier change. *Glob. Planet. Chang.* **2007**, *56*, 101–110. [CrossRef]

 *remote sensing*

Article

Satellite ASTER Mineral Mapping the Provenance of the Loess Used by the Ming to Build their Earthen Great Wall

Tom Cudahy [1,*], Pilong Shi [2,3], Yulia Novikova [1] and Bihong Fu [2]

[1] C3DMM Pty Ltd., 3/473 Cambridge Street Floreat, Western Australia 6014, Australia; ynovikova@c3dmm.com

[2] Key Laboratory of Digital Earth Science, Aerospace Information Research Institute, Chinese Academy of Sciences, Beijing 100094, China; shipl@aircas.ac.cn (P.S.); fubh@aircas.ac.cn (B.F.)

[3] Hengyang Base of International Centre on Space Technologies for Natural and Cultural Heritage under the Auspices of UNESCO, Hengyang 421002, China

* Correspondence: thomas.cudahy@c3dmm.com

Received: 30 October 2019; Accepted: 10 January 2020; Published: 14 January 2020

Abstract: The earthen border wall (Great Wall) built by the Ming is largely made of wind-blown loess. However, does the composition of this loess change along the length of the wall in response to variations in regional sediment transport pathways and impacting on the wall's erosional durability? To date, defining these sediment transport pathways has been a challenge because of the paucity of spatially-comprehensive, compositional information. Here, we show that satellite ASTER mineral maps, combined with field sample measurements along a 1200 km section of the Ming's earthen wall, reveal both the compositional heterogeneity of loess as well as the complexity of the sediment transport pathways of individual loess components, including: (i) quartz sand from Cretaceous sandstones in the Gobi Desert; (ii) gypsum from evaporative lakes in the Tengger Desert; (iii) kaolinite from Devonian Molasse in the Qilian Shan; and (iv) chlorite and muscovite from meta-volcanic rocks exposed across the Alashan Block. Sediment transport pathways involve a combination of colluvial, aeolian and fluvial (ephemeral and permanent) processes shaped by the topography. ASTER enabled mapping of compositional gradients related to two pathways, namely: (i) quartz sand driven by aeolian saltation in concert with the Yellow River; and (ii) clay and fine silt travelling large distances (>500 km) by long-term wind suspension. The most intact section of wall is found along the Hexi Corridor, which is poor in quartz sand and rich in (kaolinitic) clay and fine-silt, driven by wind-shielding by the Alashan Block. We also found evidence that the Ming: (i) mined loess from close by the wall (<1 km); (ii) targeted loess richer in finer fractions; and (iii) routinely applied a Ca-rich additive (probably lime).

Keywords: ASTER; mineral mapping; earthen Great Wall; loess; Ming Dynasty; sediment transport pathways; mineral system; erosion; deposition; aeolian; fluvial

---

1. Introduction

The Ming Dynasty's program of building an earthen border wall began following a decisive battle that forced the Ordos Mongols north of the Yellow River in 1473 [1]. The initial plan was to rapidly build an ~800 km border wall across the Ordos Plateau to separate sand dune-fields in the north from agriculturally productive lands in the south [2] (Figure 1a). The Ming were able to build this wall to a height of 5–8 m (Figure 1d,e) in just three months using their "hangtu" method of ramming earth between formwork [1,3]. This technique resulted in a characteristic 15–30 cm horizontal layering [4] and contrasts with that built by earlier dynasties where a thinner (7–10 cm) layering of earth was often sandwiched between horizons of rock and/or vegetation [5]. The completion of the wall across the Ordos Plateau soon delivered military success [1], giving impetus for eastward continuation towards Beijing and westward along the Hexi Corridor towards Jiayuguan (Figure 1a).

This rapid construction of the Ming's border wall required easy access to suitable earthen building materials. Even though wind-generated loess (dominated by angular, silt-size grains) deposits are readily eroded [6,7], the Ming selected loess for wall construction because of its pervasive development at/near the surface across the Ordos Plateau, especially south of the Mu Us Desert (Figures 1a and 2c). However, it is not known where the Ming mined their loess for wall construction. That is, was it sourced close by the wall (<1 km away) or from distant (>100 km), centralized, quarries?

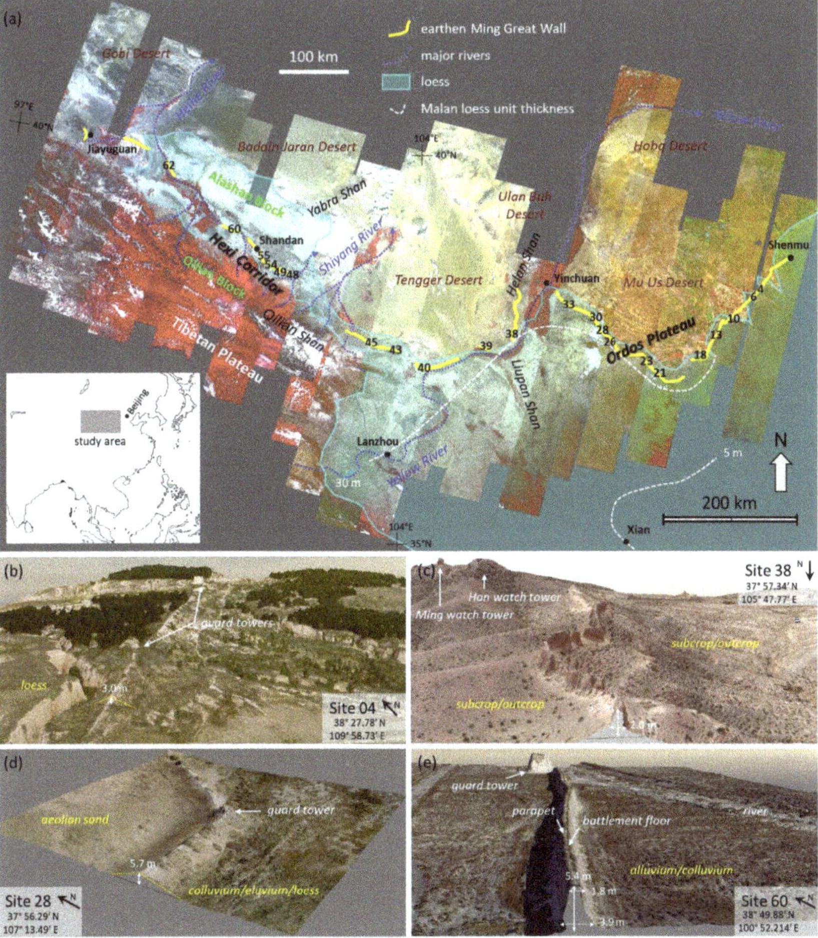

Figure 1. (**a**) Satellite ASTER false-color mosaic (Band 3: red; Band 2: green; Band 1: blue) spanning the Ordos Plateau and Hexi Corridor from Shenmu in the east to Jiayuguan in the west and including approximately 1200 km of the earthen Ming Great Wall (yellow line). The geological blocks, mountains, deserts, rivers, loess extent, thickness of the late-Pleistocene Malan loess unit [8], towns/cities, and field sample sites are also shown. (**b–e**) Oblique views of rendered digital elevation models of selected field sites generated from drone natural color imagery. The nature of local surface materials is indicated as well as several features of the wall.

Loess is typically buff colored, structureless, and flat-lying [8], although its thickness varies regionally. Over the Ordos Plateau, loess is up to 400 m thick [8] being the result of 20 million years of dust activity over a relatively stable platform. In contrast, loess deposits along the Hexi Corridor are much younger (Holocene) and often only 1 m thick [9]. This contrast in thickness is potentially driven by repeated uplift and erosion along the northeast margin of the Tibetan Plateau [10] and/or possibly the effects of climate-related variations in vegetation cover [9]. The most recent loess deposits include: the Late Pleistocene (11,700 to ~1,000,000 years) Malan Loess Unit, which decreases in thickness from Yinchuan (~30 m thick) to Xian (~5 m thick) (Figure 1a) [8]; and Holocene (0–11,700 years) layers, which are up to 8 m thick [11,12] and continue to accumulate today [13].

Mineral composition is a factor in the erosive potential of earthen materials. For example, minerals such as kaolinite ($Al_2Si_2O_5(OH)_4$) and muscovite ($KAl_2(AlSi_3O_{10})(FOH)_2$) do not expand on wetting in contrast to montmorillonite ($(Na,Ca)_{0.33}(Al,Mg)_2(Si_4O_{10})(OH)_2 \cdot nH_2O$) and illite ($(K,H_3O)(Al,Mg,Fe)_2(Si,Al)_4O_{10}((OH)_2,(H_2O))$), which swell when wet and contract when dry causing desiccation cracking and therefore are more prone to structural disintegration [14]. The cohesive strength of earthen material is also enhanced by the proportion of clay-sized material (<2 μm) because of increased levels of hydrogen bonding between water molecules and the surfaces of these grains [15], which typically comprise minerals such as kaolinite, illite, and montmorillonite. Coarser particles such as sand (63–2000 μm), which often comprises quartz (SiO_2), do not share this propensity for hydrogen bonding [15,16].

Silicate minerals can also react with calcite ($CaCO_3$), lime ($Ca(OH)_2$), and/or quicklime (CaO) to form cementing agents such as Ca-silicates (e.g., hydrogarnets) and (recrystallized) calcite [17] as part of a "pozzolanic" reaction. Loess is an effective pozzolan as it comprises $SiO_2 + Al_2O_3 + Fe_2O_3 \geq 70$ wt.%. The Ming Dynasty natural scientist, Song Yingxing, described how their "tabia" or "binding material" method used lime mixed with soil and sand [18].

These considerations help explain why the rate of erosional deterioration of earthen structures at Jiaohe in northwest China is related to the mineralogy of the building materials [19]. The question is, does the composition of the loess used by the Ming change along the length of the wall? If so, then what processes are responsible for this compositional change? Finally, have changes in composition impacted on the erosional durability of the earthen wall?

In theory, dust particles with a diameter <70 μm are capable of transportation by wind suspension [20]. The <20 μm fraction (i.e., clay and fine-silt) can potentially remain in wind-suspension for several days, travelling hundreds to thousands of kilometers and is termed long-term wind suspension [20]. In contrast, the 20–70 μm fraction (i.e., coarse-silt) can remain in wind-suspension for minutes to hours, travelling distances of meters to kilometers, and is termed short-term wind suspension [20]. Grains >70 μm in size (i.e., sand) are not carried by wind-suspension but instead bounce or saltate across the land surface, travelling only meters per wind event [20]. Factors determining where this range of particle sizes can accumulate as deposits of loess include: prevailing wind direction; distance from source; and topographic and/or vegetation traps.

The large volumes of sediment required to generate the loess deposits across the Ordos Plateau and Hexi Corridor have been attributed to the combination of cold deserts juxtaposed against rapidly uplifting mountainous regions [21,22]. That is, rocks in high areas are eroded and reduced in particulate size by glacial grinding and cold weathering processes before being transported in fluvial-suspension by snowmelt-fed rivers into neighboring desert basins that act as sediment sinks to be later re-worked by wind activity. However, identifying which mountainous source areas, rivers, deserts, and/or winds were critical in loess development has proved to be much more problematic. Mountainous regions proposed as the primary source of particulate materials include: the Qilian Shan [22]; Gobi Altay [22]; Tianshan; Alashan Block; and Tibetan Plateau [23,24]. Researchers have also identified alluvial fans as a secondary source of the eroded grains, including those along the Hexi Corridor [10] and Gobi Altay Mountains [22]. Proposed sandy or stony desert secondary sources include: the Taklamakan [25]; Qaidam Basin [23]; Gobi [22,26]; Mu Us; Hobq; Ulan Buh; Tengger; and Badain Jaran [22,26–29]. Rivers include: Yellow River [25,30,31] and ephemeral rivers along the Hexi Corridor, such as the Shiyang and Heihe [32] that drain into neighboring desert basins (Figure 1a). The prevailing wind patterns alternate between westerly to northerly flows of the East Asian Winter Monsoon (EAWM) versus southerly flows of the East Asia Summer Monsoon (EASM) [26].

To date, determining the primary/secondary sources of loess material has relied on point-sample measurements of: grain size [8,33,34]; major elements [35,36]; trace elements [35]; rare earth elements [25,35]; zircon U-Pb geochronology [37,38]; scanning electron microscope energy dispersive spectrometers [27]; luminescence [39]; electron magnetic spin [40,41]; oxygen isotopes [42]; Nd and Sr isotopes [43,44]; magnetics [45]; and mineralogy [32,46–49]. However, interpreting transport pathways using such sparse point-sample data collected from heterogeneous landscapes is thwart with difficulties. Indeed, the most recent *Loessfest* meeting [50] concluded "there is a need to develop new methods and approaches for quantitative paleoenvironmental and paleoclimatic reconstructions".

Geologists have been tackling a similar challenge in their targeting of economic mineral deposits by collecting spatially-comprehensive data, such as geophysical [51] and mineral mapping imagery [52], to better elucidate target fluid transport pathways that extend from source rocks to sites of potential (metal) deposition. This so-called "mineral-system" approach [53] requires both: (i) the definition of essential ingredients and their related mappable criteria; and (ii) the assembly of regional geoscientific data in order to create maps that target these mappable criteria.

Given that "the nature of the material is paramount, and the formation of the material is paramount. And this means the formation of the actual units, the actual particles [= mineral grains] which comprise the deposit of loess" [54], we propose a "loess system" based on a mappable suite of essential mineral components. The critical "loess system" mapping tool we used here for tracking the provenance of the loess sourced by the Ming to build their earthen wall is the satellite-borne, Advanced Spaceborne Thermal Emission and Reflection Radiometer (ASTER) sensor [55]. ASTER was designed to map land surface composition, albeit at moderate spectral resolution. It was launched in December 1999 and has since acquired multiple coverages of the Earth's land surface at <83° latitude [56].

Based on the principles of mineral spectroscopy [57], we processed the calibrated ASTER imagery [58] to generate a suite of thematic mineral mapping products at ~30 m pixel resolution for an area of ~60,000 km^2 spanning the 1200 km section of Ming earthen wall between Shenmu and Jiayuguan (Figure 1a). We focused here on five essential mineral components of loess, namely quartz-sand, white-mica (muscovite and illite), chlorite (as well as carbonate), gypsum and kaolinite, as these are measurable at ASTER's spectral resolution [58–61]. Field data (optical spectra as well as other geochemical, mineralogical, and particle size measurements) from 21 sites along the Ming earthen wall were collected to provide both validation of the satellite maps as well as information about the compositional homogeneity/heterogeneity of the wall building material and any relationships with local (<100 m) surface materials.

From this multi-scale, spatially-comprehensive framework, we obtained a more detailed perspective of the compositional variability of the loess. This compositional information then provided the basis for solving the provenance of the loess used by the Ming to build their earthen wall, which has two essential parts: (i) Where did the Ming mine their loess, i.e., was it from local sources (<1 km away) or from centralized quarries many kilometers away? (ii) Where did the particles of loess originate? With regards to the second, do specific minerals partition within particle size ranges, reflecting different transport processes? In addition, has compositional heterogeneity impacted on the wall's erosional state as this could assist in wall's future preservation? Finally, we underline the value of using the Ming earthen wall validating our satellite ASTER mineral maps because it provides a 1200 km east–west transect across the entire study area.

2. Materials and Methods

2.1. Field Recognition of the Ming Earthen Border Wall

Several features helped us establish whether a section of wall was built by the Ming or by an earlier Dynasty. The key criteria for it to be Ming related include:

- Horizontally-layered (10–30 cm), loess-dominated building material (earlier walls are often associated with abundant rock fragments);
- Regularly spaced guard towers (Figure 1b–e) also built using the same materials/methods;
- Aprons of the same types of fragmented, usually glazed stoneware and porcelain containers found scattered around these guard towers; and
- The relatively well preserved (height) nature of the earthen wall (Figure 1c), especially given that the Ming wall was the last completed in Chinese dynastic history.

2.2. Field Sampling

The Great Wall is protected by Chinese law, which is administered at the province level. Field sampling for this study thus was carefully designed to minimize any impact. Measurements types included: remote imaging, e.g., drone survey or field LIDAR; surface contact, e.g., field portable spectrometers and pXRF; and those requiring a physical sample, e.g., particle size and XRD. The field samples taken were <100 g each and were sampled as follows: (i) one sample from either side of the wall ideally from recently fallen material; (ii) surface samples 15 m to the north and south of the wall; and (iii) surface samples 45 m to the north and south of the wall. Field-portable X-ray fluorescence (pXRF) measurements were taken at each sample point. For some sites, additional samples and in situ measurements were taken, such as from within eroding layers (green arrows in Figure 2).

2.3. 3D Model Generation

A high resolution (<5 cm) 3D surface model, ortho-mosaic image, and digital elevation model (DEM) were generated for each field site (e.g., Figure 1b–e), using drone digital imagery (where allowed by government authorities) and observed at five different look angles (1 nadir and 4 oblique). Camera calibration, photo aligning, dense point-cloud building, mesh and texture building, DEM, and ortho-mosaic building were processed with photogrammetric software called Agisoft Photoscan/Metashape [61].

2.4. Assessment of the Wall's Erosional Condition

Two independent methods were used to assess the erosional status of the Ming earthen wall. The first involved visual assessment of the wall in the field using a five-level classification. The scores were reassessed later in the office using site photos and other information. A 5/5 classification was given to wall of >4 m height (above base-line), steep-sided (~80°) wall sides (e.g., Figure 2d), a flat 2–4 m wide wall-top that sometimes includes remnant brick or stone walkway lining, as well as battlements such as a parapet (Figure 1e). A 1/5 classification was given to wall characterized by a convex, narrow (<1 m wide) wall top and a lack of steep-sided walls (Figure 1b), which in places could be the result of a build-up of wind-blown sediment (Figure 1d).

The wall classification determined using the 3D model, ortho-mosaic image, and DSM generated from the drone photogrammetric involved digitally calculating the wall's geometry and projected volume per linear meter through any scree and/or sand cover.

2.5. Particle Size

Particle size measurements of wall samples were only conducted using a Malvern Mastersizer 2000 at the Institute of Geographic Sciences and Natural Resources Research, Chinese Academy of

Sciences in Beijing. The whole samples were pre-treated with a weak hydrochloric acid solution to remove organic particles. Nineteen particle size ranges were measured for each sample.

2.6. Field Portable XRF

The pXRF measurements of in situ wall and background surface materials were measured using an Olympus InnovX pXRF instrument. Measured samples were flat and air dried where possible. Thirty-five of the detectable elements were recorded, although particular attention was given to Si, Al, Fe, Ca, Mg, S, and Cl.

(g)

pXRF samples	Si/(Si+Al+Fe)	Ca	S	Cl	ASD gypsum
#1	4	0.68	6.5	3.1	0.0	y
#2	4	0.69	4.8	1.3	0.1	y
#3	1	0.72	4.3	0.9	0.0	y
#4	1	0.69	4.5	0.2	0.3	n
#5	5	0.69	5.0	0.2	1.5	n
#6	5	0.71	7.9	1.7	0.0	y
#7	2	0.69	6.3	0.2	0.7	n

Figure 2. Field site photos and data. (**a**) Twenty-centimeter-wide layering of the Ming earthen wall (cyan arrows) at Site 48 (38°19.33′N; 101°58.25′E). (**b**) Abundant white clay-carbonate fragments within "hangtu" construction layers at Site 28 (37°56.29′N; 107°13.49′E). (**c**) A recently exposed cutting <1 km from Site 28 exposing a ~1 m thick surface layer of loess developed over an eroded red and white horizontal layered clay-rich paleosol with white clay-carbonate fragments. (**d**) South-side view of the well-preserved earthen wall at Site 48 (38°19.33′N; 101°58.25′E). Preferentially eroding "hangtu" layers are shown by green arrows while sub-vertical rills are shown by yellow arrows. Field sampling points for pXRF and ASD measurements are numbered (#). (**e**) Close-up view at Site 48 showing isolated, sub-rounded grey siltstone pebbles and cobbles in a beige colored loess matrix. (**f**) A road cutting along the Shandan section (38°27.40′N; 101°27.53′E) showing a ~1 m thick surface layer of loess developed over grey, sub-rounded, siltstone pebbles, and cobbles similar to those evident in (**e**). (**g**) A table of field pXRF chemistry and ASD gypsum information of selected sample points (#) in (**d,i**). (**h**) Preferential erosion along multiple "hangtu" layers at Site 33 (38°16.34′N; 106°32.18′E). (**i**) Recent collapse of the earthen wall at Site 38 (37°57.34′N; 105°47.78′E) related to undercutting along a ~50 cm wide erosional zone (#7) rich in halite. (**j**) Visual classification of the earthen wall's erosional status where: 1 = little remaining of the original wall above the exposed land surface; and 5 = well-preserved wall, including the remnants of battlements. The magenta box highlights the well-preserved Shandan section.

2.7. Laboratory XRD

Laboratory mineralogical analyses were conducted on a Bruker D4 Endeavour XRD instrument. Whole samples were prepared as random powders (nominally 80% <75 μm) and back-mounted into circular holders. Samples were typically scanned from 2° to 70°. First pass mineral identification was facilitated using the XPLOT software program [62]. This software search-matches peak positions with respect to the ICDD mineral standards library [63]. Quartz was identified in the diffraction patterns and used as an internal standard to correct for any instrumental shifts in 2θ position. The same software was also used to measure mineral-targeted peak heights and full-width-half-heights (FWHH). For example, the FWHH of the 10 Å peak was used to identify muscovite (FWHH < 0.2) versus illite (FWHH > 0.2).

2.8. Field and Laboratory Spectral Measurements

A single beam, Analytical Spectral Devices (ASD) FieldSpec Pro spectrometer [64] was used to measure the 350–2500 nm bi-directional reflectance of field samples. A 100% reflectance Spectralon panel [65] was used as the reference standard, with both the target and reference illuminated in series, either: (1) under the same optical geometry (off the specular angle) with a 1000 W Quartz Halogen light source; or (2) using the contact-probe attachment, which has its own illumination source.

The emissivity of field samples was measured using a portable FTIR (Fourier transform infrared) spectroradiometer Model 102 designed and built by Designs and Prototypes [66]. This instrument measures emissivity in the 3–5 and 8–14 μm wavelength regions at approximately 6 wavenumber resolution. The area sensed is approximately 20 mm diameter or less. Measurements of hot and cold blackbodies establish calibration to radiance at sensor. A correction for background or sky down welling irradiance is implemented using a brass reference plate to retrieve absolute surface radiance. The method for temperature-emissivity separation and extraction of surface emissivity involved an assumption for an emissivity of ~1.0 at the Christiansen frequency.

2.9. Satellite ASTER Imagery

Orthorectified, radiance-at-sensor (L1T) ASTER images of the study area were sourced from NASA's *Earthdata* web portal [67]. Approximately 250 images from 25 overlapping paths were selected on the basis of cloud cover (<10%) and season (ideally late summer). However, for some areas, the only available images in the archive were compromised by cloud cover, snow, and/or green vegetation, which potentially impacted on the accuracy of the subsequent image cross-calibration.

Pre-processing the 250 satellite ASTER L1T images into a cross-calibrated mosaics involved the following steps conducted using ENVI™ software: (i) spatial resampling of the VNIR and TIR module data to 30 m (nearest the nearest neighbor sampling); (ii) merging of all three wavelength modules for each image into a single 30 m pixel resolution file; (iii) masking any pixels without a full complement of VNIR, SWIR, and TIR radiance-at-sensor data; (iv) finding pairs of overlapping images followed by manual selection of ~25 invariant targets for each pair and spanning a wide range of radiance values from which linear regressions were established (using EXCEL™) and the gains and offsets calculated and used to adjust one image to the same levels as the adjoining image—this procedure was conducted for all single images and/or paths of temporally coincident images till a single cross-calibrated mosaic was established; (v) a linear, across-image correction (de-ramp) for a systematic calibration error most evident in ASTER Band 5 (see Appendix A) was implemented; and (vi) an offset for each of Bands 1–9 was calculated using a dark-point estimation method developed by the lead author (not published) and then corrected.

Even using this pre-processing strategy, residual calibration errors between images/paths remain in the final mosaic (examples highlighted by white block arrows in Figures 4a, 5a, 6a and 7a). Note, however, that these apparent calibration errors do not persist for the entire length of a given satellite path and much less so for those areas close by where the invariant targets were selected. That is, poorly exposed areas with extensive snow or dynamic green vegetation cover and often associated with higher topographic elevation often show these overlapping image mismatches. There are also errors caused by local variations in atmospheric conditions, especially water vapor and aerosols. These localized errors could be reduced through additional cross-calibration of the aberrant images.

2.10. ASTER Spectral Mineral Indices

We used spectral indices, such as band ratios, to enhance the often-subtle mineralogical signal by cancelling extraneous effects that often dominate the signal. In the VNIR-SWIR, these extraneous "multiplicative" effects include topographic illumination and surface scattering while in the TIR it is kinetic temperature [52]. Care is essential in ensuring that any additive effects are first corrected before attempting to use normalization-based compositional indices [60]. The ASTER spectral indices used here are based on those developed in other studies [58,60,68].

The TIR-based ASTER Gypsum Index (GI) [68], which is sensitive to the presence and abundance of gypsum and targets a related reststrahlen feature near 8600 nm, is calculated as follows:

$$GI = (B_{10} + B_{12})/B_{11} \tag{1}$$

This ratio is inverse to the commonly used quartz index [58,68].

The TIR-based ASTER Silica Index (SI^3) [69,70], which is sensitive to the amount of coarse (>100 µm particulate size) silicate minerals such as quartz [71,72], targets silicate reststrahlen features between 8000 and 9000 nm [73] and is calculated as follows:

$$SI^3 = B_{13}/(B_{10} + B_{11} + B_{12}) \tag{2}$$

The SWIR-based ASTER *AlOH abundance* index (2200D) [60], which is sensitive to the content of dioctahedral minerals, such as kaolinite, white mica (e.g., illite, muscovite, phengite, and lepidolite), and montmorillonite, and targets related absorption at 2200 nm [57], is calculated as follows:

$$2200D = (B_5 + B_7)/B_6 \tag{3}$$

However, the 2200D index is also sensitive to changes in dioctahedral mineral composition with kaolinite generating a lower response while white mica generates a higher response. It is also complicated by the nature of scattering interactions with other mineral grains, whether they be opaque phases such as graphite [74] or transparent ones such as quartz [75,76].

The SWIR-based ASTER *AlOH composition* index (2165D) [70], which targets changes in the geometry of the 2200 nm absorbing minerals, especially kaolinite (left-asymmetric) to white mica (right-asymmetric), is calculated as follows:

$$2165D = (B_7 + B_8)/(B_5 + B_6) \tag{4}$$

The SWIR-based ASTER *MgOH abundance* index (2330D) [70], which is sensitive to a broad suite of trioctahedral silicate (e.g., chlorite, amphibole, talc, and serpentine) and carbonate (e.g., calcite and dolomite) minerals and targets related absorption/s between 2250 and 2380 nm, is calculated as follows:

$$2330D = ((B_5 + B_6 + B_9)/B_8) \tag{5}$$

The above spectral indices were implemented on both the satellite ASTER and field ASD and microFTIR data, with the latter two first convolved to simulate ASTER band responses.

2.11. Vegetation Unmixing of ASTER Mineral Indices

The seasonally-variable effects of green vegetation on the ASTER spectral mineral indices were compensated using a linear unmixing approach [70,77,78]. This first step involves estimating the abundance of green vegetation (GV) as follows:

$$GV = (B_3/B_2)^{1/3} \qquad\qquad (6)$$

An orthogonal mixing relationship between GV and the target mineral index is then assessed via a 2D scattergram before scaling the ranges of the two input indices between 0 and 1. It is then a simple task to add or subtract GV and the target mineral index, depending on the geometry of the data cloud [70]. Removing the effects of dry vegetation can also be implemented [52,70] but was not attempted here.

2.12. Validation of the ASTER Mineral Indices Using the Field Spectral Data

The statistical comparison of the field versus satellite results was based on selecting three regions of interest (ROI) of 5–50 pixels for each field site in the ASTER imagery: (i) north of the Wall; (ii) south of the Wall; and (iii) along the trace of the Wall (1–3 pixels wide, i.e., <100 m). Where possible, pixels were excluded from the ROI where they were considered to be compromised by snow, cloud, vegetation, creeks/rivers, roads, and other manmade infrastructure including earlier Walls. These ASTER ROI data (Figures 4b, 5b, 6b and 7b) were then used for visual comparison with the associated field data (Figures 4c, 5c, 6c and 7c). Note that we do not imply that the 15–90 m pixel resolution of ASTER can be used to directly map the composition of the wall, which has a width ~6 m (not including any erosional scree slope that can extend the surface expression of the wall's loess material by up to 30 m width). Instead, our methodology relies on recognizing similar compositional patterns along the 1200 km length of the Ming earthen wall between the ASTER ROIs and the field data, which include field sample points located 15 m and 45 m on either side of the wall as well as those from the wall.

2.13. Interpreting Mineral Transport Pathways

The assumptions used in interpreting the loess system mineral sediment transport pathways include:

- The loess-related surface materials sensed by the ASTER satellite sensor are either Holocene or possibly late Pleistocene in age.
- Sediment dispersal (transport) of a given mineral type from its source generates a related decreasing compositional gradient, especially downslope or along flat topography.
- Reversals in compositional gradients can be caused by sediment "sinks" such as topographic lows or banking-up against topographic highs.
- The spatial pattern and distance travelled by a particular sediment type is dependent on its particle size and the nature (energy) of the transport process, namely:

  - colluvial transport is short (<50 km) and located adjacent to topographic highs;
  - fluvial transport is spatially restricted, e.g., within floodplains and (dry) river beds (wadis);
  - aeolian transport by saltation is confined to connected lowlands; and
  - depending on wind energy and particle size, aeolian transport by suspension can rise above topography and cross-cut fluvial transport networks.

- The relative timing of transport events, whether they be related to ephemeral rivers or aeolian sand saltation flow, can be assessed by their cross-cutting (compositional) nature.
- Sand dune patterns have largely developed since the last ice age, i.e., post late-Pleistocene, and thus are a potential indicator of prevailing wind direction/s.

- Longitudinal sand dunes, which are large amplitude (1–10 km) and often vegetation-stabilized, are formed by prevailing winds operating parallel to the dunes.
- Transverse dunes, which are small amplitude (~100 m) and typically free of vegetation, are formed by winds operating orthogonal to the dunes.

3. Results

3.1. Earthen Wall Building Materials

All 21 field sites (Figure 1a) showed that the Ming's "hangtu" method was used to build the earthen border wall, including the characteristic 15–30 cm thick, sub-horizontal layering of loess (Figure 2a). We often found sporadic (<15% by volume) pebbles and cobbles (1–10 cm in diameter; Figure 2b,e) in the wall, especially in areas of thin (<1 m) loess cover. The composition and form of these clasts were similar to local (<100 m) rock fragments exposed on the ground surface or exhumed by nearby building works and road cuttings (Figure 2c,f, respectively). We found loess-type material nearby (<100 m) for all but Sites 38 and 43 (Figure 1a,c), which were built directly on rocky ground in higher relief terrain.

3.2. Earthen Wall Status

Exposed, steep-sided walls often show a variety of erosional characteristics [79]. These include layer-parallel, erosional embayments with widths ranging from 3 to 60 cm (green arrows, Figure 2). The smaller ones (1–3 cm width) are often associated with the 15–30 cm "hangtu" layering (Figure 2a) while the larger ones (5–60 cm) often span a number of these layers (Figure 2h,i) and are usually located <2 m above the wall's base (Figure 2d,h,i) where not backfilled by aeolian sand (Figure 1d). These embayments can become so recessed that they eventually cause the collapse of the overhanging wall (Figure 2i). Sub-vertical rills (yellow arrows in Figure 2d,h) also contribute to this erosional decay.

The condition of the remnant Ming earthen wall was gauged both visually in the field and digitally using a high-resolution digital elevation model (DEM) generated from drone-acquired, stereo, visible imagery (Figure 1b–e). These different methods yielded similar results though only the visual assessments are presented here. The results (Figure 2j) show a trend of better-preserved wall westward, except for Site 62, which is largely covered by sand dunes. The best-preserved part of the earthen wall is the ~120 km long "Shandan section" (Sites 48–60 and highlighted by a magenta box in Figure 2j).

3.3. Loess Particle Size

The particle size of wall material (43 samples from 21 sites) was measured for 19 particle size bins spanning the range from 1 to 1000 µm. The results (Figure 3a) show tight standard deviations (STD ~2%) for all bin sizes ≤100 µm. The volumetrically-dominant bin-size is 20–50 µm (25 ± 7%), i.e., coarse-silt fraction (yellow box, Figure 3a). Combining all of the silt-size bins (2–63 µm) accounts for ~54% of the loess material, whereas sand (>50 µm) represents ~38% and clay (<2 µm) ~8%. This dominance of silt-size grains and left-skewness in particle size distribution (Figure 3a) are characteristic indicators for loess [80]. There is also inverse correlation ($R^2 = 0.74$) between the combined clay and fine-silt fractions (<20 µm) versus the combined sand fractions (>50 µm), given the coefficient of determination (R^2) for 21 samples is significant at the 90% confidence level where $R^2 > 0.57$. However, this excludes the 20–50 µm coarse-silt fraction, which yields no improvement when combined with either the finer or coarser fractions. In addition to the main peak at 20–50 µm, there are minor peaks spanning the 0–8 µm (centered at 1–2 µm) and 8–14 µm (centered at 8–10 µm) ranges.

To better understand the spatial pattern of these particle size data, we grouped the particle size bins into four types: (i) 0–3 µm (~clay); (ii) 3–20 µm (~fine-silt); (iii) 20–50 µm (~coarse-silt); and (iv) 50–1000 µm (~sand). Figure 3b shows the coarse-silt group has a relatively flat trend (~25%) along the 1200 km length of the wall. In contrast, the clay and fine-silt groups show smooth-changing curves that vary by a factor of ~2 and are well-modeled using fourth-order polynomials, with minima located near Site 28 and maxima located near Sites 55 (major) and 10 (minor). The 50–1000 μm group is also well-modeled using a fourth-order polynomial and is broadly opposite in pattern to the 1–3 μm and 3–20 μm data, although it also reveals a linear upward trend from Sites 4 (~35%) to 39 (~55%) before rapidly dropping to a minimum at Site 55 (~10%). The well-preserved "Shandan" section of earthen wall (Figure 2j) is associated with this minimum in sand content, where it becomes less than the combined finer fractions. Normalizing all the <50 μm fractions by the combined 50–2000 μm fraction highlights this sand-poor nature of the well-preserved Shandan section of the Ming earthen wall (magenta box in Figure 3c).

To assess possible local (<300 km) sources of the different loess dust components, we measured for each of the 19 particle size bins the distance between a given sample site and three potential source regions: (i) sand dune fields of the Badain Jaran, Tengger, and Mu Us Deserts; (ii) evaporative lakes at the outflow zone of the Shiyang River in the Tengger desert; and (iii) alluvial fans along the northern edge of the Qilian Shan (Figure 1a). Using fitted linear-functions, the results (Figure 3d) show: (i) clay is inversely correlated with the distance from the alluvial fans and to a lesser degree the evaporative lakes and positively correlated with the dune fields; (ii) fine-silt is positively correlated with the sand dunes and to a lesser degree inversely correlated with the alluvial fans and evaporative lakes; (iii) sand is inversely related with the dune fields, albeit weakly; and (iv) coarse-silt shows no relationship. These correlations are improved when fitted with higher order functions. For example, when using a second-order polynomial: (i) clay versus alluvial fans increases up to $R^2 = 0.74$, with the related minima coinciding with the location of the Yellow River; (ii) fine-silt versus sand dunes increase up to $R^2 = 0.73$, with the related minima located ~30 km to the south of the dune-fields; and (iii) sand versus deserts increases up to $R^2 = 0.43$ with the related maximum located ~30 km to the south of the dune-fields. Coarse-silt shows no apparent improvement.

3.4. Field XRF Chemistry—Wall and Background

The pXRF measurements of both wall and background samples show the following spatial patterns: (i) the Si/(Si + Al + Fe) ratio (Figure 3e) approximates a sinusoidal shape for both datasets with their peaks positioned near Site 20 and their inflexion points positioned between Sites 33 and 38 (i.e., where the Yellow River crosses); (ii) the Si/(Si + Al + Fe) ratios for background samples (red diamonds) are ~15% higher than their associated wall samples (blue dots) for localities nearby (<30 km) sandy deserts, i.e., Sites 10–33, 39, and 62; (iii) wall and background samples show a similar trend for Fe and Mg contents which linearly increase westward from 2% to 4% (red dots and green diamonds, respectively, in Figure 3f,g), although wall samples have less variability, (iv) Ca contents are more variable though both wall and background samples show a weak, linear, increasing trend westward (blue triangles in Figure 3f,g); (v) Ca contents of wall samples are on average 2–3% higher than their corresponding background samples (6.0% ± 1.6% versus 3.8 ± 2.2%, respectively); (v) there is a zone between Sites 28 and 48 in wall samples where Ca contents are up ~4% above the general trend of ~6%, which also corresponds to the zone of elevated S; (vi) S contents are typically <0.01% (i.e., below detection limits) for both wall and background samples accept west of Site 28 and especially for a zone between Sites 28 and 48 where S levels can reach up to 5% (yellow circles in Figure 3f,g); and (vii) Cl levels are typically below detection limits (~0.03%) except for wall samples (and one background sample at Site 28) between Sites 28 and 62 where contents are up to 3% (magenta squares in Figure 3f,g).

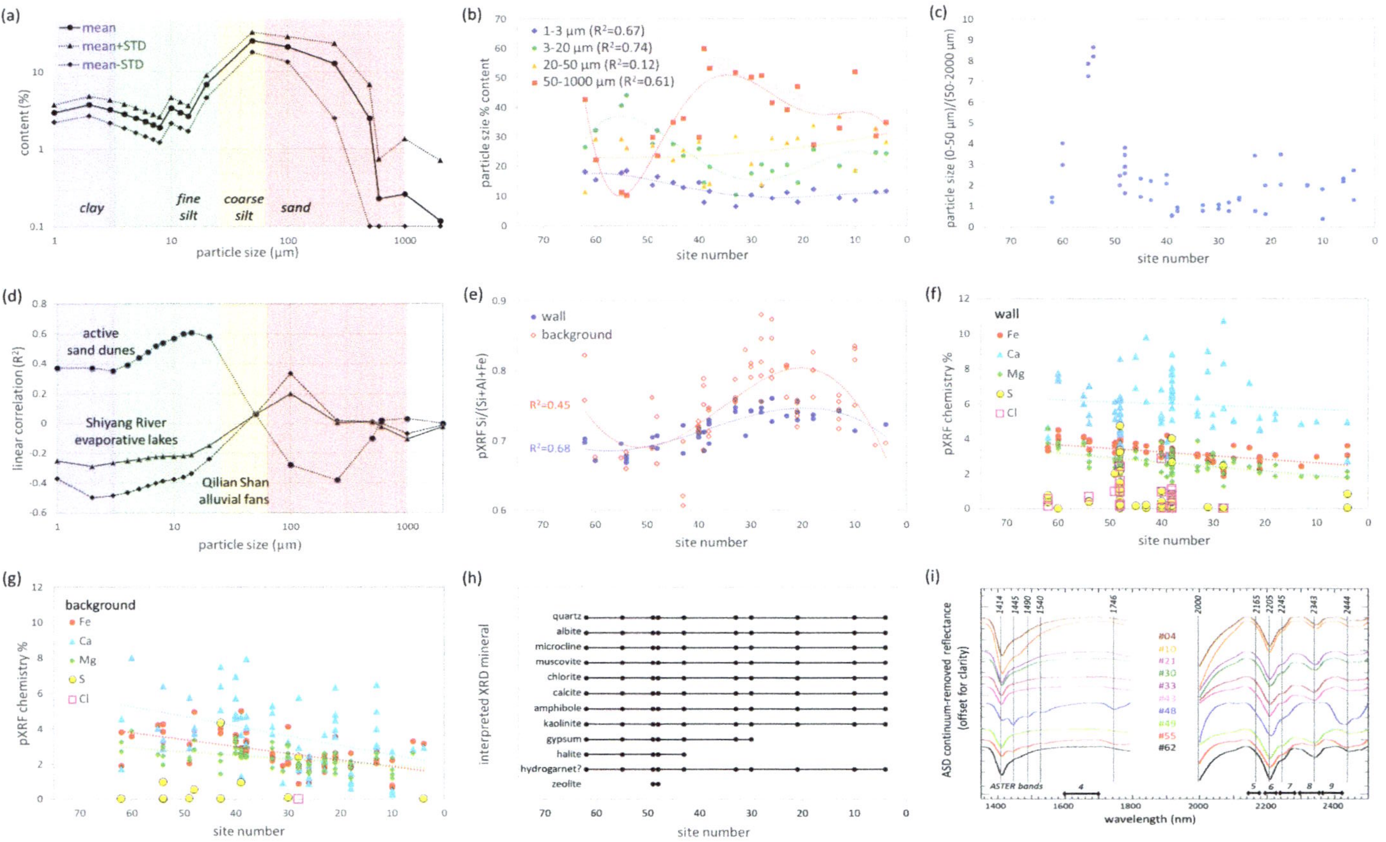
(a)
mean
mean+STD
mean-STD
content (%)
clay
fine silt
coarse silt
sand
particle size (µm)

(b)
1-3 µm (R²=0.67)
3-20 µm (R²=0.74)
20-50 µm (R²=0.12)
50-1000 µm (R²=0.61)
particle szie % content
site number

(c)
particle size (0-50 µm)/(50-2000 µm)
site number

(d)
active sand dunes
Shiyang River evaporative lakes
Qilian Shan alluvial fans
linear correlation (R²)
particle size (µm)

(e)
wall
background
R²=0.45
R²=0.68
pXRF Si/(Si+Al+Fe)
site number

(f)
wall
Fe
Ca
Mg
S
Cl
pXRF chemistry %
site number

(g)
background
Fe
Ca
Mg
S
Cl
pXRF chemistry %
site number

(h)
quartz
albite
microcline
muscovite
chlorite
calcite
amphibole
kaolinite
gypsum
halite
hydrogarnet?
zeolite
interpreted XRD mineral
site number

(i)
ASD continuum-removed reflectance (offset for clarity)
1414
1445
1490
1540
1746
2000
2165
2205
2245
2343
2444
#04
#10
#21
#30
#33
#43
#48
#49
#55
#62
ASTER bands
4
5
6
7
8
9
wavelength (nm)

Figure 3. Laboratory analyses of field samples. (**a**) Scattergram of the mean ($n = 43$) and standard deviation (STD) for each of the 19 particle size bins. Four populations are highlighted: 0.1–3.0 μm (blue box) with a peak at 2 μm and called "clay"; 3–20 μm (green box) with a peak at 10 μm and called "fine-silt"; 20–50 (yellow box) with a peak at 50 μm called "coarse silt"; and 50–2000 μm (red box) and called "sand". (**b**) Scattergram of the west-to-east pattern in mean particle size content for each of the four groupings. (**c**) Scattergram of the west-to-east spatial trend in the combined clay and fine-silt fractions (<20 μm) normalized with respect to sand (>50 μm) content. (**d**) Scattergram of the particle-size bin versus the fitted linear correlation (R^2, $n = 43$) between a sample site's given particle size content and its distance to one of three potential source areas, i.e., sand dunes, Tengger Desert evaporative lakes, and Qilian Shan alluvial fans. (**e**) Scattergram of the west-to-east pattern for field site portable XRF measurements of Si/(Si + Al + Fe) for both wall and background samples. (**f**) Scattergram of the west-to-east pattern of field site portable XRF measurements of Fe, Ca, Mg, S, and Cl of wall samples. (**g**) Scattergram of the west-to-east spatial trend of field site portable XRF measurements of Fe, Ca, Mg, S, and Cl of background samples. (**h**) East-to-west pattern of minerals detected using XRD. A magenta box highlights the well-preserved "Shandan" wall section. (**i**) Selected field ASD spectra. Diagnostic mineral absorptions are highlighted as well as the ASTER band-passes.

At a site-scale, the field pXRF measurements show considerable heterogeneity and differences between chemical elements down a given vertical section of the wall. For example, at Site 48 (Figure 2d), S decreases from 3.1% at ~2 m above ground (#1 in Figure 2d,g) to 0.2% at ~30 cm above ground (#5 in Figure 2d,g). Similarly, Cl also shows considerable variability. For example, at Sites 48 (Figure 2d) and 38 (Figure 2i), Cl measurements from within (e.g., #2, #5, and #7) and immediately adjacent (<15 cm, e.g., #4) to wide (15–50 cm), sub-horizontal embayments (green arrows in Figure 2d,h,i) yield high (0.1–3.1%) values (Figure 2g). Away from these eroded layers (e.g., #1, #3, and #6), Cl values are below pXRF detection limits. In contrast, the Si/(Si + Al + Fe) ratio remains constant throughout.

We interpret these pXRF as follows: (i) the similarity between the Si/(Si + Al + Fe) chemistry (Figure 3e) and particle size distribution (Figure 3b,c) is driven by a change in composition across the spectrum of particle sizes, namely Si-rich (i.e., quartz) coarser fractions to Al-rich (i.e., white mica, kaolinite) and/or Fe-rich (e.g., chlorite) finer fractions; (ii) Fe and Mg are related to the mineral chlorite [$(Fe^{2+},Mg,).5Al_2Si_3O_{10}(OH)_8$], which was sourced from the west (up gradient); (iii) the similar patterns for wall versus background samples for Si/(Si + Al + Fe) (Figure 3e) as well as Fe, Mg, and S (Figure 3f,g) are evidence that the Ming locally (~100 m) sourced the loess to build their earthen wall; (iv) the lower Si/(Si + Al + Fe) values of wall samples versus their associated background surface samples nearby sand deserts is evidence that the Ming had developed methods to assess and selectively mine loess poor in quartz-sand content; (v) the low variability of Fe for wall (1 STD = 0.6%) versus background samples (1 STD = 1.0%) is evidence that the Ming used visual color for grading the quality of the loess, given that Fe is contained in either chlorite (green/brown) or related weathering products, namely hematite (red) or goethite (yellow); (vi) the elevated S contents west of Site 28 in both wall and background samples (yellow circles in Figure 3f,g) is related to gypsum ($CaSO_4.2OH$), which is aeolian in origin and derived from evaporative lakes in the Tengger Desert; (vii) the relative lack of Cl in background surface samples but common occurrence in wall materials west of Site 28 can be explained by aeolian halite (Na.Cl) derived from a nearby source (e.g., evaporative lakes in the Tengger Desert) that is readily leached from the surface into the groundwater where it then can be mobilized, presumably by capillary action, up into specific "hangtu" layers of the wall; and (viii) the consistently higher Ca content of wall samples versus background samples is evidence that the Ming routinely applied a Ca-rich additive, most likely lime.

3.5. XRD Mineralogy—Selected Wall Samples

Ten wall samples were measured using laboratory X-ray diffraction (XRD), namely Sites 4, 10, 21, 30, 33, 43, 48, 49, 55, and 62. The mineralogy interpreted from these whole-sample analyses include: quartz, albite, microcline, muscovite, chlorite (chlinochlore and chamosite), kaolinite, calcite (no dolomite), amphibole (actinolite), gypsum, halite, poorly crystalline, hydrated clay (~montmorillonite and/or interstratified clay), a zeolite, and possibly a hydrogarnet. We determined muscovite rather than illite in these bulk sample XRD analyses given the 10 Å peak-width at half-height-maximum (FWHM) is <0.2 Å. The zeolite mineral, laumontite [$Ca(AlSi_2O_6)_2 \cdot 4H_2O$]), was identified by peaks at 9.46 Å and 6.83 Å (Sites 48 and 49). Similar to the pXRF Ca, Cl, and S results (Figure 3e), gypsum, halite, and laumontite were only detected by XRD west of Site 28 (Figure 3h). All samples show a number of other minor but difficult to interpret XRD peaks, including one at 2.64 Å, which is possibly related to a Ca-bearing hydrogarnet.

3.6. Field and Satellite VNIR-SWIR Results

Selected field ASD spectra of wall loess samples taken from the same sites chosen for the XRD analyses (Figure 3h) are presented in Figure 3i. Absorption by water vapor (and surface water) is masked out between 1800 and 2000 nm. These wall loess samples are all intimate mixtures of minerals, including: muscovite (2200, 2350, and 2450 nm); illite (2200, 2350, 2450 nm, and well-developed water related shoulder near 2000 nm); chlorite (2000, 2245, and 2330 nm); and gypsum (1445, 1490, 1540, 1746, 2215, and 2410 nm). Kaolinite (2165 and 2205 nm) and calcite (2335 nm) are more difficult to recognize, although kaolinite is identified because of a characteristic sharp absorption locked in at 2205 nm (~18 nm FWHH, whereas the muscovite and illite absorption is >30 nm width and can vary in wavelength [52]) and a depressed shoulder near 2165 nm, as shown by ASD spectra from Sites 43–55 (excluding Site 48). This subtlety reflects the fact that kaolinite represents only a minor component of the loess compared with minerals such as muscovite.

The mineral information targeted using ASTER's moderate spectral-resolution capabilities (Figure 3h) included: (i) the presence of gypsum; (ii) the proportion of quartz sand versus clay (called sand-clay index or SI^3); (iii) the content of silicate minerals such as kaolinite and white mica (called 2200D); (iv) the proportion of kaolinite versus white mica and/or montmorillonite (called 2165D); and (iv) the combined content of chlorite, carbonate and/or amphibole (called 2330D). We validated these ASTER mineral maps using coincident ROIs of the field ASD data convolved to ASTER responses. Full-spectral resolution field ASD were also used where required to more accurately interpret the composition of the mineral components.

3.6.1. Gypsum

The full spectral resolution field ASD data show diagnostic gypsum absorptions at 1446, 1490 and 1540 nm for wall samples collected from Sites 40, 48 (Figure 3i), 55 (Figure 3i), 60, and 62 and background surface samples collected at Site 43 only. Thus, the ASD, XRD, and pXRF data all indicate a similar pattern where gypsum is present in a zone spanning Sites 28–62.

The ASTER gypsum index [61] was clipped to identify pixels most likely to contain gypsum. The resultant map (white filled polygons in Figure 4a) reveals numerous, large (up to 50 km wide) occurrences of gypsum associated with evaporative lakes. Most of these ephemeral lakes are found in the Tengger Desert ("A"), with less in the Badain Jaran Desert and no apparent gypsum occurrences are evident in the Mu Us and Hobq Deserts, at least at this map-scale (Figure 4a).

We propose that northerly to westerly winds (light yellow arrows in Figure 4a) transported gypsum-laden dust as long-term suspension from evaporative lakes in the Tengger and Badain Jaran Deserts to areas <300 km away. We also suggest that this gypsum-laden dust was able to cross topographic highs of the Helan Shan and the Alashan Block (Figures 1a and 4a).

3.6.2. Sand-clay Index

The SI^3 index is sensitive to the abundance of sand rich in quartz (and other silicates such as feldspars) relative to clay materials (size and composition) and is driven by the wavelength and intensity of the silicate reststrahlen feature near 8.6 μm [70,81]. The satellite and field SI^3 data show similar sigmoidal patterns (Figure 5a,b, respectively) that are well-modeled using third-order polynomials and are consistent with the pXRF Si/(Si + Al + Fe) results (Figure 3e). All show inflexion points positioned between Sites 33 and 39 (i.e., approximate position of the Yellow River), with quartz-sand-rich materials to the east and clay-rich materials to the west. Similar to the particle size results, which show that the loess along the Shandan section is poor in sand (Figure 3c), the associated field SI^3 values are also at their lowest, i.e., least amount of quartz sand (Figure 4c).

The ASTER SI^3 map (Figure 4a) depicts areas rich in quartz-sand as warmer tones while areas rich in clays are cool tones. The two areas most abundant in quartz-sand are parts of the Gobi Desert (near "C") and the Hobq Desert ("D"). From the limited spatial coverage selected for our study area, we recognize SI^3 gradients extending away from these enriched zones as follows. The area of highest SI^3 values in the Gobi Desert (near "C") is located over terrain comprising exposed Mesozoic and Cainozoic sedimentary rocks, including Cretaceous sandstones. From here, the SI^3 progressively decreases east-southeastward (i.e., the direction of quartz-sand transport) along a linear ~500 km long trajectory ("C" to "A"), paralleling (restricted by) the highlands of the Alashan Block to the south ("H"). In the process, this SI^3 signature seamlessly crosses the >40 km wide wadi of the Heihe River (Figure 1a). That is, this aeolian activity is sufficiently frequent to obscure the potentially cross-cutting effects of fluvial sediment transport.

Continuing eastward from the Heihe River wadi, the SI^3 signature extends down-gradient for ~100 km across dune-fields of the Badain Jaran Desert, which comprise linear to complex dune patterns consistent with an easterly flow of quartz-sand mobilized by saltation. The SI^3 gradient then begins to increase for ~100 km across similar dune-fields concomitant with a rise in topography (100 m), reaching a maximum height along the Yabra Shan (Figure 1a). Beyond here, the pathway of quartz sand transport becomes more diffuse as it enters the lowland expanse of the Tengger and Ulan Buh Deserts (Figure 1a). The SI^3 map also provides evidence for the input of quartz sand into this desert pathway from the Qilian Shan ("F") and Alashan Block ("H"), although we suggest this sediment was mobilized downslope chiefly by fluvial and/or colluvial processes (dark red arrows in Figure 4a).

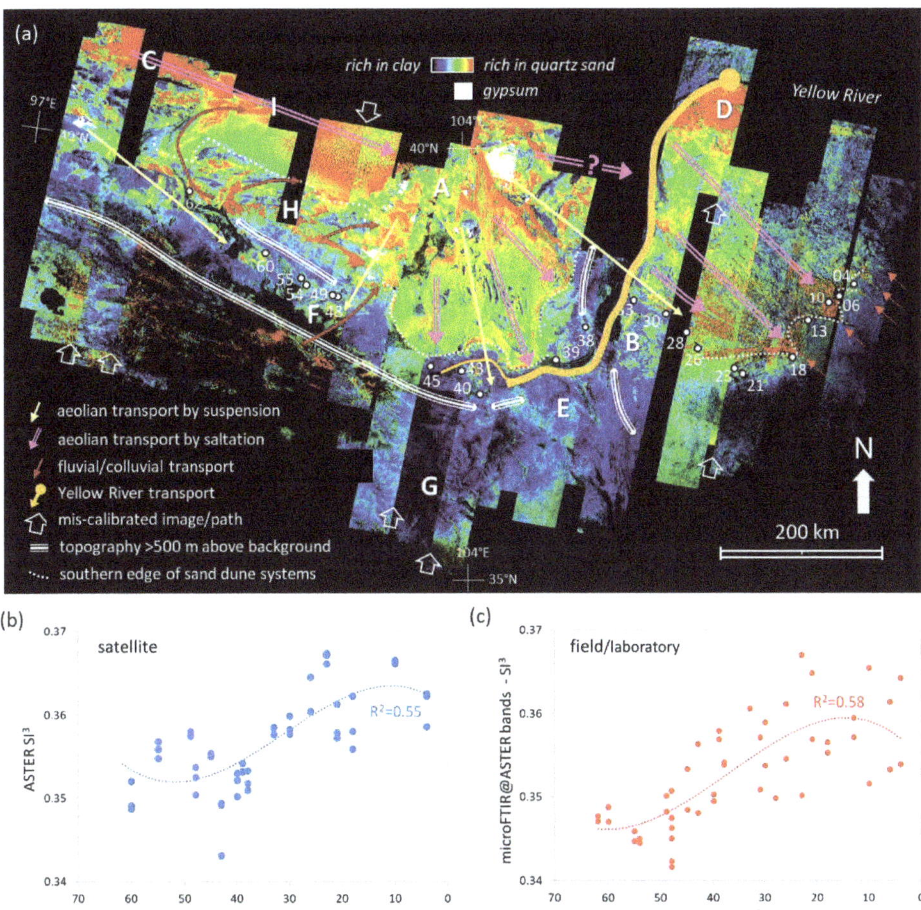

Figure 4. (a) ASTER SI^3 base-map depicting the proportion of quartz sand (warm colors) to clay (cool colors) material overlain with a threshold mask of the ASTER Gypsum Index (white). (b) Scattergram of the west-to-east spatial trend of satellite ASTER SI^3 responses collected from three (north, south, and centered over the wall) ROIs (~50 pixels each) of the field sites. (c) Scattergram of the west-to-east spatial trend of field SI^3 responses of wall samples only. A magenta box highlights the well-preserved "Shandan" wall section.

The SI^3 gradients across the Tengger Desert are more variable, which is consistent with the complex pattern of linear dune systems, both longitudinal and transverse. The net result is a saltation movement of aeolian quartz sand (magenta arrows) that terminates against the Yellow River or its ephemeral tributaries (e.g., between Sites 39 and 45, Figure 4a). Importantly, the ASTER SI^3 values markedly step-down along this boundary, which is clear evidence for aeolian saltation flow. That is, quartz sand has not been transported by wind-suspension across the river to be deposited some distance across on the opposite bank near, e.g., near "E".

Thus, the entire volume of quartz sand flowing by aeolian saltation through the Tengger Desert, and presumably Ulan Buh Desert, is consumed by the Yellow River before being transported downstream (thick orange arrow). This is consistent with field studies which have found a pronounced increase in the sand sediment load along this section of the Yellow River [81–84] and corresponds to a change in flow regime from a deep, meandering channel upstream to a shallow, braided system downstream that extends along the margin of the Hobq Desert. It then returns to a single, deep channel, presumably because the river's load of sand has been deposited.

The SI^3 pattern across the Ordos Plateau is less complex, driven largely by southeastward saltation of quartz sand (magenta arrows) paralleling longitudinal dunes in the southern half of the Mu Us Desert. The highest SI^3 values are located in a ~50 km wide, lens-shaped zone spanning vegetation-free, transverse dunes (<100 m amplitude) of the Hobq Desert ("D" in Figure 4a). Southeastward from here the SI^3 gradient decreases for ~100 km values before gradually increasing over the Mu Us Desert concomitant with the development of longitudinal dunes. A series of quartz sand rich incursions or cusps (red arrows) extend for >30 km southeastward into the zone of loess. These cusps appear not to be constrained by topography as they are located over both shallow valleys and low hills. In contrast with the sharp southern boundary of the Tengger Desert, the southern margin of the Mu Us Desert (white dotted line in Figure 4a) shows a diffuse SI^3 transition with the adjacent loess deposits across a ~50 km zone.

There is no indication from the SI^3 map (Figure 4a) that the upper catchment of the Yellow River (west of "G") in northeast part of the Tibetan Plateau was a significant source of quartz sand. This contrasts with the Qilian Shan near "F", which has both exposed rocks and associated downslope fans and alluvial plains of the Shiyang River with characteristically high SI^3 values.

3.6.3. Al-Clay Content Index

Both the field and satellite 2200D data show similar weak patterns which are modeled here using fifth-order polynomials. These show a narrow low centered near Sites 55 and 56 (Figure 5b,c), which is coincident with the zone poor in (quartz) sand size material (Figure 3c). This association helps to explain this apparent 2200D low as the lesser amount of optically-transparent quartz grains reduces the overall optical thickness, resulting in less opportunity for electromagnetic radiation to interact with clay mineral particles [75,76]. This optical transparency effect also helps explain why the quartz-rich dune-fields (Figure 5a) have moderate to high 2200D values (warmer tones) even though they generally contain less clay mineral content compared with loess (Figure 3b,e and Figure 5b,c). One exception, however, is the western part of the Badain Jaran Desert ("I"), which has a low 2200D response that seamlessly merges with the wadi of the Heihe River and not the nearby Gobi Desert ("C") or the Alashan Block ("H"). Instead, we trace the source rocks up the Heihe River to areas of low 2200D signature along the western part of the Qilian Shan, i.e., south of Sites 54–62. Similarly, the higher 2200D response of the eastern half of the Badain Jaran Desert can be traced to sediment eroded from the Alashan Block (dark red arrows in Figure 5a).

Further east along the Qilian Shan, there are rock exposures near "F" (Figure 5a) that generate the highest 2200D responses across the study region. This high response then persists downslope across associated alluvial fans and then along the Shiyang River wadi before merging seamlessly into the Tengger Desert (dark red arrow). In contrast, the upper reaches of the Yellow River catchment (near "G") comprise a relatively low 2200D signature, although it does increase downstream near "E", where it seamlessly merges with the 2200D signature of the Tengger Desert. This signature can be traced southward across the Yellow River and over loess deposits for >100 km ("E"), although appears to be constrained (given image miscalibration error) by the eastward extension of the Qilian Shan, where topography is >500 m above background (triple white lines). This apparent lack of 2200D contrast between dune-fields and loess is also evident across the Ordos Plateau between the Mu Us Desert and loess to its south (between Sites 4 and 26 in Figure 5a).

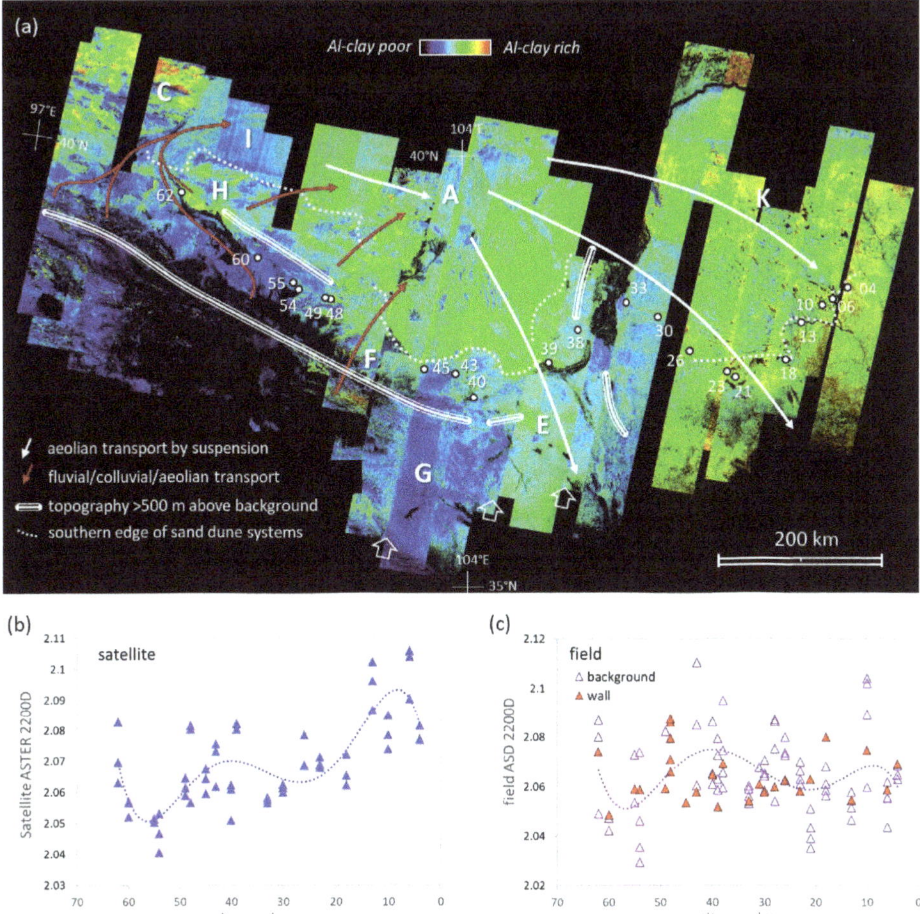

Figure 5. (**a**) Satellite ASTER 2200D index base-map depicting the amount of Al-bearing minerals (i.e., muscovite/illite, kaolinite) with greater abundances in warmer tones. (**b**) Scattergram of the west-to-east spatial trend of satellite ASTER 2200D responses of ROIs collected from three (north, south, and centered over the wall) ROIs (~50 pixels each) of the field sites. (**c**) Scattergram of the west-to-east spatial trend of field 2200D responses of both wall (filled red triangles) and background (open red triangles) samples. Fifth-order polynomials are fitted to the satellite and background field sample data. A magenta box highlights the well-preserved "Shandan" wall section.

3.6.4. Kaolin—White Mica Index

The ASTER 2165D is sensitive to the composition of dioctahedral clay minerals with kaolinite generating a higher response and white mica a lower response, although this interpretation is most appropriate for those areas sharing a similar 2200D response. Both the field and satellite ASTER 2165D data show a similar sigmoidal (dextral offset) pattern (Figure 6b,c, respectively), with the Yellow River valley (Sites 33–39) marking the inflexion point. That is, there are higher levels of kaolinite relative to white mica (high 2165D response) in the zone between Sites 40 and 48 in background samples, whereas there are higher levels of white mica to kaolinite in the zone between Sites 26 and 33. This pattern is broadly consistent with the height of the kaolinite related XRD peak at 7.169 Å (green squares in Figure 6c). Wall samples along the well-preserved Shandan wall section tend to be rich in kaolinite, although the associated background samples are more variable (Figure 6c). This is related to the fact that background surface samples were not always available from exposed loess but instead either alluvium (e.g., Sites 49, 54, and 55) or sand dunes (e.g., Site 62).

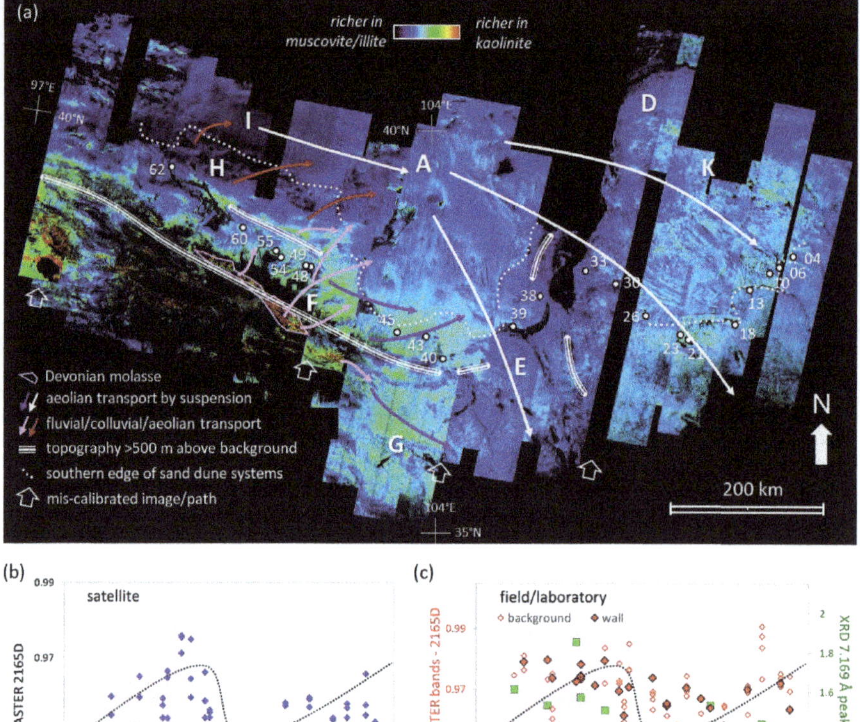

Figure 6. (**a**) Satellite ASTER 2165D index base-map depicting the proportion of kaolin (warmer tones) versus white mica (cooler tones). (**b**) Scattergram of the west-to-east spatial trend of satellite ASTER 2165D ROI responses. (**c**) Scattergram of the west-to-east spatial trend of field 2165D responses of both wall (filled red circles) and background (open red circles) samples. These are underlain by laboratory XRD results of the 7.169 Å kaolinite peak height (green boxes) for selected wall samples. A magenta box highlights the well-preserved "Shandan" wall section.

The satellite 2165D map (Figure 6a) shows areas richer in kaolinite as warmer tones. Regionally, kaolinite is more abundant over the Tibetan Plateau with some of the highest 2165D responses associated with Devonian molasse in the Qilian Shan (purple polygon). These exposures have associated downslope colluvial/alluvial aprons (light purple arrows in Figure 6a). However, aeolian transport by long-term westerly wind suspension (dark purple arrows in Figure 6a) is also required to account for the 2165D gradient that extends for up to 300 km eastward across to the southern parts of the Tengger Desert and the upper Yellow River catchment ("G").

The lowest 2165D responses (cooler tones, i.e., rich in white mica) are found over the Badain Jaran Desert (I) and the Alashan Block ("H"). We propose that the eroding geology of the Alashan Block is a (the) primary source of white mica with derived sediments initially transported downslope by colluvial/alluvial processes before entering the aeolian sediment transport pathway along the Badain Jaran and then ultimately the Tengger and Ulan Buh Deserts. The associated low 2165D response then crosses southward across the Yellow River into loess deposits near E as well as eastward to the Ordos Plateau. However, the 2165D gradient then increases eastward, which could be a function of increased weathering related to higher rainfall in this region resulting in the formation of kaolinite after minerals such as chlorite and feldspar [85]. Given this apparent, mildly elevated level of kaolinite across the Ordos Plateau is caused by weathering, the dominant white mica component is likely transported by long-term wind suspension uplifted from the Badain Jaran and Tengger Deserts (white arrows) with related particles originating from source rocks exposed across the Alashan Block ("H").

3.6.5. Chlorite/Carbonate/Amphibole Content

We identified chlorite as the dominant 2330D absorbing mineral from the field ASD spectra based on absorptions at around 2000, 2250, and 2230 nm. Both the satellite and field 2330D data show a similar sigmoidal (sinistral offset) pattern (Figure 7b,c, respectively), that is opposite to that observed with the 2165D data (Figure 6b,c). However, all share a similar position for their inflexion point, which is located over Yellow River valley (Sites 33–40). Wall samples along the well-preserved Shandan section are relatively poor in chlorite and less variable compared with their associated background samples (Figure 7c), which often comprise colluvial/alluvial material (e.g., Sites 49, 54, and 55).

The satellite 2330D map (Figure 7a) shows areas richer in chlorite as warmer tones. The highest 2330D responses are found on the Alashan Block ("H") from which stems a seamless, decreasing gradient spanning the eastern part of the Badain Jaran Desert and then across the Tengger Desert ("A") before dispersing across the Yellow River southeastward towards "E" and eastward over the Ordos Plateau. The lack of a 2330D north–south gradient across the Ordos Plateau is evidence that the source of these aeolian-borne trioctahedral minerals were not local but from a distant region. The low 2330D signatures that characterize the Tibetan Plateau, loess deposits in the upper reaches of the Yellow River ("G"), the Qilian Shan, Hexi Corridor, and Heihe and Shiyang Rivers (Figures 1a and 7a) indicate these areas are not intrinsic to the chlorite (and carbonate and other trioctahedral minerals) mineral transport pathway. We thus interpret the 2330D transport pathway from source to loess deposition as initially being downslope, colluvial-fluvial movement from source rocks exposed across the Alashan Block (dark red arrows) before being entrained by westerly to northwesterly wind long-term suspension (white arrows) that deliver chlorite to distant (>500 km) areas including loess deposits of the Ordos Plateau (Figure 1a).

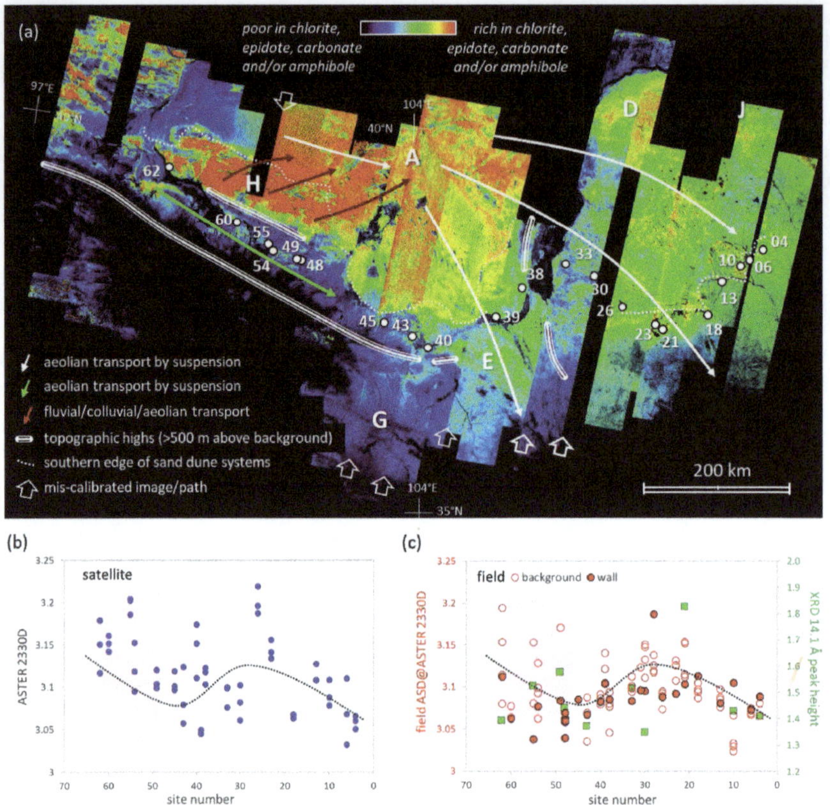

Figure 7. (**a**) Satellite ASTER 2330D mosaic, which is sensitive to the content of chlorite. (**b**) Scattergram of the west-to-east spatial trend of satellite ASTER 2330D ROI responses. (**c**) Scattergram of the west-to-east spatial trend of field 2330D responses of both wall (filled red circles) and background (open red circles) samples. These are plotted with laboratory XRD results of the 14.1 Å chlorite peak height (green boxes) for selected wall samples. A magenta box marks the well-preserved "Shandan" wall section.

4. Discussion

We have presented here a range of evidence that shows the Ming earthen wall changes in composition along its 1200 km length between Shenmu and Jiayuguan (Figure 1a). These compositional changes are apparent in field sample measurements of wall samples and nearby surface background materials (Figures 3b,c,e–h, 4c, 5c, 6c, and 7c) and are consistent with the ASTER surface mineral maps and related ROI data (Figures 4a,b, 5a,b, 6a,b, and 7a,b). All show that the loess used by the Ming to build their earthen wall is compositionally heterogeneous. This has implications for improved understanding of: (i) where the Ming mined loess to build their earthen wall; (ii) the origin of the loess materials; and (iii) the importance of mineral composition in determining the wall's long-term erosional durability.

In addition to the similarity in composition (mineralogy and chemistry) between wall and background (<50 m away from the wall) materials (Figures 3e–g, 4b,c, 5b,c, 6b,c, and 7b,d), there is also the similarity in the nature of rock fragments (where available) at a given site. We cite these as key evidences for the Ming sourcing their earthen wall building material from nearby (<500 m), surficial (<30 m depth) loess deposits of presumably Holocene or possibly late Pleistocene (i.e., Malan loess units) age. We conjecture that, where available, the Ming's inclusion of up to 20% local rock fragments into the wall building material (Figure 2b,e) simply served as an expedient additive rather than being designed for improving the wall's structural performance, unlike previous Dynasties such as the Han who incorporated layers of rock into their wall design, i.e., were a mixture of both grain (=rock) supported as well as matrix (=loess) supported.

We further propose that the Ming established at least three methods to improve the quality of the loess for their earthen wall construction. The first is that they routinely gauged the sand content of the loess and, if required, selectively mined deeper loess layers richer in silt and/or clay content if required (Figure 3e). This would have been particularly important across the Ordos Plateau where incursions of saltating sand grains regularly occurred because of a lack of physical barriers such as vegetation, mountains, and permanent rivers. At the same time, it was fortuitous for the Ming that the limited depth of loess across the Hexi Corridor (~1 m layer, Figure 2f) was of high-quality building material rich in finer fractions (Figures 3c,e and 4c), else they would have had to either transport loess from mines located a great distance away or found an alternative locally-available building material, much of which has a high content of rock. Similarly, we propose that the Ming used color, which is related to the content of iron oxide and/or chlorite, to assess the quality or at least maintain consistency of the mined loess (Figure 3f).

The third method we propose is that the Ming applied a Ca-rich material, possibly calcite but more likely lime. This is based on the consistent ~2% higher levels in Ca content of wall versus background materials across the study area (Figure 3f,g, respectively). The lack of a similar pattern for both Mg and S indicates that neither dolomite [$(Ca,Mg)CO_3$] nor gypsum was involved. Lime was well known to the Ming having been used in construction back to the Shang Dynasty (1700–1027 BC) [2]. In theory, a 2% lime application rate is sufficient to reach the "lime fixation point" where ions are absorbed by clay minerals, which increases the unconfined compressive strength of the earthen material [86]. After this point, alkali activation can cause the (re)precipitation of calcite as part of a pozzolanic type reaction [54]. However, nano-fibers of secondary calcite are also found in natural loess material [48] such that (re)crystalized calcite in the earthen wall material is not itself conclusive evidence for lime addition by the Ming.

The advanced stages of pozzolanic reaction generate a variety of Ca-bearing silicate and/or alumino-silicate minerals, such as ettringite, tobermorite, zeolites, and hydrogarnets [17,86]. From our limited XRD data, we detected the zeolite mineral, laumontite, at Sites 48 and 49 (Figure 3h). Given that zeolites can also form by subaerial dissolution of silicate minerals in a range of geologic environments at near-surface conditions [87], its limited distribution is unlikely to be the result of systematic application of lime by the Ming. Instead, we suggest its overlapping pattern with gypsum and halite (Figure 3h) is the result of aeolian transport from nearby evaporative lakes (Figure 4a). From the available data, we are not confident that hydrogarnet or any other Ca-bearing alumino-silicate cementing minerals are present in the wall material. Thus, even though we conclude that the Ming added ~2% Ca, most probably lime, to their loess building material, this may only have had the intended effect of improving its compressive strength rather than generating a binding cement.

Previous studies have been unable to agree on the sediment transport pathways that generated the vast loess deposits across north-central China [21–50]. We suggest that this lack of consensus is a function both of the complexity of the loess system and the limitations of the analytic tools and sampling strategies used to date. Similar to the parable of the elephant and the six blind men, only when one can see the bigger picture is it possible to begin to resolve the relationship of the parts. We argue that our use of satellite ASTER mineral maps has assisted in untangling this complexity by enabling a view of the "whole", at least for a few mineral components of the loess. To that end, Table 1 lists the essential ingredients and mappable criteria [53] we conclude are valuable for mapping the "loess system" using ASTER's limited but valuable spectral/spatial/radiometric resolution.

Using the ASTER mineral maps (Figures 4a, 5a, 6a, and 7a), we identify at least two sediment transport pathways, both of which source mineral grains from a variety of exposed rocks. These two pathways or "loess subsystems" are driven by their grainsize energy potential: (i) sand; and (ii) clay to fine-silt (Table 1).

Table 1. A mineral-based "loess system" sediment transport model using ASTER satellite imagery.

	Sand Sub-System	Clay and Fine Silt Sub-System
Essential components	(i) quartz	(ii) white mica (muscovite and/or illite) (iii) chlorite (iv) kaolinite (v) gypsum
ASTER mappable criteria	(i) ASTER SI^3 index	(ii) ASTER 2200D and 2165D indices (iii) ASTER 2200D and 2165D indices (iv) ASTER 2300D index (v) ASTER SI^3 index
Particle size	(i) sand	(ii)–(iv) fine-silt to clay (v) fine silt (?)
Primary source	(i) Mesozoic and Cainozoic sedimentary rocks exposed in the Gobi Desert	(ii) pre-Mesozoic metamorphic rocks of the Alashan Block (iii) pre-Mesozoic metamorphic rocks of the Alashan Block (iv) Devonian molasse exposed in the Qilian Shan (v) evaporative lakes of the Tengger Desert
Transport mechanisms	• aeolian saltation by westerly winds west of the Yellow River • Fluvial erosion alongside the Tengger and Hobq Deserts and deposition alongside the Hobq Desert. • aeolian saltation by northerly winds east of the Yellow River	• aeolian long-term suspension by westerly to northwesterly winds
Detectable gradient size	• <200 km	• >500 km
Pathway constraints	• connected (downslope or flat) lowlands • topography <100 m elevation above background	• topography <2000 m elevation above background
Sediment traps/reservoirs	• eastern Badain Jaran Desert (topographic upslope) • Hobq Desert (Yellow River braided system bedload deposition) • Mu Us Desert (topographic upslope)	(ii) and (iii) eastern Badain Jaran and Tengger Deserts (iv) alluvial fans off the Qilian Shan (ii)–(v) low energy, long-term suspension dust fallout areas, topographically shielded from saltation flow of sand

The transport pathway for the quartz sand fraction (>50 µm), which accounts for ~40% of the loess material (Figure 3a,b), is initially driven by colluvial/alluvial downslope movement from source rocks in the Gobi Desert, Alashan Block, and Qilian Shan (dark red arrows in Figure 4a). There could also be a significant amount of quartz sand originating from rocks to the north and/or west of the limits of the current study area. The quartz sand then travels by aeolian saltation flow along lowlands of the Badain Jaran, Tengger, and Ulan Buh Deserts (magenta arrows in Figure 4a) forming arrays of multi-scale linear dunes systems. Most of this quartz sand is then consumed by the Yellow River and delivered downstream to where it is then deposited along the banks of the western Mu Us and Hobq Deserts (thick orange arrow). It then becomes available for transport by northwesterly wind saltation flow before finally depositing along the northern margin (<100 km wide) of the loess plateau. Related ASTER SI^3 gradients change over distances of <300 km determined by distance from "source" and topography. A gentle rise of even 1:1000 appears to retard saltation flow causing a bank-up (deposition) of quartz sand, resulting in a reversal of the SI^3 gradient. An example is the eastern part of the Badain Jaran Desert (between "I" and "A"), which has the highest star dunes (~500 m high) in the world. A greater rise in topography to 1:100 appears to inhibit the saltation flow of quartz sand, as demonstrated by the western flank of the Helan Shan Figures 1a and 4a). These observations help explain why the loess developed along the Hexi Corridor is relatively poor in quartz sand because of shielding by mountains of the Alashan Block from quartz sand moving by north-northwesterly wind saltation along the Badain Jaran Desert (Figure 5a).

The transport pathway for the combined clay and fine-silt fraction (<20 μm), which accounts for ~30% of the loess material (Figure 3a,b), is characterized by aeolian long-term suspension with related sediment travelling for many hundreds of kilometers, often at high elevation (>2000 m). We identified several rock sources of these finer particles in our study area using the ASTER indices (Table 1): (i) gypsum (SI^3 index) from evaporative lakes in the Tengger Desert (Figure 4a); (iii) kaolinite (2200D and 2165D indices) from Devonian Molasse in the Qilian Shan (Figures 5a and 6a); and (iv) chlorite (2330D index) and muscovite (2200D and 2165D indices) from the Alashan Block (Figures 5a, 6a, and 7a). Previous mineralogical work across the Chinese Loess Plateau on the Malan Loess Unit [88,89] found that the dominant minerals of the fine-silt fraction comprise quartz (~30%), muscovite (~27%) and chlorite (~22%) while the clay fraction comprises illite (~46%), kaolinite (~16%), and chlorite (~15%). This mineralogical distinction according to particle size is consistent with our spectral results. That is, both muscovite and chlorite share the same sediment transport pathway, originating primarily from source rocks in the Alashan Block by colluvial/fluvial processes (dark red arrows in Figures 5a, 6a, and 7a) before being uplifted by westerly to northwesterly wind long-term suspension (white arrows in Figures 5a, 6a, and 7a) to eventually be deposited in areas >500 km away, including the Chinese Loess Plateau. This shared transport pathway is expressed as a continuous down-gradient in the ASTER 2165D (dark blue to cyan in Figure 6a) and 2330D (red to yellow to green in Figure 7a) maps. It is also expressed as a southeastward decrease in chlorite and muscovite content of the loess developed across the Ordos region [34,46,48]. These observations are consistent with both muscovite and chlorite being significant components of the fine-silt fraction.

We trace kaolinite using the ASTER 2200D and 2165D maps back to source rocks in the Tibetan Plateau, including Devonian Molasse exposed along the Qilian Shan (Figures 5a and 6a). However, this kaolinite comprises a range of particle sizes given that it shows a 2165D gradient spanning <500 km north/east from "F" and with characteristics of both downslope colluvial/alluvial transport, associated with fans and ephemeral drainage networks (mauve arrows in Figure 6a), and aeolian dispersal across the southern parts of the Tengger Desert and western loess plateau (dark purple arrows in Figure 6a). Note that this "primary" kaolinite is different to the "weathering" kaolinite in the southeastern part of the study area, which has been identified by other workers [13,34,85] and suggested to be related to increasing rainfall.

In addition to the "sand" and "clay to fine-silt" subsystems described in Table 1, we also acknowledge the existence of other sediment transport systems contributing to the development of loess: (i) "clay"; and (ii) "coarse-silt". The coarse-silt fraction (20–50 μm), which accounts for ~30% of the loess material, shows no coherent spatial pattern along the length of the Ming earthen wall (Figure 3b). It also does not yield any statistical association with other particle size fractions or targeted source areas (Figure 3d). Given that U-Pb studies [24,30,31,37], which measure zircon grains of >30 μm size (i.e., coarse silt size), have consistently concluded that the Yellow River is sourcing a significant amount of material from the Tibetan Plateau (and adjoining Qaidam Basin), we suggest that the Yellow River is transporting the bulk of this coarse-silt sediment through the study region. The mineralogy of this coarse-silt likely includes plagioclase, alkali feldspar, and amphibole [31,88]. Similar to the sand fraction, a significant part of this coarse-silt material is deposited alongside the Hobq Desert before mobilization southward by aeolian saltation where it is finally incorporated into loess. The thickest part of the Malan Loess Unit (Figure 1a) reflects this south(east)ward flow of coarse-silt (and sand) originating from the banks of the Yellow River during the late Pleistocene and most likely through the Holocene [8].

Clay-size grains, which can travel distances of 1000s of kilometers by long-term wind suspension [8,47,89], are also readily transported by fluvial suspension. The Yellow River, which carries the largest volume of sediment of any river on the Earth [90], owes its distinctive beige-yellow color to a mixture of suspended clay minerals and iron oxides. Much of this suspended clay material is deposited into marine environments >1000 km to the east of the study area. Drill cores sampled from the Bohai and Yellow Seas [90] show that the clay-size mineralogy comprises: illite (average: ~60%), smectite (average: ~15%), chlorite (average: ~15%), and kaolinite (average: ~10%). This dominance of illite in the clay fraction is mirrored in the Malan loess unit [8] found across the Ordos region (Figure 1a). Illite formation is unlikely to occur through regional weathering because at surface temperatures the crystallization of illite requires a specific range of Si and K activities usually found in saline environments [91]. That is, illite is more likely to have formed in evaporative basins, such as parts of the Tengger, Badain Jaran, and Taklamakan Deserts [25,92,93]. The Taklamakan Desert has been identified as a key source of clay size material transported by long-term westerly wind suspension to the Ordos region [33,40,41,93].

The Ordos Plateau region thus represents a depositional "cross-road" for at least four sediment transport pathways. This contrasts with the Hexi Corridor which lacks significant contributions of sand associated with the Badain Jaran aeolian saltation pathway as well as coarse-silt associated with the Yellow River pathway. This has resulted in the loess developed along the Hexi Corridor being relatively abundant in clay and fine silt (Figure 3c) and with a composition relatively poor in quartz (Figures 3e and 4c) and richer in kaolinite (Figure 6c) because of its proximity to kaolinitic source rocks in the Qilian Shan ("F" in Figures 5a and 6a). These factors have likely driven the relative erosional robustness of the Ming's earthen wall along the Shandan section (Figure 2j), because of greater opportunity for hydrogen bonding between water molecules and the surfaces of the finer mineral grains [16].

Groundwater penetration along specific "hangtu" layers continues to generate a cycle of crystallization and dissolution of salts, resulting in erosion and undercutting of the wall (Figure 3i) [94]. We observed in the field that when irrigated farming abutted the wall, then the erosion immediately above was dramatically enhanced compared with the wall <10 m away from the edge of the cropped field (e.g., at Site 49, 38°19.952′N; 101°54.259′E). Thus, the control of groundwater movement is crucial in the continued preservation of the earthen wall.

ASTER has proven valuable for mapping and understanding the complexity of mineral-composition patterns not apparent in point-sample field data alone. In so doing, ASTER has enabled us to build a "loess system" model for at least two sediment transport pathways. However, ASTER's modest spectral resolution (compared with hyperspectral systems) has impacted what loess mineral information can be targeted. Higher spectral resolution systems with tens to hundreds of spectral bands should in theory provide more detailed compositional information and thus improve the accuracy and detail of this loess system analysis. For example, the Tschermak chemical composition of both muscovite and chlorite can be measured and traced back to more specific rock sources [52].

NASA's hyperspectral VNIR-SWIR EMIT imaging system, which is scheduled for operation starting 2020 on board the International Space Station (ISS) [95], is one candidate, especially given that it has the task of mapping the surface mineralogy of dust generating regions of the Earth to assist in global climate models. It is a shame that ASTER was not also tasked to help deliver this important information, especially given its 20-year archive of multi-temporal, global land surface imagery. Interestingly, EMIT will operate on the ISS alongside Japan's hyperspectral VNIR-SWIR imaging system called HISUI [96], which is focused on mineral and energy resource applications.

These two hyperspectral VNIR-SWIR imaging systems will essentially be acquiring similar mineral compositional information from the same platform and at the same time, albeit for different end-users. However, unlike ASTER, these two systems do not possess spectral bands that cover the TIR, which is essential for mapping quartz sand information (as well as carbonates and sulfates). This gap in wavelength coverage could be augmented by NASA's ECOSTRESS [97], which is currently operating on the ISS.

ASTER's publicly available, global data archive presents us with the future opportunity to extend our mapping and understanding of loess-related mineral transport pathways beyond our current study area (Figure 1a), especially northwards across the Ulan Buh and Gobi Deserts, westwards towards the Taklamakan Desert, and southwestwards over Tibetan Plateau and Qaidam Basin. It also provides the wider geoscience community with the unprecedented opportunity to map and understand a range of other earth science challenges, including temporal monitoring of soil loss and the related process of desertification [98–101].

5. Conclusions

The main conclusions from this study include:

- The Ming earthen wall provided a valuable 1200 km transect for validating a >600,000 km^2 mosaic of satellite ASTER mineral maps, with both showing similar patterns for quartz sand, muscovite-kaolinite, and chlorite content.
- The composition (mineralogy, particle size, and chemistry) of loess used by the Ming to build their earthen wall across the Ordos Plateau and Hexi Corridor is heterogeneous.
- The ASTER mineral maps enable the tracking of sediment transport pathways of loess related minerals not detected in previous studies relying on point-sample data.
- These pathways help explain both the compositional variation of the loess along the Ming earthen wall as well as the wall's erosional robustness.
- Two sediment transport pathways are well mapped using ASTER, namely:

  - Quartz sand is sourced from exposed rocks (e.g., Cretaceous sandstones) in the Gobi stony desert, Alashan Block, and Qilian Shan. This sand travelled by west-northwesterly wind saltation along lowlands of the Badain Jaran, Tengger, and Ulan Buh Deserts before being consumed by the Yellow River. It was then transported downstream where it is was deposited along the margins of the Mu Us and Hobq Deserts, where it is finally moved by northwesterly wind saltation across to the Loess Plateau.
  - Clay and fine-silt are relatively rich in either muscovite and chlorite, which are sourced from metavolcanics and associated sediments of the Alashan block, or kaolinite, which is largely sourced from Devonian molasse exposed in the Qilian Shan. Initial movement of these minerals is via downslope colluvial/alluvial processes before eventual uplift from alluvial fans and wadis by westerly to northerly wind long-term suspension. These fine mineral grains are then deposited 100 s to 1000 s of kilometers away across fields of loess deposition.

- The well-preserved Shandan section of the earthen wall along the Hexi Corridor is associated with loess poor in quartz sand and an increase, albeit minor, in kaolinite content.
- We also propose that the Ming established a number of methods for building their earthen wall, including:

  - Locally sourcing loess (not from distant, centralized mines);
  - Gauging the amount of sand content relative to clay and fine silt so that better-quality loess layers (finer fractions) could be mined; and
  - Adding a Ca material, possibly calcite but more likely lime, to either improve the compressive strength of the loess or to generate a cement as part of a pozzolanic reaction.

Author Contributions: Conceptualization, T.C., P.S., and B.F.; Data curation, T.C., P.S., and Y.N.; Formal analysis, T.C., P.S., and Y.N.; Funding acquisition, P.S. and B.F.; Investigation, T.C., P.S., and Y.N.; Methodology, T.C., P.S., and Y.N.; Project administration, T.C., P.S., and B.F.; Resources, T.C. and P.S.; Supervision, T.C.; Validation, T.C., P.S., and Y.N.; Visualization, T.C.; Writing—original draft, T.C.; Writing—review and editing, T.C., P.S., Y.N., and B.F. All authors have read and agreed to the published version of the manuscript.

Funding: This research was funded by the Chinese Academy of Sciences President's International Fellowship Initiative (Grant No. 2017VTA0001), Remote Sensing Geological Survey of Key Earth Belts (Grant No. DD20190536), Strategic Priority Research Program of the Chinese Academy of Sciences (Grant No. XDA19030502), and C3DMM Pty Ltd.

Acknowledgments: We thank: (i) the Chinese Academy of Sciences, Institute of Remote Sensing and Digital Earth for access to laboratory and office space in Beijing during the fieldwork program as well as access to a range of field and laboratory equipment used in this study; (ii) Jimin Sun from the Institute of Geology and Geophysics, Chinese Academy of Sciences regards discussions on loess formation and the Ming earthen wall; (iii) Shixiang Li from the Ningxia Cultural Heritage Administration for project discussions in the field; (iv) Ji Wei from the Chinese Research Academy of Environmental Sciences for arranging access to a Malvern Mastersizer 2000 instrument; and (v) Mike Verrall from CSIRO Mineral Resources for access to a Bruker D4 Endeavour XRD instrument and Xplot software.

Conflicts of Interest: The authors declare no conflicts of interest.

Appendix A Systematic Detector-Array Calibration Error in ASTER Band 5

The process of generating "seamless", compositionally-accurate mosaics from ~250 ASTER images spanning the Ming earthen wall study region required cross-calibration of 26 overlapping satellite paths. This is because, without applying this cross-calibration step to the as-received L1T ASTER data [67], subsequent band normalization techniques (e.g., band ratios) to generate compositional information would result in mismatches between adjacent images/paths, especially for the VNIR-SWIR bands (Figure A1a,c). Much of this apparent error is related to atmospheric effects, both aerosol scattering (additive effect) in the VNIR and absorption attenuation by gases such as water vapor in the SWIR (multiplicative effect). However, instrument related error is also likely, especially an additive SWIR component [60] related to uncorrected residuals of the so-called "cross-talk" effect [102].

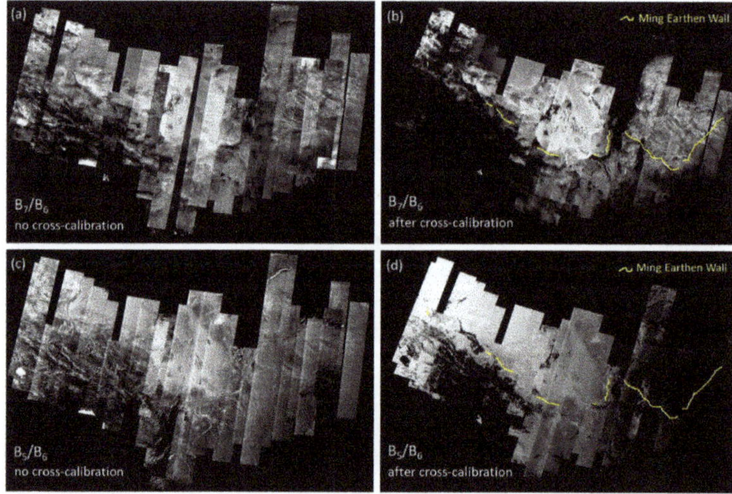

Figure A1. (**a**) ASTER B_7/B_6 mosaic generated from L1T data, i.e., no image cross-calibration applied; (**b**) ASTER B_7/B_6 mosaic after image cross-calibration (and transformation from universal transverse Mercator (UTM) to geographic latitude-longitude); (**c**) ASTER B_5/B_6 mosaic generated from L1T data, i.e., no image cross-calibration applied; and (**d**) ASTER B_5/B_6 mosaic after image cross-calibration (and geometric transformation from UTM to geographic latitude-longitude).

To solve these miscalibrations between images/paths, we collected VNIR-SWIR band statistics for 20–35 coincident invariant targets (or regions of interest (ROIs)) from each opposing pair of image overlaps, with each ROI comprising between 10 and 200 pixels. We then gauged whether there existed significant linear correlations for the coincident ROI band means sampled from each pair of opposing paths (which always proved to be the case). The resultant gains and offsets from these linear correlations were then used to transform (cross-calibrate) one image path to the other. This was repeated 25 times until the entire mosaic was complete. This cross-calibration method was effective for most of the VNIR-SWIR bands, as demonstrated by the B_7/B_6 ratio (Figure A1b). That is, previous mismatches obvious in the uncorrected L1T ratio product (Figure A1a) are absent in the cross-calibrated L1T ratio product (Figure A1b). However, ratios involving ASTER Band 5 generated a ramp effect across the mosaic, as shown by ASTER B_5/B_6 (Figure A1d). This ramp effect is also apparent in color composites of the cross-calibrated ASTER bands that include Band 5 (Figure A2a).

To better understand the nature of this ramp effect, we extracted pixel values from a transect across the mosaic (red line in Figure A3a) for Bands 5–7 (Figure A3c) and then generated ratios for B_7/B_6 and B_5/B_6 (Figure A2d). In contrast with the B_7/B_6 ratio (orange data points), which shows an overall flat trend across the transect, the B_5/B_6 ratio (purple data points) shows a linear slope, except for a segment of pixels from 34,000 to 37,000, which corresponds to an image path not cross-calibrated using coincident ROIs as there was no available image overlap. These results indicate that: (i) ASTER Band 5 has a systematic calibration error; and (ii) our method of using coincident ROIs across image overlaps was susceptible to this error.

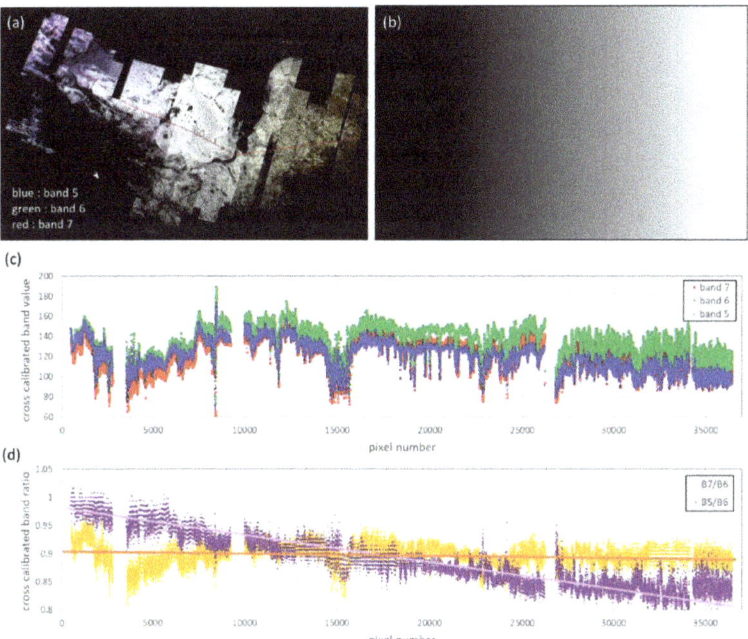

Figure A2. (**a**) Cross-calibrated ASTER color composite of SWIR Bands R:7 G:6 B:5. A transect from which pixel values were extracted is shown by a red line. (**b**) A linear ramp model. (**c**) Pixel values of cross-calibrated Bands 5 (blue), 6 (green) and 7 (red) for the transect shown in (**a**). (**d**) The same data as in (**b**) but with Bands 5 and 7 each normalized with respect to Band 6. Note the linear trend in the B_5/B_6 ratio data (mauve line), which was used for the design of the model.

To better understand the nature of this systematic ASTER Band 5 calibration error, a random suite of full-image-width segments from different paths and dates were sampled from the B_5/B_7 mosaic (Figure A3a). These normalized segments all show the same systematic noise not evident in the raw Bands 5 and 7. All segments show: (i) a broad (<200 pixels wide) bright band on the left side of each image swath (red line); (ii) a pair of narrow (~20 pixels) dark bands (yellow arrow); and (iii) a broad shoulder (blue dotted lines). The mean B_5/B_7 values for six transects taken from these six images (Figure A3b) highlights this systematic line across different dates of imagery. This mean is clearly not flat across the image swath. The implication is that cross-calibration using ROIs collected from either side of a given image swath will be compromised by this systematic non-flatness. Indeed, ROIs collected from the 150 pixel-wide column on the left highlighted by the red bar will be ~2% higher than those ROIs collected from the 150 pixel-wide column on the right highlighted by the green bar. This difference drove the Band 5 miscalibration using coincident ROIs sourced from the as-received L1T data.

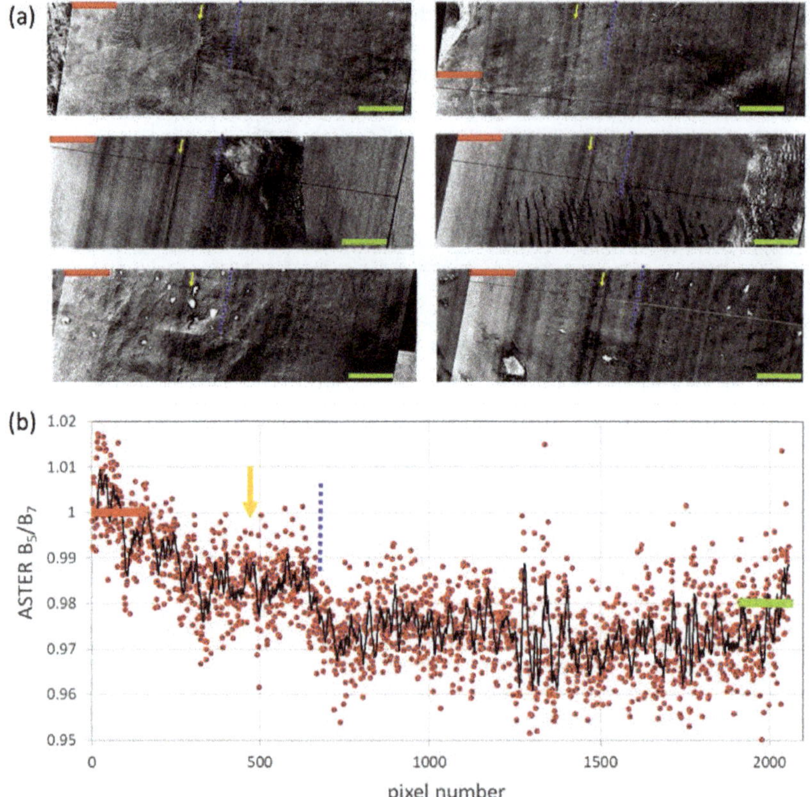

Figure A3. (**a**) A selection of six full-image-width segments from across the non-cross-calibrated ASTER B_5/B_7 mosaic with each showing the similar systematic column noise, including: a broad (<200 pixels wide) bright band on the left side of each image (red line); a pair of narrow (~20 pixels) dark bands (yellow arrow); and a broad shoulder (blue dotted lines). (**b**) Scattergram of the ASTER scene pixel number versus the mean B_5/B_7 values for six transects taken from (**a**). The 10-point moving average is shown as a black line.

We solved this ASTER Band 5 detector array calibration issue by constructing a scaled linear-ramp model across the entire mosaic using ENVI™ software (Figure A2b), which was then used to normalize the cross-calibrated Band 5 (Figure 4a). This de-ramped Band 5 (Figure A4b) was then normalized with the cross-calibrated Band 6 to generate a B_5/B_6 ratio that no longer shows the ramp effect (Figure A4d). This Band 5 de-ramp process proved to be essential for all ASTER compositional products involving Band 5, such as 2200D (Figure 5a) and 2165D (Figure 6a).

Figure A4. (**a**) Band 5 after cross-calibration but before applying the linear ramp in (**b**). (**b**) Band 5 after cross-calibration and applying the linear ramp in (**b**). (**c**) The B_5/B_6 ratio after cross-calibration but before applying the linear ramp in (**b**). (**d**) The B_5/B_6 ratio after cross-calibration and applying the linear ramp in (**b**).

In theory, this systematic ASTER Band 5 calibration error could be solved at the stage of L0 corrections where the data have not yet been rotated from their original detector array vertical column configuration, similar to that developed for satellite Hyperion push-broom sensor [103]. Implementation of an improved ASTER Band 5 detector array calibration file would solve both this east–west bias in image cross-calibration using coincident ROIs (Figure A1d) as well as remove the systematic striping observed in the current ASTER ratio products, such as those highlighted by yellow arrows in Figure 3a. Note that the image cross-calibration method used to generate the Australian ASTER mineral maps [104] was not prone to this error, as the statistics were calculated on a per-scene basis (not image overlaps) and weighted against "global" statistics.

References

1. Lovell, J. *The Great Wall: China against the World, 1000 BC-AD 2000*; Grove: New York, NY, USA, 2006.
2. Han, Z.Q. Change of the Mu Us sandy land and its relationship with the cultivation in the Ming Dynasty. *Soc. Sci. China* **2003**, *23*, 191–204.
3. Carran, D.; Hughes, J.; Leslie, A.; Kennedy, C. A Short History of the Use of Lime as a Building Material Beyond Europe and North America. *Int. J. Archit. Herit.* **2012**, *6*, 117–146. [CrossRef]
4. Jaquin, P.A.; Augarde, C.; Gerrard, C.M. A chronological description of the spatial development of rammed earth techniques. *Int. J. Archit. Herit. Conserv. Anal. Restor.* **2008**, *2*, 377–400. [CrossRef]

5.   Zewen, L.; Wilson, D.; Drege, J.-P.; Delahaye, H.; Wenbao, D.; Gernet, J. *The Great Wall*; McGraw-Hill: New York, NY, USA, 1984; ISBN 10: 0070707456.

6.   Shi, H.; Shao, M. Soil and water loss from the Loess Plateau in China. *J. Arid Environ.* **2000**, *45*, 9–20. [CrossRef]

7.   Li, P.; Vanapalli, S.; Li, T. Review of collapse triggering mechanism of collapsible soils due to wetting. *J. Rock Mech. Geotech. Eng.* **2016**, *8*, 256–274. [CrossRef]

8.   Porter, S.C. Chinese loess record of monsoon climate during the last glacial–interglacial cycle. *Earth-Sci. Rev.* **2001**, *54*, 115–128. [CrossRef]

9.   Küster, Y.; Hetzel, R.; Krbetschek, M.; Tao, M. Holocene loess sedimentation along the Qilian Shan (China): Significance for understanding the processes and timing of loess deposition. *Quat. Sci. Rev.* **2006**, *25*, 114–125. [CrossRef]

10. Derbyshire, E.; Meng, X.; Kemp, R.A. Provenance, transport and characteristics of modern aeolian dust in western Gansu Province, China, and interpretation of the Quaternary loess record. *J. Arid Environ.* **1998**, *39*, 497–516. [CrossRef]

11. Anwar, T.; Kravchinsky, V.A.; Zhang, R.; Koukhar, L.P.; Yang, L.; Yue, L. Holocene climatic evolution at the Chinese Loess Plateau: Testing sensitivity to the global warming-cooling events. *J. Asian Earth Sci.* **2018**, *166*, 223–232. [CrossRef]

12. Lai, Z.-P.; Wintle, A.G. Locating the boundary between the Pleistocene and the Holocene in Chinese loess using luminescence. *Holocene* **2006**, *16*, 893–899. [CrossRef]

13. Mahaney, W.C.; Hancock, R.G.V.; Zhang, L. Stratigraphy and paleosols in the Sale terrace loess section, northwest China. *Catena* **1990**, *17*, 357–367. [CrossRef]

14. Morgan, R.P.C. *Soil Erosion and Conservation*; Longman Group UK Ltd.: Harlow Essex, UK, 2005; 298p.

15. Ikari, M.J.; Kopf, A.J. Cohesive strength of clay-rich sediment. *Geophys. Res. Lett.* **2011**, *38*, L16309. [CrossRef]

16. Trask, P.D.; Close, J.E.H. Effect of clay content on strength of soils. *Coast. Eng. Proc.* **1957**, *1*, 50. [CrossRef]

17. Babu, N.; Poulose, E. Effect of lime on soil properties: A review. *Int. Res. J. Eng. Technol.* **2018**, *5*, 2395-0056.

18. Pheng, L.S. Construction of dwellings and structures in ancient China. *Struct. Surv.* **2001**, *19*, 262–274. [CrossRef]

19. Shao, M.; Li, L.; Wang, S.; Wang, E.; Li, Z. Deterioration mechanisms of building materials of Jiaohe ruins in China. *J. Cult. Herit.* **2013**, *14*, 38–44. [CrossRef]

20. Pye, K. *Aeolian Dust and Dust Deposits*; Academic Press: London, UK, 1987.

21. Smalley, I.J.; Krinsley, D.H. Loess deposits associated with deserts. *Catena* **1978**, *5*, 53–66. [CrossRef]

22. Sun, J. Provenance of loess material and formation of loess deposits on the Chinese Loess Plateau. *Earth Planet. Sci. Lett.* **2002**, *203*, 845–859. [CrossRef]

23. Pullen, A.; Kapp, P.; McCallister, A.T.; Chang, H.; Gehrels, G.E.; Garzione, C.N.; Heermance, R.V.; Ding, L. Qaidam Basin and northern Tibetan Plateau as dust sources for the Chinese Loess Plateau and paleoclimatic implications. *Geology* **2011**, *39*, 1031–1034. [CrossRef]

24. Nie, J.; Stevens, T.; Rittner, M.; Stockli, D.; Garzanti, E.; Limonta, M.; Bird, A.; Andó, S.; Vermeesch, P.; Saylor, J.; et al. Loess Plateau storage of Northeastern Tibetan Plateau-derived Yellow River sediment. *Nat. Commun.* **2015**, *6*, 8511. [CrossRef]

25. Liu, C.Q.; Masuda, A.; Okada, A.; Yabuki, S.; Fan, Z.L. Isotope geochemistry of Quaternary deposits from the arid lands in northern China. *Earth Planet. Sci. Lett.* **1994**, *127*, 25–38. [CrossRef]

26. Sun, Y.; Tada, R.; Chen, J.; Liu, Q.; Toyoda, S.; Tani, A.; Ji, J.; Isozaki, Y. Tracing the provenance of fine-grained dust deposited on the central Chinese Loess Plateau. *Geophys. Res. Lett.* **2007**, *35*, L01804. [CrossRef]

27. Nie, J.; Peng, W. Automated SEM–EDS heavy mineral analysis reveals no provenance shift between glacial loess and interglacial paleosol on the Chinese Loess Plateau. *Aeolian Res.* **2014**, *13*, 71–75. [CrossRef]

28. Hu, F.; Yang, X. Geochemical and geomorphological evidence for the provenance of aeolian deposits in the Badain Jaran Desert, northwestern China. *Quat. Sci. Rev.* **2016**, *131*, 179–192. [CrossRef]

29. Wang, X.; Cai, D.; Sun, J.; Lu, H.; Liu, W.; Qiang, M.; Cheng, H.; Che, H.; Hua, T.; Zhang, C. Contributions of modern Gobi Desert to the Badain Jaran Desert and the Chinese Loess Plateau. *Sci. Rep.* **2019**, *9*, 985. [CrossRef]

30. Stevens, T.; Carter, A.; Watson, T.P.; Vermeesch, P.; Andò, S.; Bird, A.F.; Lu, H.; Garzanti, E.; Cottam, M.A.; Sevastjanova, I. Genetic linkage between the Yellow River, the Mu Us desert and the Chinese Loess Plateau. *Quat. Sci. Rev.* **2013**, *78*, 355–368. [CrossRef]

31. Bird, A.; Stevens, T.; Rittner, M.; Vermeesch, P.; Carter, A.; Andò, S.; Garzanti, E.; Lu, H.; Nie, J.; Zeng, L.; et al. Quaternary dust source variation across the Chinese Loess Plateau. *Palaeogeogr. Palaeoclimatol. Palaeoecol.* **2015**, *435*, 254–264. [CrossRef]

32. Wang, F.; Sun, D.; Chen, F.; Bloemendal, J.; Guo, F.; Li, Z.; Zhang, Y.; Li, B.; Wang, X. Formation and evolution of the Badain Jaran Desert, North China, as revealed by a drill core from the desert centre and by geological survey. *Palaeogeogr. Palaeoclimatol. Palaeoecol.* **2015**, *426*, 139–158. [CrossRef]

33. Shi, Z.; Liu, X. Distinguishing the provenance of fine-grained eolian dust over the Chinese Loess Plateau from a modelling perspective. *Tellus* **2011**, *63*, 959–970. [CrossRef]

34. Huang, C.Q.; Zhao, W.; Li, F.Y.; Tan, W.F.; Wang, M.K. Mineralogical and pedogenetic evidence for palaeoenvironmental variations during the Holocene on the Loess Plateau, China. *Catena* **2012**, *96*, 49–56. [CrossRef]

35. Ding, Z.L.; Sun, J.M.; Yang, S.L.; Liu, T.S. Geochemistry of the Pliocene red clay formation in the Chinese Loess Plateau and implications for its origin, source provenance and paleoclimate change. *Geochim. Cosmochim. Acta* **2001**, *65*, 901–913. [CrossRef]

36. Zhao, W.; Liu, L.; Chen, J.; Ji, J. Geochemical characterization of major elements in desert sediments and implications for the Chinese loess source. *Sci. China Earth Sci.* **2019**, *62*, 1428–1440. [CrossRef]

37. Licht, A.; Pullen, A.; Kapp, P.; Abell, J.; Giesler, N. Eolian cannibalism: Reworked loess and fluvial sediment as the main sources of the Chinese Loess Plateau. *Geol. Soc. Am. Bull.* **2016**, *128*, 944–956. [CrossRef]

38. Sun, J.; Ding, Z.; Xi, X.; Sun, M.; Windley, B.F. Detrital zircon evidence for the ternary sources of the Chinese Loess Plateau. *J. Asian Earth Sci.* **2018**, *155*, 21–34. [CrossRef]

39. Lü, T.; Sun, J. Luminescence sensitivities of quartz grains from eolian deposits in northern China and their implications for provenance. *Quat. Res.* **2011**, *76*, 181–189. [CrossRef]

40. Sun, Y.; Chen, H.; Tada, R.; Weiss, D.; Lin, M.; Toyoda, S.; Yan, Y.; Isozaki, Y. ESR signal intensity and crystallinity of quartz from Gobi and sandy deserts in East Asia and implication for tracing Asian dust provenance. *Geochem. Geophys. Geosyst.* **2013**, *14*, 2615–2627. [CrossRef]

41. Ma, L.; Sun, Y.; Tada, R.; Yan, Y.; Chen, H.; Lin, M.; Nagashima, K. Provenance fluctuations of aeolian deposits on the Chinese Loess Plateau since the Miocene. *Aeolian Res.* **2015**, *18*, 1–9. [CrossRef]

42. Yan, Y.; Ma, L.; Sun, Y. Tectonic and climatic controls on provenance changes of fine-grained dust on the Chinese Loess Plateau since the late Oligocene. *Geochim. Cosmochim. Acta* **2017**, *200*, 110–122. [CrossRef]

43. Chen, Z.; Li, G. Evolving sources of eolian detritus on the Chinese Loess Plateau since early Miocene: Tectonic and climatic controls. *Earth Planet. Sci. Lett.* **2013**, *371–372*, 220–225. [CrossRef]

44. Zhang, W.; Chen, J.; Li, G. Shifting material source of Chinese loess since ~2.7 Ma reflected by Sr isotopic composition. *Sci. Rep.* **2015**, *5*, 10235. [CrossRef]

45. Maher, B.A.; Mutch, T.J.; Cunningham, D. Magnetic and geochemical characteristics of Gobi Desert surface sediments: Implications for provenance of the Chinese Loess Plateau. *Geology* **2009**, *37*, 279–282. [CrossRef]

46. Zhao, L.; Ji, J.; Chen, J.; Liu, L.; Chen, Y.; Balsam, W. Variations of illite/chlorite ratio in Chinese loess sections during the last glacial and interglacial cycle: Implications for monsoon reconstruction. *Geophys. Res. Lett.* **2005**, *32*, L20718. [CrossRef]

47. Jeong, G.Y.; Hillier, S.; Kemp, R.A. Quantitative bulk and single-particle mineralogy of a thick Chinese loess–paleosol section: Implications for loess provenance and weathering. *Quat. Sci. Rev.* **2008**, *27*, 1271–1287. [CrossRef]

48. Jeong, G.Y.; Hillier, S.; Kemp, R.A. Changes in mineralogy of loess–paleosol sections across the Chinese Loess Plateau. *Quat. Res.* **2011**, *75*, 245–255. [CrossRef]

49. Formenti, P.; Schutz, L.; Balkanski, Y.; Desboeufs, K.; Elbert, M.; Kandler, K.; Petzold, A.; Scheuvens, D.; Weinbruch, S.; Zhang, D. Recent progress in understanding physical and chemical properties of African and Asian mineral dust. *Atmos. Chem. Phys.* **2011**, *11*, 8231–8256. [CrossRef]

50. Schaetzl, R.J.; Bettis, E.A., III; Crouvi, O.; Fitzsimmons, K.E.; Grimley, D.A.; Hambacf, U.; Lehmkuhl, F.; Marković, S.B.; Mason, J.A.; Owczarek, P.; et al. Approaches and challenges to the study of loess—Introduction to the LoessFest Special Issue. *Quat. Res.* **2018**, *89*, 563–618. [CrossRef]

51. McCuaig, C.T.; Beresford, S.; Hronsky, J. Translating the mineral systems approach into an effective exploration targeting system. *Ore Geol. Rev.* **2010**, *38*, 128–138. [CrossRef]

52. Cudahy, T.J. Mineral Mapping for Exploration: An Australian Journey of Evolving Spectral Sensing Technologies and Industry Collaboration. *Geosciences* **2016**, *6*, 52. [CrossRef]

53. Wyborn, L.A.; Heinrich, C.A.; Jacques, A.L. Australian Proterozoic mineral systems: Essential ingredients and mappable criteria. In Proceedings of the AusIMM Annual Conference Proceedings, Darwin, Australia, 5–9 August 1994; pp. 109–115.

54. Smalley, I.; Marshall, J.; Fitzsimmons, K.; Whalley, W.B.; Ngambi, S. Desert loess: A selection of relevant topics. *Geologos* **2019**, *25*, 91–102. [CrossRef]

55. Yamaguchi, Y.; Kahle, A.B.; Tsu, H.; Kawakami, T.; Pniel, M. Overview of Advanced Space-borne Thermal Emission and Reflection Radio. meter (ASTER). *IEEE Trans. Geosci. Remote Sens.* **1998**, *36*, 1062–1071. [CrossRef]

56. Abrams, M.; Tsu, H.; Hulley, G.; Iwao, K.; Pieri, D.; Cudahy, T.J.; Kargel, J. The Advanced Spaceborne Thermal Emission and Reflection Radiometer (ASTER) after fifteen years: Review of global products. *Int. J. Appl. Earth Obs. Geoinf.* **2015**, *38*, 292–301. [CrossRef]

57. Clark, R.N. Chapter 1: Spectroscopy of Rocks and Minerals, and Principles of Spectroscopy. In *Manual of Remote Sensing*; Rencz, A.N., Ed.; John Wiley and Sons: New York, NY, USA, 1999; Volume 3, pp. 3–58.

58. Cudahy, T.J. *Australian ASTER Geoscience Product Notes*; CSIRO Report, EP-30-07-12-44; Commonwealth Scientific and Industrial Research Organisation (CSIRO): Canberra, Australia, 2012. Available online: https://data.csiro.au/dap/landingpage?pid=csiro%3A6182 (accessed on 28 September 2019).

59. Rowan, L.C.; Mars, J.C. Lithologic mapping in the Mountain Pass, California area using Advanced Spaceborne Thermal Emission and Reflection Radiometer (ASTER) data. *Remote Sens. Environ.* **2003**, *84*, 350–366. [CrossRef]

60. Hewson, R.D.; Cudahy, T.J.; Mizukiko, S.; Mauger, A.L. Seamless geological map generation using ASTER in the Broken Hill Curnamona Province of Australia. *Remote Sens. Environ.* **2005**, *99*, 159–172. [CrossRef]

61. Agisoft Metashape Software. Available online: https://www.agisoft.com (accessed on 7 December 2019).

62. Raven, M.D.; Self, P.G. Xplot user manual, manipulation of powder X-ray diffraction data. In *CSIRO Division of Soils Technical Memorandum*; 30/1988; CSIRO: Canberra, Australia, 1988; 25p.

63. ICDD Mineral Standards Library. Available online: http://www.icdd.com (accessed on 30 October 2019).

64. Analytical Spectral Devices (ASD) FieldSpec Pro Spectrometer. Available online: https://www.malvernpanalytical.com/en/products/product-range/asd-range (accessed on 30 October 2019).

65. Spectralon Panel. Available online: http://www.labsphere.com (accessed on 30 October 2019).

66. Korb, A.R.; Dybwad, P.; Wadsworth, W.; Salisbury, J.W. Portable Fourier transform infrared spectroradiometer for field measurements of radiance and emissivity. *Appl. Opt.* **1996**, *35*, 1679–1692. [CrossRef] [PubMed]

67. Earthdata Web Portal. Available online: https://search.earthdata.nasa.gov/search (accessed on 30 October 2019).

68. Ninomiya, Y.; Fu, B.; Cudahy, T.J. Detecting lithology with Advanced Spaceborne Thermal Emission and Reflectance Radiometer (ASTER) multispectral thermal infrared "radiance-at-sensor" data. *Remote Sens. Environ.* **2005**, *99*, 127–139. [CrossRef]

69. Shi, P.; Fu, B.; Cudahy, T.J.; Guo, Q.; Xu, H.; Chen, X.; Ma, Y.; Xue, G. Desertification monitoring using the ASTER global emissivity dataset. In Proceedings of the 2017 IEEE International Geoscience and Remote Sensing Symposium (IGARSS), Fort Worth, TX, USA, 23–28 July 2017; pp. 4501–4504. [CrossRef]

70. Cudahy, T.J.; Jones, M.; Lisitsin, V.; Caccetta, M.; Collings, S.; Bateman, R. *3D Mineral Mapping of Queensland-Version 2 ASTER and Related Geoscience Products*; CSIRO Mineral Resources; EP1767; 18p, Available online: https://publications.csiro.au/rpr/download?pid=csiro:EP17697&dsid=DS3 (accessed on 6 October 2019).

71. Ninomiya, Y. Quantitative estimation of SiO_2 content in igneous rocks using thermal infrared spectra with a neural network approach. *IEEE Trans. Geosci. Remote Sens.* **1995**, *33*, 684–691. [CrossRef]

72. Hunt, G.R.; Vincent, R.K. The behaviour of spectral features in the infrared emission from particulate surfaces of various grain sizes. *J. Geophys. Res.* **1968**, *73*, 6039–6046. [CrossRef]

73. Salisbury, J.W.; Walter, L.S. Thermal infrared (2.5–13.5 m) spectroscopic remote sensing of igneous rock types on particulate planetary surfaces. *J. Geophys. Res.* **1989**, *94*, 9192–9202. [CrossRef]

74. Clark, R.N. Spectral properties of mixtures of montmorillonite and dark carbon grains: Implications for remote sensing minerals containing chemically and physically absorbed water. *J. Geophys. Res.* **1983**, *88*, 10635–10644. [CrossRef]

75. Haest, M.; Cudahy, T.; Laukamp, C.; Gregory, S. Quantitative mineralogy from visible to shortwave infrared spectroscopic data—I. Validation of mineral abundance and composition products of the Rocklea Dome channel iron deposit in Western Australia. *Econ. Geol.* **2012**, *107*, 209–228. [CrossRef]

76. Haest, M.; Cudahy, T.; Laukamp, C.; Gregory, S. Quantitative mineralogy from visible to shortwave infrared spectroscopic data—II. 3D mineralogical characterisation of the Rocklea Dome channel iron deposit, Western Australia. *Econ. Geol.* **2012**, *107*, 229–249. [CrossRef]

77. Rodger, A.R.; Cudahy, T.J. Vegetation Corrected Continuum Depths at 2.20 μm: An Approach for Hyperspectral Sensors. *Remote Sens. Environ.* **2009**, *113*, 2243–2257. [CrossRef]

78. Haest, M.; Cudahy, T.J.; Rodger, A.; Laukamp, C.; Martens, C.; Caccetta, M. Unmixing vegetation from airborne visible-near to shortwave infrared spectroscopy-based mineral maps over the Rocklea Dome (Western Australia), with a focus on iron rich palaeochannels. *Remote Sens. Environ.* **2013**, *129*, 17–31. [CrossRef]

79. Du, Y.; Chen, W.; Cui, K.; Gong, S.; Pu, T.; Fu, X. A Model characterizing deterioration at earthen sites of the Ming Great Wall in Qinghai Province, China. *Soil Mech. Found. Eng.* **2017**, *53*, 426–434. [CrossRef]

80. Allen, J.R. *Physical Processes of Sedimentation*; Earth Science Series 1; Sutton, J., Watson, J.W., Eds.; Unwin University Books: London, UK, 1970.

81. Hook, S.J.; Dmochowski, J.E.; Howard, K.A.; Rowan, L.C.; Karlstrom, K.E.; Stock, J.M. Mapping variations in weight percent silica measured from multispectral thermal infrared imagery—Examples from the Hiller Mountains, Nevada USA and Tres Virgenes-La Reforma, Baja California Sur, Mexico. *Remote Sens. Environ.* **2005**, *95*, 273–289. [CrossRef]

82. Pan, P.; Pang, H.; Zhang, D.; Guan, Q.; Wang, L.; Li, F.; Guan, W.; Cai, A.; Sun, X. Sediment grain-size characteristics and its source implication in the Ningxia–Inner Mongolia sections on the upper reaches of the Yellow River. *Geomorphology* **2015**, *246*, 255–262. [CrossRef]

83. Pan, B.; Pang, H.; Gao, H.; Garzanti, E.; Zou, Y.; Liu, X.; Li, F.; Jia, Y. Heavy-mineral analysis and provenance of Yellow River sediments around the China Loess Plateau. *J. Asian Earth Sci.* **2016**, *127*, 1–11. [CrossRef]

84. Jia, X.; Li, Y.; Wang, H. Bed sediment particle size characteristics and its sources implication in the desert reach of the Yellow River. *Environ. Earth Sci.* **2016**, *75*, 950. [CrossRef]

85. Kalm, V.E.; Rutter, N.W.; Rokosh, C.D. Clay minerals and their paleoenvironmental interpretation in the Baoji loess section, Southern Loess Plateau, China. *Catena* **1996**, *27*, 49–61. [CrossRef]

86. Bell, F.G. Lime stabilization of clay minerals and soils. *Eng. Geol.* **1996**, *42*, 223–237. [CrossRef]

87. Koporulin, V.I. Formation of laumontite in sedimentary rocks: A case study of sedimentary sequences in Russia. *Lithol. Miner. Resour.* **2013**, *48*, 122–137. [CrossRef]

88. Eden, D.N.; Qizhong, W.; Hunt, J.L.; Whitton, J.S. Mineralogical and geochemical trends across the Loess Plateau, North China. *Catena* **1994**, *21*, 73–90. [CrossRef]

89. Begét, J.E.; Keskinen, M.; Severin, K. Mineral particles from Asia found in volcanic loess on the island of Hawaii. *Sediment. Geol.* **1993**, *84*, 189–197. [CrossRef]

90. Yao, Z.; Shi, X.; Qiao, S.; Liu, Q.; Kandasamy, S.; Liu, J.; Liu, Y.; Liu, J.; Fang, X.; Gao, J.; et al. Persistent effects of the Yellow River on the Chinese marginal seas began at least ~880 ka ago. *Sci. Rep.* **2017**, *7*, 2827. [CrossRef]

91. Harder, H. Illite mineral synthesis at surface temperatures. *Chem. Geol.* **1974**, *14*, 241–253. [CrossRef]

92. Laurent, B.; Marticorena, B.; Bergametti, G.; Mei, F. Modeling mineral dust emissions from Chinese and Mongolian deserts. *Glob. Planet. Chang.* **2006**, *52*, 121–141. [CrossRef]

93. Wang, Q.; Zhuang, G.; Huang, K.; Liu, T.; Lin, Y.; Deng, C.; Fu, Q.; Fub, J.S.; Chen, J.; Zhang, W.; et al. Evolution of particulate sulfate and nitrate along the Asian dust pathway: Secondary transformation and primary pollutants via long-range transport. *Atmos. Res.* **2016**, *169*, 86–95. [CrossRef]

94. Pu, T.; Chen, W.; Du, Y.; Li, W.; Su, N. Snowfall-related deterioration behavior of the Ming Great Wall in the eastern Qinghai-Tibet. *Nat. Hazards* **2016**, *84*, 1539–1550. [CrossRef]

95. Green, R.; Mahowald, N.; Thompson, D.; Clark, R.; Ehlmann, B.; Ginoux, P.; Kalashnikova, O.; Miller, R.; Okin, G.; Painter, P.; et al. The Earth Surface Mineral Dust Source Investigation Planned for the International Space Station. *Geophys. Res. Abstr.* **2019**, *21*, EGU2019-10660.

96. HISUI. Available online: https://hyspiri.jpl.nasa.gov/downloads/2018_Workshop/day3/14_1808_HISUI_Matsunaga_04b.pdf (accessed on 4 July 2019).

97. ECOSTRESS. Available online: https://ecostress.jpl.nasa.gov/news (accessed on 4 July 2019).

98. Tollefson, J.; Gilbert, N. Rio report card. *Nature* **2012**, *486*, 20–23. [CrossRef]
99. Sommer, S.; Zucca, C.; Grainger, A.; Cherlet, M.; Zougmore, R.; Sokona, Y.; Hill, J.; Della Peruta, R.; Roehrig, J.; Wang, G. Application of indicator systems for monitoring and assessment of desertification from National to Global scale. *Land Degrad. Dev.* **2012**, *22*, 184–197. [CrossRef]
100. Vogt, J.V.; Safriel, U.; von Maltitz, G.; Sokona, Y.; Zougmore, R.; Bastin, G.; Hill, J. Monitoring and assessment of land degradation and desertification: Towards new conceptual and integrated approaches. *Land Degrad. Dev.* **2011**, *22*, 150–165. [CrossRef]
101. Cudahy, T.; Caccetta, M.; Thomas, M.; Hewson, R.; Abrams, M.; Kato, M.; Kashimura, O.; Ninomiya, Y.; Yamaguchi, Y.; Collings, S.; et al. Satellite-derived mineral mapping and monitoring of weathering, deposition and erosion. *Sci. Rep.* **2016**, *6*, 23702. [CrossRef] [PubMed]
102. Iwasaki, A.; Tonooka, H. Validation of a crosstalk correction algorithm for ASTER/SWIR. *IEEE Trans. Geosci. Remote Sens.* **2005**, *43*, 2747–2751. [CrossRef]
103. Cudahy, T.J.; Rodger, A.R.; Barry, P.S.; Mason, P.; Quigley, M.A.; Folkman, M.A.; Pearlman, J. Assessment of the stability of the Hyperion SWIR module for hyperspectral mineral mapping using multi-date images from Mount Fitton, Australia. In Proceedings of the IEEE 2002 International Conference on Geoscience and Remote Sensing, Toronto, ON, Canada, 24–28 June 2002.
104. Caccetta, M.; Collings, S.; Cudahy, T.J. A calibration method for continental scale mapping using ASTER imagery. *Remote Sens. Environ.* **2012**, *139*, 306–317. [CrossRef]

Article

Radiometric Degradation Curves for the ASTER VNIR Processing Using Vicarious and Lunar Calibrations

Satoshi Tsuchida [1,*], Hirokazu Yamamoto [1], Toru Kouyama [2], Kenta Obata [1,3],
Fumihiro Sakuma [4], Tetsushi Tachikawa [4], Akihide Kamei [5], Kohei Arai [6],
Jeffrey S. Czapla-Myers [7], Stuart F. Biggar [7] and Kurtis J. Thome [8]

[1] Geological Survey of Japan, National Institute of Advanced Industrial Science and Technology, Central 7,
    Higashi 1-1-1, Tsukuba 305-8567, Japan; hirokazu.yamamoto@aist.go.jp (H.Y.); obata@ist.aichi-pu.ac.jp (K.O.)
[2] Artificial Intelligence Research Center, National Institute of Advanced Industrial Science and Technology,
    2-4-7 Aomi, Koto-ku, Tokyo 135-0064, Japan; t.kouyama@aist.go.jp
[3] Department of Information Science and Technology, Aichi Prefectural University, 1522-3 Ibaragabasama,
    Nagaute, Aichi 480-1198, Japan
[4] Research and Development Division, Japan Space Systems, Tokyo, 3-5-8 Shibakoen, Minato-ku,
    Tokyo 105-0011, Japan; fsakuma@jcom.home.ne.jp (F.S.); Tachikawa-tetsushi@jspacesystems.or.jp (T.T.)
[5] National Institute for Environmental Studies, Tsukuba 305-8506, Japan; kamei.akihide@nies.go.jp
[6] Saga University, Saga City 840-8502, Japan; arai@is.saga-u.ac.jp
[7] Remote Sensing Group, College of Optical Sciences, University of Arizona, Tucson, Arizona,
    1630 E University Blvd, Tucson, AZ 85721, USA; jscm@optics.arizona.edu (J.S.C.-M.);
    biggar@optics.arizona.edu (S.F.B.)
[8] NASA Goddard Space Flight Center, Greenbelt, MD 20771, USA; kurtis.thome@nasa.gov
*   Correspondence: s.tsuchida@aist.go.jp; Tel.: +81-29-861-3947

Received: 16 December 2019; Accepted: 26 January 2020; Published: 29 January 2020

Abstract: The Advanced Spaceborne Thermal Emission and Reflection Radiometer (ASTER) onboard
Terra platform, which was launched in 1999, has three separate subsystems: a visible and near-infrared
(VNIR) radiometer, a shortwave-infrared radiometer, and a thermal-infrared radiometer. The ASTER
VNIR bands have been radiometrically corrected for approximately 14 years by the sensor degradation
curves estimated from the onboard calibrator according to the original calibration plan. However,
this calibration by the onboard calibrator encountered a problem; specifically, it is inconsistent with
the results of vicarious calibration and cross calibration. Therefore, the ASTER VNIR processing
was applied by the radiometric degradation curves calculated from the results of three calibration
approaches, i.e., the onboard calibrator, the vicarious calibration, and the cross calibration since
February 2014. Even though the current degradation curves were revised, the inter-band and lunar
calibrations show some inconsistencies owing to the different traceability in the bands by different
calibration approaches. In this study, the current degradation curves and their problems are explained,
and the new curves that are derived from the vicarious calibration with lunar calibration are discussed.
The new degradation curves that have the same traceability in the bands will be used for future
ASTER VNIR processing.

Keywords: ASTER; vicarious calibration; lunar calibration; radiometric calibration; VNIR

---

1. Introduction

The Advanced Spaceborne Thermal Emission and Reflection Radiometer (ASTER) onboard Terra
platform, which was launched in 1999, has three separate subsystems: a visible and near-infrared (VNIR)
radiometer, a shortwave-infrared (SWIR) radiometer, and a thermal-infrared (TIR) radiometer [1].

The detector temperature of the SWIR radiometer started to increase owing to the degradation of the detector cooling system, which resulted in the progressive deterioration of product quality in 2007, and low-quality SWIR images have been acquired since May 2008 [2]. However, the VNIR and TIR radiometers have been providing good images for approximately 20 years, and several calibration approaches have been applied to the radiometers [3–17].

The preflight calibration utilized standard large integrating spheres whose radiance levels were traceable to primary standard fixed-point blackbodies [3]. The VNIR and SWIR radiometers have onboard calibrators that consist of two halogen lamps with photodiode monitors for the onboard calibration [4]. For VNIR, the vicarious calibration using the reflectance-based approach has been conducted by three separate groups [5]. In April 2003 and August 2017, ASTER observed the Moon (and the deep space) to conduct a radiometric calibration (i.e., lunar calibration) that can measure the temporal variation in the sensor sensitivity of the VNIR bands sufficiently accurately (better than 0.1%) [6]. Numerous studies have reported the cross calibration of ASTER and Moderate Resolution Imaging Spectroradiometer (MODIS) VNIR bands [7–11], but the trends in the cross-calibration results are not necessarily consistent [11]. Inter-band radiometric calibration, which can be used to evaluate inter-band consistency, was applied for the ASTER VNIR bands; then, the radiometric degradation curves of ASTER Bands 2 and 3N with a reference to Band 1, which is well calibrated compared with other bands, were calculated [12].

ASTER data were radiometrically corrected by the L1 processing using the radiometric calibration coefficients (RCC) in the radiometric database (DB) [13], which has the information of the degradation curves. For the VNIR radiometer, its degradation curves in the radiometric DB ver. 1 to 3 have been estimated from the onboard calibrator [14] according the original calibration plan [15] for approximately 14 years since Terra's launch. However, the degradation curves produced by onboard calibration are inconsistent with the results obtained by vicarious and cross calibrations [16]. Therefore, the degradation curves calculated from the results of three calibration approaches (i.e., onboard, vicarious, and cross calibrations) have been applied to the latest version of the radiometric DB (ver. 4) on the L1 processing since February 2014 [17]. Even though the degradation curves have been revised in the radiometric DB ver. 4, the inter-band and lunar calibrations have reported a few inconsistencies [6,12], which is attributed to the different traceability of each band with a different calibration approach.

In this study, the current degradation curves and their problems are explained, and the new curves, which are derived from vicarious and lunar calibrations, are discussed. The new degradation curves that have the same traceability in the bands have been selected to be used for the next radiometric DB ver. 5 in the ASTER VNIR processing.

2. Vicarious and Lunar Calibrations for ASTER VNIR

2.1. ASTER VNIR

The ASTER VNIR radiometer is the multiband sensor with a 15 m spatial resolution and a 60 km swath width; the instrument has three bands (Bands 1, 2, and 3N) with a nadir view and one band (Band 3B) with a backward view. Bands 3N and 3B were used for the topographic interpretation and the digital elevation model (DEM) mapping.

The center wavelength of Bands 1, 2, and 3 are 0.56, 0.66, and 0.81 micrometers, respectively [3]. This VNIR radiometer has three gain modes (high, normal, and low gains). The values of normal and low gains in all bands are 1.0, 0.75, respectively. The value of high gain in Band 1 is 2.5, and the value of high gain in Bands 2 and 3 is 2.0 [1].

2.2. Vicarious Calibration Using the Reflectance-Based Method

Vicarious calibration using the reflectance-based method [18,19] was performed for ASTER VNIR by three separate groups [i.e., Saga University, the University of Arizona, and the National Institute of Advanced Industrial Science and Technology (AIST)]. The method is used by all three groups with differences in the equipment used in the data collection, sampling strategies, and the radiative transfer code results. Although there were differences in the method used by the groups, the results produced by the three groups were in good agreement between each other for four years after the launch [5].

In this study, the RCCs for all ASTER VNIR bands (Bands 1, 2, 3N, and 3B) were calculated from the vicarious calibration by the AIST method [5,12,20–22]. The AIST group performed a field campaign for the vicarious calibration for ASTER at six sites in the United States and two sites in Australia. We used the data from all eight sites (the United States six and Australian two sites) to obtain the degradation curves for the current radiometric DB ver. 4. However, to have a more clear traceability with almost the same quality, the new degradation curves for the next ver. 5 of the radiometric DB were derived from the vicarious calibration at only three sites of Ivanpah Playa (35.57N, 115.40W) in California, USA, Railroad Valley Playa (38.50N, 115.69W), and Alkali Lake (37.85N, 117.41W) in Nevada, USA.

At the three sites, which are horizontal planes in clay-dominated dry lakes in semi-arid areas, the field campaigns were carried out with a rectangular target of 90 × 80 m in size and approximately 900 measurements of the surface reflectance factor of the target for one band for each field campaign. However, except these three sites, the field campaigns were performed for the current radiometric DB ver.4 without the abovementioned conditions of ground surface and/or measurements, owing to the convenience of each site.

The solar model for this work is based on the World Radiation Center model, because this model was standard for the Terra sensors and was selected by the ASTER science team [5]. After the Terra lunch, many newer solar models have been proposed, and one of the recent ASTER products, ASTER L1T generated by the United States Geological Survey [23], used a new solar model; however, the effect of this switch of the model is estimated at <0.3% for all VNIR bands [24].

Figure 1 shows the RCCs vs. days since launch (DSL) in ASTER VNIR bands derived from this vicarious calibration from 2000 to 2017.

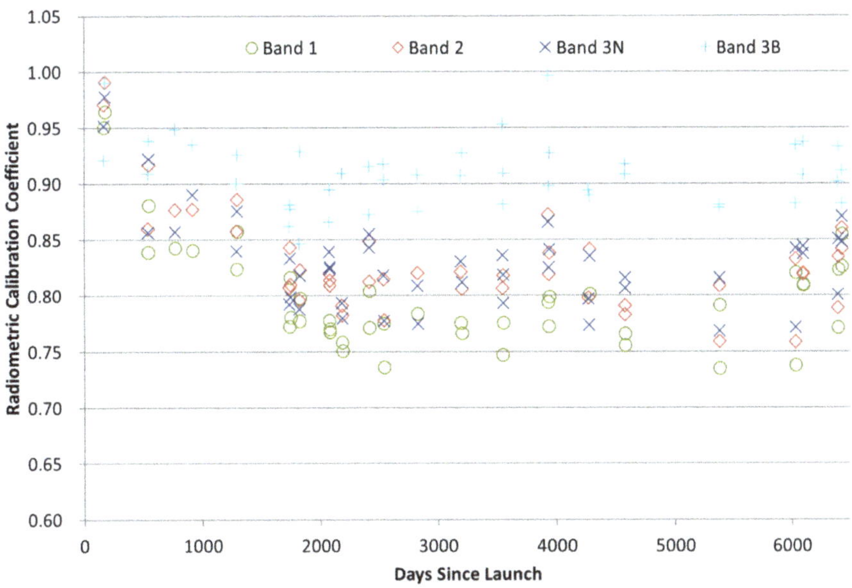

Figure 1. Radiometric calibration coefficients (RCCs) derived from the vicarious calibration of the reflectance-based method carried out from 2000 to 2017.

2.3. Lunar Calibration with the Spectral Profiler Model

Lunar calibration is one of the radiometric calibrations using the Moon that is conducted by comparing the observed Moon brightness with the expected brightness calculated from the lunar surface reflectance model [25–27]. Although the absolute accuracy of the lunar calibration method is considered to be insufficient (5–10%), the lunar calibration, temporal variation of relative sensor sensitivity can be accurately measured on the order of 0.1% [6].

In April 2003 (DSL = 1213) and August 2017 (DSL = 6440), ASTER observed the Moon and the deep space to perform lunar calibration with the Robotic Lunar Observatory [25] and spectral profiler (SP) [26,27] models, and the calibration results of the two models were consistent with each other within the error range [6].

In this study, we use the relative sensor sensitivity degradation (RD) from 2003 to 2017, and the RD value is the degradation ratio of RCC (DSL = 6440)/RCC (DSL = 1213). The RD_L is the RD value derived from the lunar calibration with the SP model. The values of RD_L in Bands 1, 2, 3N, and 3B are shown in Table 1 [6].

Table 1. Relative sensor sensitivity degradation (RD_L), which is the degradation ratio of RCCs for VNIR bands from 2003 (DSL = 1213) to 2017 (DSL = 1213) by lunar calibration using the SP model [6].

Band	Relative Sensor Sensitivity Degradation (RD_L) Degradation Ratio: RCC (6440)/RCC (1213)
1	0.969
2	0.948
3N	0.942
3B	0.968

3. Radiometric Degradation Curves in the Current Radiometric DB and Its Problems

ASTER data were radiometrically corrected by the L1 processing using the RCC in the Radiometric DB [13], which has the information of the degradation curves. For the VNIR radiometer, the degradation curves in the old radiometric DBs ver. 1 to 3 have been estimated from the onboard calibrator [14] according the original calibration plan [15] for approximately 14 years since the launch. However, the degradation curves produced by onboard calibration are inconsistent with the results from vicarious and cross calibrations [16], especially for Bands 1 and 2.

The radiometric degradation curves in the current radiometric DB ver. 4 have been used for L1 processing since February 2014, and the best curve obtained using a different estimation method of each band was selected as of 2014 by the ASTER science team [17]. However, a problem has occurred using a different traceability of each band as described below.

3.1. Radiometric Degradation Curves

The dotted lines in Figure 2 indicate the current degradation curves in the radiometric DB ver. 4 for ASTER VNIR bands estimated from the results of three calibration approaches (i.e., onboard, vicarious, and cross calibrations) from 2000 to 2012. The equations and their parameters are shown in Tables 2 and 3.

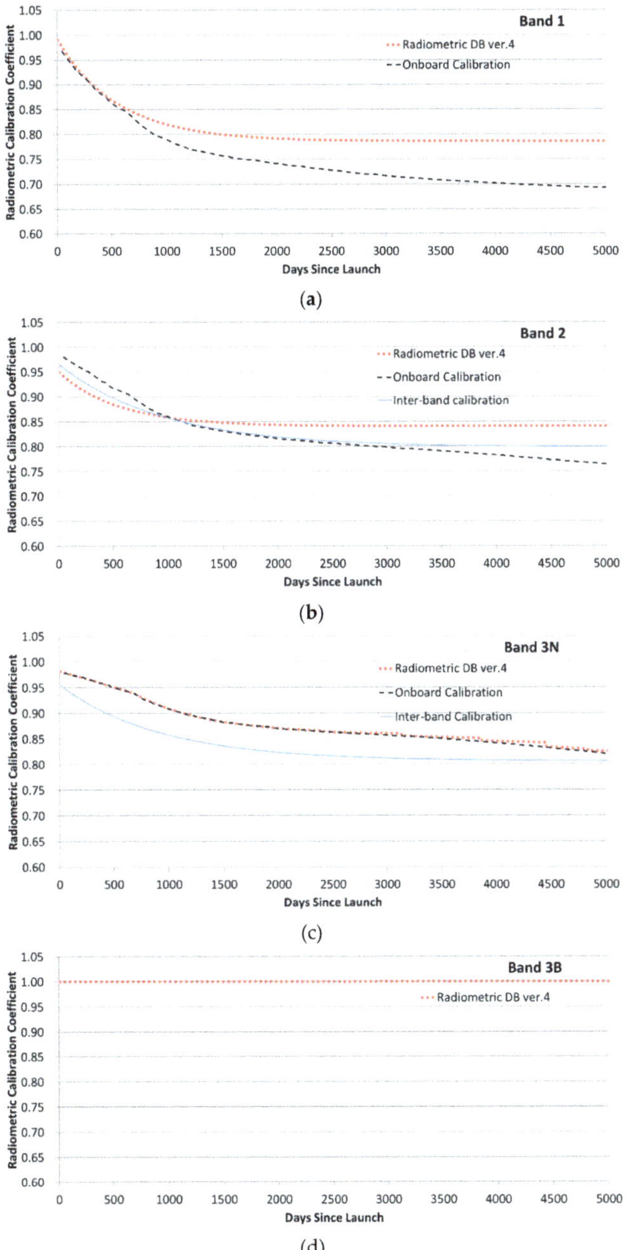

Figure 2. Degradation curves of ASTER VNIR bands in the current radiometric DB ver. 4. Bands (**a**) 1, (**b**) 2, (**c**) 3N, and (**d**) 3B. Dotted lines correspond to the current degradation curve in the current radiometric DB estimated from the results of three calibration approaches (i.e., onboard, vicarious, and cross calibrations) from 2000 to 2012. Dashed lines indicate the degradation curves derived from the onboard calibration, which is almost the same as that in the radiometric DB ver. 3. The solid lines indicate the degradation curves estimated from the inter-band calibration for ASTER VNIR Bands 2 and 3N.

Table 2. Equations and parameters of the current radiometric degradation curves in the radiometric DB ver. 4 for ASTER VNIR bands.

Band	Duration $(d$: DSL)	Equation $(R(d)$: RCC)	Coefficients		
			a_0	a_1	a_2
1	whole duration	$R(d) = a_0 \, (1.0 - a_1) \exp(-a_2 \, d) + a_0 \, a_1$	0.990	0.794	0.00181
2	whole duration	$R(d) = a_0 \, (1.0 - a_1) \exp(-a_2 \, d) + a_0 \, a_1$	0.949	0.886	0.00181
3N	$d < 673$	$R(d) = a_0 + a_1 \, d + a_2 \, d^2$	0.9817	-5.726×10^{-5}	-9.360×10^{-9}
	$673 \leq d < 4825$	See Table 3			
	$4825 \leq d$	$R(d) = a_0$	0.8259	-	-
3B	whole duration	$R(d) = a_0$	1.000	-	-

Table 3. Equations and parameters of the current radiometric degradation curves in the radiometric DB ver. 4 for ASTER Band 3N.

Band	Duration $(d$: DSL)	Equation $(R(d)$: RCC)	Coefficients		
			a_0	a_1	a_2
3N	$673 \leq d < 2394$	$R(d) = a_1 \, \exp(-a_2 \, d) + a_0$	0.8599	0.2163	0.0014974
	$2394 \leq d < 3123$		0.8590	0.5750	0.0019668
	$3123 \leq d < 3857$		0.8428	0.1054	0.0006679
	$3857 \leq d < 4450$		0.8310	0.1176	0.0005303
	$4450 \leq d < 4825$		0.7086	0.2051	0.0001096

The degradation curve of Band 1 in the radiometric DB ver. 4 [dotted line in Figure 2a] was estimated from the results of three calibration approaches, but the vicarious calibration influences the degradation curve the most among the approaches. The degradation curve is expressed by the following exponential equation in the degradation model for the contamination and/or corrosion [28].

$$R(d) = a_0 c = a_0 \times T / T_0 = a_0 (1.0 - a_1) \exp(-a_2 \, d) + a_0 \, a_1 \tag{1}$$

where $R(d)$ represents the radiometric calibration coefficient (RCC) of the DSL d, c is the scaled calibration coefficient, T represents the transmittance of DSL d in the radiometer optics (lenses, mirrors and etc.), and T_0 represents the transmittance T when $d = 0$. The degradation parameters, a_0, a_1, and a_2 represent the RCC at launch, the minimum (saturated) transmittance of the contamination/corrosion layer of the sensor lenses, and the degradation rate, respectively. Band 1 has $a_0 = 0.990$, $a_1 = 0.794$, and $a_2 = 0.00181$ coefficients (Table 2).

The a_0 coefficient, which equals the $R(0)$, was extrapolated by the RCC trend (i.e., the a_0 coefficient was estimated from best-fit with the distribution of RCCs) in the period of $0 < d \leq 673$ by the onboard calibration [4]. That was because the RCC trend in this period by the onboard calibration shows smooth curve [dashed line in Figure 2a] and was consistent with the trend by the vicarious and cross calibrations.

The a_1 and a_2 coefficients were estimated from the distribution of the RCCs of the vicarious calibration [16] and the RCCs of the cross calibration [29,30] by the least squares method. The RCCs of the cross calibration were obtained from image data with high-gain mode and solar zenith angle limited to 45° or less.

The degradation curve of Band 2 in the radiometric DB ver. 4 [dotted line in Figure 2b] was estimated from the RCCs in the vicarious and cross calibrations, and the curve is also expressed by Equation (1). The coefficients of degradation curve in Band 2 are $a_0 = 0.949$, $a_1 = 0.886$, and $a_2 = 0.00181$ (Table 2). However, Band 2 had a problem with the gain ratio at the time of determining the value of these coefficients; the radiance obtained from the Band 2 image of the high-gain mode shows a 7–11% difference with the normal-gain mode [7,29,30]. In most ASTER images used for the vicarious calibration, the gain modes of Bands 1, 2, 3N, and 3B are set to High, High, Normal, and

Normal, respectively. However, the normal-gain mode was used for all ASTER bands in the onboard calibration. Therefore, the large difference (almost 10%) existed in the RCCs between vicarious and onboard calibrations. To determine the degradation coefficients in the radiometric DB ver. 4, the onboard calibration result was not applied to Band 2. It was difficult to use both calibrations together owing to the abovementioned large difference, and the recent RCC from the onboard calibration seemed too small to apply the coefficients. Because the distribution of RCCs of vicarious calibration and cross calibration is highly scattered, it is not desirable to obtain the coefficient of the degradation curve only from this distribution. Therefore, the a_2 coefficient that determines the shape of the curve, which is the degradation rate, was assumed to be the same as that of Band 1. The $a_{0\text{-high}}$ coefficient, which equals the $R(0)$ in the high-gain mode, was estimated from the best-fit with the distribution of the RCCs by vicarious and cross calibrations using the image in the high-gain mode. The a_0 coefficient, which equals the $R(0)$ in the normal-gain mode, was calculated from $a_{0\text{-high}}$ and the gain ratio r_g of the normal-gain and the high-gain mode (i.e., $a_0 = a_{0\text{-normal}} = a_{0\text{-high}} \times r_g$). The value of r_g was investigated from the two sets of the cross calibration, MODIS Band 1 vs. ASTER Band 2 with the normal-gain mode and MODIS Band 1 vs. ASTER Band 2 with the high-gain mode [29]. The values of $a_{0\text{-high}}$ and r_g were 0.87 and 1.091, respectively; thus $a_0 = a_{0\text{-normal}} = a_{0\text{-high}} \times r_g = 0.87 \times 1.091 = 0.949$. Lastly, a_1 was estimated from the distribution of the RCCs of the vicarious calibration and the RCCs of the cross calibration by the least squares method. The RCCs of the cross calibration was obtained from image data with a high-gain mode and solar zenith angle limited to 45° or less. The cause of this gain ratio problem is that the ratio after launch changed by almost 10% from the preflight calibration ratio, which was determined from the analysis of the onboard electrical calibration [31]. On the basis of this analysis, in the radiometric DB ver. 4, the Band 2 images in the high-gain mode were radiometrically corrected by +8% [17].

The degradation curve of Band 3N in the radiometric DB ver. 4 [dotted line in Figure 2c] was mainly estimated from the RCCs in the onboard calibration, and this curve is expressed by the following equations [4,14]:

$$R(d) = a_0 + a_1 d + a_2 d^2 \text{ in } d < 673$$
$$R(d) = a_1 \exp(-a_2\, d) + a_0 \text{ in } 673 \leq d < 4825 \qquad (2)$$
$$R(d) = a_0 \text{ in } 4825 \leq d$$

where the values of coefficients (i.e., a_0, a_1, and a_2) of the degradation curve are shown in Tables 2 and 3.

Band 3N does not have a large difference in the RCCs between onboard and vicarious calibrations but has a small difference in the degradation trend. The degradation trend by the onboard calibration shows a gradual downward slope in all periods (from the launch to the present) [dashed line in Figure 2c]. The trend obtained by the vicarious calibration also shows a downward slope, but the recent trend does not appear to have a slope (Figure 1). Therefore, the degradation curve in $d < 4825$ was obtained from the onboard calibration, as with the radiometric DB ver. 3. It was assumed that the curve in $4825 \leq d$ did not degrade, and the RCCs were then set to be a constant.

Band 3B does not have a correction for the degradation [dotted line in Figure 2d], and the equation is as follows:

$$R(d) = a_0 = 1 \qquad (3)$$

In the vicarious calibration, the degradation of Band 3B was observed. However, this degradation is small compared with that of the other bands (Bands 1, 2, and 3N), and Band 3B does not have an onboard calibrator. Therefore, the radiometric DB ver. 4 assumed that Band 3B had no degradation, as with the radiometric DB ver. 3.

3.2. Problems with the Current Radiometric DB

The degradation curves were revised in this current radiometric DB ver. 4, but the inter-band and lunar calibrations have reported problems [6,12], which are probably caused by the different traceability of each band with the different calibration approach.

Inter-band radiometric calibration can be used to evaluate the inter-band consistency. The solid lines in Figure 2b,c are the degradation curves of Bands 2 and 3N calculated from the degradation curve of reference band, Band 1, using the inter-band calibration technique. In this study, Band 1 is set to a reference band, because the three calibration approaches (i.e., onboard, vicarious, and cross calibrations) exhibited good agreement between each other for the Band 1 degradation curve estimation, in the early stage (from launch day to approximately day 673), and Bands 2 an 3N do not have good agreement among three approaches [12]. The degradation curves of Bands 2 and 3N from this inter-band calibration have inconsistencies with the curves from the radiometric DB ver. 4 and the onboard calibration in Figure 2.

The inter-band inconsistency owing to the difference in traceability can also be read on the basis of the lunar calibration results. Figure 3 shows the comparison of the RD in VNIR bands from 2003 to 2017 derived from the RCCs in the radiometric DB ver. 4 and three different calibration methods (i.e., lunar, inter-band, and onboard calibrations) [6]. RD_4, RD_L, RD_I, and RD_O are the RD values derived from the radiometric DB ver.4, lunar calibration, inter-band calibration, and onboard calibration, respectively.

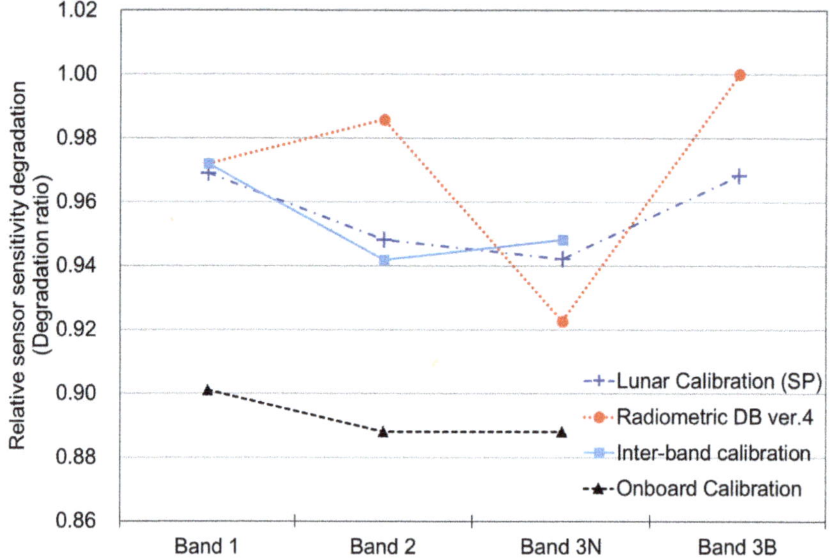

Figure 3. Comparison of the relative sensor sensitivity degradation (RD) in ASTER-VNIR bands from 2003 to 2017 derived from the RCCs in the radiometric DB ver. 4 and three different calibration methods (i.e., lunar, inter-band, and onboard calibrations) [6].

The RD_4 of Band 1 in the radiometric DB ver. 4 is consistent with RD_L in the lunar calibration, but for Bands 2 and 3B, the RD_4 of ver. 4 is larger than RD_L of the lunar calibration; however, for Band 3N, the RD_4 of ver. 4 is smaller than RD_L of the lunar calibration. However, the RD_I in the inter-band calibration is consistent with the RD_L in the lunar calibration for all three bands, and RD_O in the onboard calibration shows lower values compared with RD_L in the lunar calibration for all bands.

On the basis of the abovementioned points, it was necessary to improve the degradation curve in the current radiometric DB.

4. Radiometric Degradation Curves in the Next Radiometric DB

In the next radiometric DB ver. 5 for ASTER VNIR L1 processing, new radiometric degradation curves with the same traceability among the bands are necessary to obtain good inter-band consistency.

4.1. Selection of the Calibration Approach

In the ASTER science team, several calibration approaches (i.e., preflight, onboard, vicarious, cross, lunar, and inter-band calibrations) have been performed and discussed for the ASTER VNIR radiometers. Vicarious and lunar calibration approaches have been selected [32], and the radiometric degradation curves for ASTER VNIR processing have been derived from the vicarious and lunar calibrations. The basic ideas (i.e., reason and philosophy) of this selection are explained as follows.

For the onboard calibration, the uncertainty of the onboard calibrator typically increases with time in the harsh environment of space [5,33]. In ASTER, as described above, the RCCs of Band 1 obtained from the onboard calibration tended to deviate from the RCCs of other calibration approaches after 673 days since the launch, and Band 3B did not have an onboard calibrator. Therefore, we did not use the ASTER onboard calibrator, which has been in space for a long period of time, to obtain the degradation curves for the next radiometric DB.

Recently, numerous studies have reported the cross calibration of ASTER and Terra-MODIS VNIR bands [7–11], but the trends in the cross-calibration results are not necessarily consistent. This inconsistency may be attributed to the differences between the versions of the radiometric DBs used in the ASTER radiometric calibration [11]. The different corrections of spectral and spatial effects between ASTER and MODIS can also cause inconsistencies. At this point in time, it is difficult to obtain the degradation curves of the ASTER VNIR bands from the cross calibration.

The inter-band calibration appears to be a promising approach, because it was consistent with vicarious and lunar calibrations, as shown in the previous section. However, the inter-band calibration in the current analysis method cannot be applied between bands with different view-angles and observation times, i.e., between nadir-view bands (Bands 1, 2, and 3N) and backward-view bands (Band 3B). Therefore, the degradation curve of Band 3B cannot be obtained from inter-band calibration with reference to Band 1, and the same traceability cannot be given to all bands.

The degradation curve of the sensor cannot be obtained from only the preflight calibration, and its RCC can only be used at the launch. The assumption that the sensor is not damaged at the launch is needed in order to apply the RCC of the preflight calibration to the RCC of the degradation curve at the launch. From the point of view of the onboard and vicarious calibrations in Figures 1 and 2, it appears that there was a small damage at the launch for ASTER VNIR, and thus, the RCC from the preflight calibration is not used.

Therefore, the remaining calibration approaches (i.e., vicarious and lunar calibrations) are selected. Vicarious calibration is an excellent approach to obtain an absolute value of the RCC when the sensor is performing normal observation (observing the earth). Although the absolute accuracy of the lunar calibration is considered to be insufficient (5–10%), the relative accuracy of lunar calibration (i.e., temporal variation of relative sensor sensitivity) is extremely high (the order of 0.1%) [6]. Vicarious calibration in ASTER VNIR has been carried out many times throughout the ASTER operation period; however, lunar calibration has only been conducted twice, in April 2003 (DSL = 1213) and August 2017 (DSL = 6440). Therefore, the base of the degradation curve is obtained from the vicarious calibration, and the result of the lunar calibration with high accuracy against relative changes (relative degradation) is used as the constraint condition.

4.2. Radiometric Degradation Curves

Because the uncertainty of the vicarious calibration using the reflectance-based method is not small (Appendix A), it is difficult to correctly obtain the curve parameters of the degradation from the RCC distribution of the vicarious calibration. However, if the degradation stops and the RCC value can be regarded as a constant (or can be regarded to have a linear distribution), a relatively accurate average (or regression line) of RCCs can be obtained even from the vicarious calibration.

In Figure 1, after 3000 DSL, the degradation stops in all bands, and the RCC is almost constant. Thus, the RCC value after 3000 DSL can be set by the average of the RCCs of vicarious calibration. On the other hand, before 3000 DSL, we decided to obtain the degradation curve parameters by the least squares method from the distribution of the RCCs of the vicarious calibration, and the result (the value of relative degradation) from the lunar calibration is used as a constraint for the estimation of the degradation curve parameters as described below.

After 3000 DSL, the degradation is thought to have stopped; the RCC is assumed to be constant and is expressed by the following equation:

$$R(d) = a_0 = x \text{ in } 3000 < d \tag{4}$$

where x is the average of the RCCs derived from the vicarious calibration over 3000 DSL. For the period before 3000 DSL, the same degradation model [28] as Equation (1) is adopted and expressed by the following equation:

$$R(d) = a_0\, c = a_0 \times T/T_0 = a_0\,(1.0 - a_1)\,\exp(-a_2\,d) + a_0\,a_1 \text{ in } 0 \le d \le 3000 \tag{5}$$

where $R(d)$ represents the radiometric calibration coefficient (RCC) of the DSL d, c is the scaled calibration coefficient, T represents the transmittance of DSL d in the radiometer optics (lens, mirror and etc.), and T_0 represents the transmittance T when $d=0$. The degradation parameters a_0, a_1, and a_2 represent the RCC at launch, the minimum (saturated) transmittance of the contamination/corrosion layer of the sensor lenses, and the degradation rate, respectively. The degradation coefficients a_0, a_1, and a_2 are obtained by the least squares method from the RCCs of the vicarious calibration in Figure 1, and the following two constraints are given:

$$R(3000) = a_0\,(1.0 - a_1)\,\exp(-3000 \times a_2) + a_0\,a_1 = x \tag{6}$$

$$\frac{R(6440)}{R(1213)} = \frac{x}{a_0\,(1.0 - a_1)\,\exp(-1213 \times a_2) + a_0\,a_1} = y \tag{7}$$

where y is the value of RD_L in Table 1 calculated from the lunar calibration. Equation (6) is used to connect (at 3000 DSL) Equation (4) of the straight line (i.e., constant value) to Equation (5) of the curve; Equation (7) is used to reflect the result of the lunar calibration.

The radiometric degradation curves obtained from the abovementioned equations are shown in Figure 4, and their coefficients are shown in Table 4.

Table 4. The equations and parameters of the new radiometric degradation curves in the radiometric DB ver. 5 for ASTER VNIR bands.

Band	Duration (d: DSL)	Equation (R(d): RCC)	Coefficients		
			a_0	a_1	a_2
1	$0 \le d \le 3000$	$R(d) = a_0\,(1.0 - a_1)\,\exp(-a_2\,d) + a_0\,a_1$	1.017	0.7730	0.001791
	$3000 < d$	$R(d) = a_0$	0.7869	-	-
2	$0 \le d \le 3000$	$R(d) = a_0\,(1.0 - a_1)\,\exp(-a_2\,d) + a_0\,a_1$	1.008	0.8016	0.001114
	$3000 < d$	$R(d) = a_0$	0.8152	-	-
3N	$0 \le d \le 3000$	$R(d) = a_0\,(1.0 - a_1)\,\exp(-a_2\,d) + a_0\,a_1$	0.9849	0.8192	0.0008238
	$3000 < d$	$R(d) = a_0$	0.8218	-	-
3B	$0 \le d \le 3000$	$R(d) = a_0\,(1.0 - a_1)\,\exp(-a_2\,d) + a_0\,a_1$	0.9762	0.9009	0.0003670
	$3000 < d$	$R(d) = a_0$	0.9116	-	-

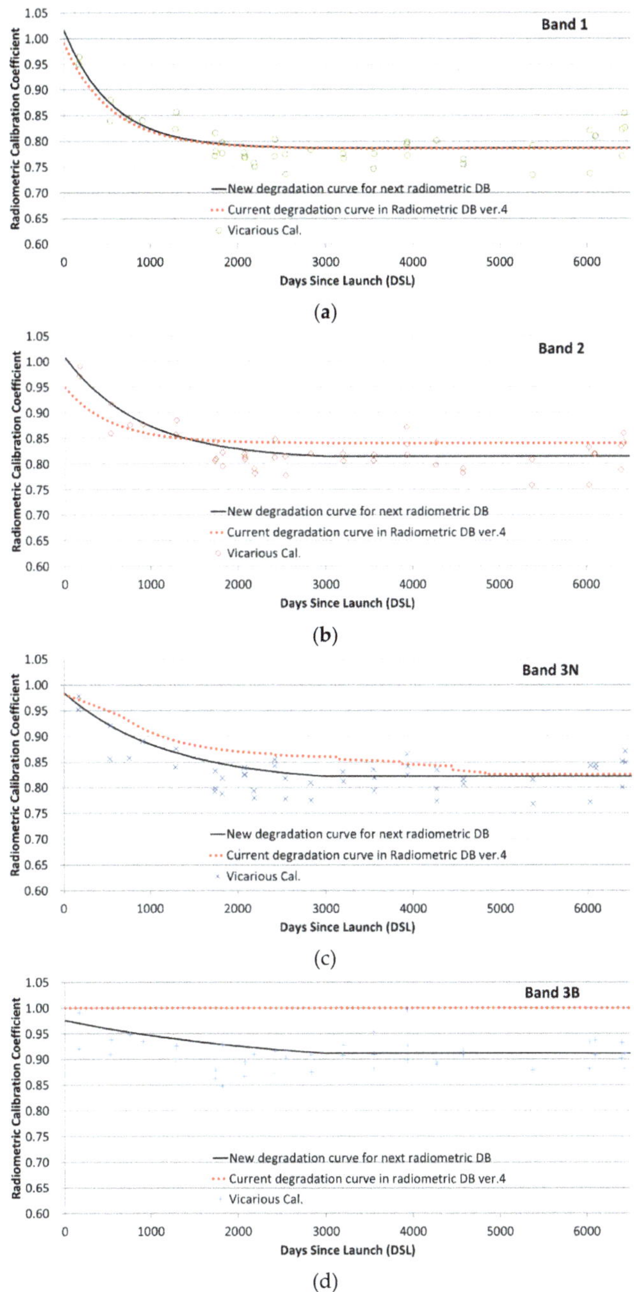

Figure 4. New degradation curves of ASTER VNIR bands for the next radiometric DB ver. 5. Bands (**a**) 1, (**b**) 2, (**c**) 3N, and (**d**) 3B. Solid lines correspond to the new degradation curves for the next radiometric DB estimated from the vicarious calibration data from 2000 to 2017 with the lunar calibration on April 14, 2003 (DSL = 1213), and August 5, 2017 (DSL = 6440). Dotted lines correspond to the current degradation curves in the current radiometric DB ver. 4. The marks show the RCCs from the vicarious calibration.

5. Discussion

The shape of the newly obtained degradation curves in Figure 4 for the next radiometric DB is similar between bands. This approach considerably differs from that used in the current radiometric DB ver. 4. The shape of new degradation curves is similar to the curve shape of the inter-band calibration, and the inter-band consistency of new curve shows almost same trend as the inter-band calibration within the uncertainty of the inter-band calibration (Appendix B). In addition, the shape of new degradation curves is also different from the curve shape of the onboard calibration in Figure 2. The onboard calibrator was in the outer space for a long period of time after the launch, and its degradation may have occurred. In all bands, the RCC value at the launch (the RCC value when $d = 0$) obtained from the new degradation curve slightly differs from 1.00 (by approximately 1–3%). This may indicate that the sensor experienced some changes at the launch, but the difference was small; thus, this discrepancy may be within the uncertainty in the degradation curve.

The new degradation curve of Band 1 is almost the same as that in the current ver. 4. There is only a slight difference in the RCC value at launch. Because Band 1 degraded more rapidly than the other bands, the degradation curve can be accurately obtained from the vicarious calibration, even if there is no constrain from the result of the lunar calibration.

The new degradation curve of Band 2 is very different from that in the current ver. 4. The degradation curve in the current ver. 4 has been calculated from the data in the old radiometric DB ver. 3. The problem of electrical calibration for the gain ratio [31] was not improved in the old radiometric DB. Therefore, the degradation curve of ver. 4 was obtained on the basis of several assumptions, and some of the assumptions may have been inappropriate.

For Band 3N, because the degradation in the current ver. 4 is based on the onboard calibration, the new curve obtained from vicarious and lunar calibrations also considerably differs from the current version. However, after 4825 DSL, the consistency between the current and new curves is higher because the vicarious calibration is also used as a reference for the current curve.

Band 3B in the current ver. 4 was set to have no degradation. However, for the first time, its degradation curve is obtained in this study. In this new degradation curve, approximately 10% degradation exists.

The new degradation curves for the next radiometric DB have the same traceability in the bands, can represent the distribution of the RCCs from the vicarious calibration, and are highly consistent with the lunar calibration.

6. Conclusions

In 2014, the radiometric correction of ASTER VNIR bands was once improved by the current radiometric degradation curves in the radiometric DB ver. 4 [17], compared to the previous DBs ver. 1 to 3. However, the calibration approach applied for each band is different; thus, the traceability is complicated. Therefore, the problem with the radiometric DB ver. 4 was indicated by inter-band and lunar calibrations.

For the next radiometric DB ver. 5, vicarious and lunar calibrations were selected, and the base of the degradation curve was obtained from the vicarious calibration, and the result of the lunar calibration with high accuracy against relative changes (relative degradation) was used as the constraint condition. The shape and format of the calibration function of the newly obtained degradation curves for the next radiometric DB are now similar between bands.

The new degradation curve for Band 1 is almost the same as that in the current radiometric DB ver. 4; however, other bands have curves that differ from the current version. Specifically, for Band 3B, there is no degradation in the current ver. 4, but the degradation of nearly 10% exists in the new degradation curve.

In this study, for the next update of the radiometric DB, new favored degradation curves were obtained. The favored curves have the same traceability in the bands, can represent the distribution of the RCCs from the vicarious calibration, and are consistent with inter-band and the lunar calibrations.

However, in order to obtain a better degradation curve in the future, further analysis of other calibration approaches, e.g., cross calibration and pseudo-invariant calibration had better to be conducted (Appendix C).

Such long-term calibration activities for one sensor are rare, and we expect that the experience and knowledge will be useful for the calibration of other sensors in the future.

Author Contributions: Conceptualization, S.T., T.K., K.O., and K.J.T.; methodology for new degradation curve, S.T., H.Y., T.T., and K.J.T.; methodology for the current degradation curve, S.T., K.A., S.F.B., J.S.C.-M., K.J.T., F.S., H.Y., and T.T.; analysis for the AIST vicarious calibration, S.T., A.K., and H.Y.; analysis for the lunar calibration, T.K.; analysis for the inter-band calibration, K.O.; analysis for the onboard calibrator, F.S.; writing—original draft preparation, S.T.; writing—review and editing, S.T., H.Y., K.O. All authors have read and agreed to the published version of the manuscript.

Funding: This research was supported partly by the Ministry of Economy, Trading and Industry of Japan (METI) and National Aeronautics and Space Administration (NASA).

Acknowledgments: The authors appreciate Yasushi Yamaguchi and Michael Abrams for their special support for the ASTER-VNIR calibration activity, and to all members in ASTER science team. The authors also appreciate the Bureau of Land Management offices in Tonopah, NV, and Needles, CA, for their help in accessing the test sites in California and Nevada, and the numerous people for assisting with the field data collections. The authors thank Koki Iwao for providing information of ASTER L1 processing and Yukiko Doyama for assisting with field equipment preparation in AIST. The authors also thank Satoru Yamamoto who provided the onboard calibration trend.

Conflicts of Interest: The authors declare no conflict of interest.

Appendix A

In this study, we have used the legacy vicarious calibration approach based on the reflectance-based method that was established in 1990s, and the uncertainty of this approach is not small, approximately 5%, [34] which is calculated by the analysis of the error sources for one field campaign (or one data). However, the uncertainty of the degradation curve $R(d)$ derived from multiple field campaigns (or multiple data) shows smaller than the uncertainty from one field campaign as described below.

The value u_c of the uncertainty in the degradation curve $R(d)$ is expressed by the following equation [35]:

$$u_c = \sqrt{u_r^2 + u_s^2} \tag{A1}$$

where u_r and u_s represent the uncertainty of the degradation curve $R(d)$ for the random and systematic components respectively. The u_r value is written by the deviation of the $R(d)$ as a regression curve of the RCC derived from the vicarious calibration.

$$u_r = \sqrt{\frac{\sum_{i=1}^{n}(R(d_i) - R_i)^2}{n(n-p)}} \tag{A2}$$

where d_i represents the DSL of the i-th field data (or field campaign), R_i is the RCC of the DSL d_i, n is the number of the field data, and the p is the number of the parameters for the function $R(d)$. On the other hand, the error sources of the reference panel and solar irradiance can be set at least as the systematic components. Therefore, the u_s value is calculated from the error sources of the reference panel (2.0%) [34] and solar irradiance (0.3% or less) [24]. Table A1 shows these uncertainties calculated above equations for ASTER VNIR in this study, and the uncertainty u_c is estimated to be approximately 2% or more.

Many of the error sources of the vicarious calibration are assumed to be the random components, however if there are unknown error sources of the systematic components except the reference panel and solar irradiance, the u_c value should become larger.

Table A1. The uncertainties of $R(d)$ calculated from the multiple field data for ASTER VNIR bands.

Uncertainty of $R(d)$	Duration (d:DSL)	Band 1	Band 2	Band 3N	Band 3B
u_r	$0 \leq d \leq 3000$	0.0053	0.0056	0.0076	0.0089
	$3000 < d$	0.0072	0.0065	0.0064	0.0067
u_s	Independent to d	0.020	0.020	0.020	0.020
u_c	$0 \leq d \leq 3000$	0.021	0.021	0.022	0.022
	$3000 < d$	0.021	0.021	0.021	0.021

Appendix B

The shape of newly obtained degradation curves is similar to the curve shape of the inter-band calibration, but the shape of the current degradation curve of Band 2 and 3N in the radiometric DB ver.4 is different from it (Figure A1a–c). The RCC difference of Band 1, Band 2, and Band 3N between new degradation curve and the degradation curve from the inter-band calibration are 0.003, 0.016, and 0.018 respectively, which are average values in the period of 0–6500 DSL.

To compare the inter-band consistency among three kinds of degradation curve, Figure A2a–c shows the RCC ratio between bands for each degradation curve. The curve shape of the RCC ratio of new degradation curve is similar to the shape of the RCC ratio curve from the inter-band calibration, but the curve shape of the RCC ratio of current degradation curve is different from the curve from the inter-band calibration. The percent difference of the RCC ratio curve in Band 1 / Band 3N, Band 2 / Band 1, and Band 3N / Band 2 between new degradation curve and the degradation curve from the inter-band calibration are 1.8%, 1.6%, and 0.4% respectively, which are average values in the period of 0–6500 DSL. The uncertainties of the inter-band calibration for ASTER VNIR bands are 3.0, 2.6, and 2.5, respectively, for Bands 1 and 3N, Bands 1 and 2, and Bands 2 and 3N [12]. The inter-band consistency of new curve shows almost same trend as the inter-band calibration within the uncertainty of the inter-band calibration.

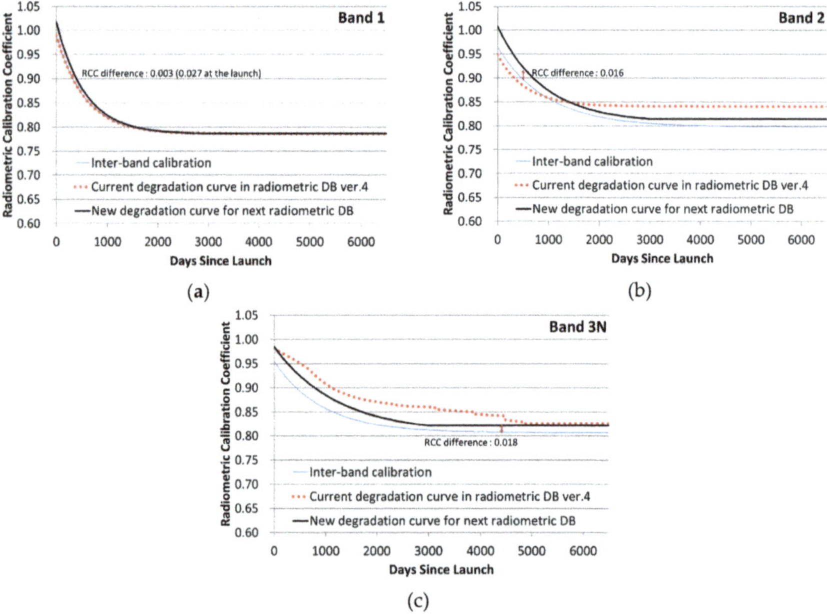

Figure A1. Comparison of the shape of the degradation curve calculated from the inter-band calibration, the current radiometric DB ver.4, and the next radiometric DB. Bands (**a**) 1, (**b**) 2, and (**c**) 3N.

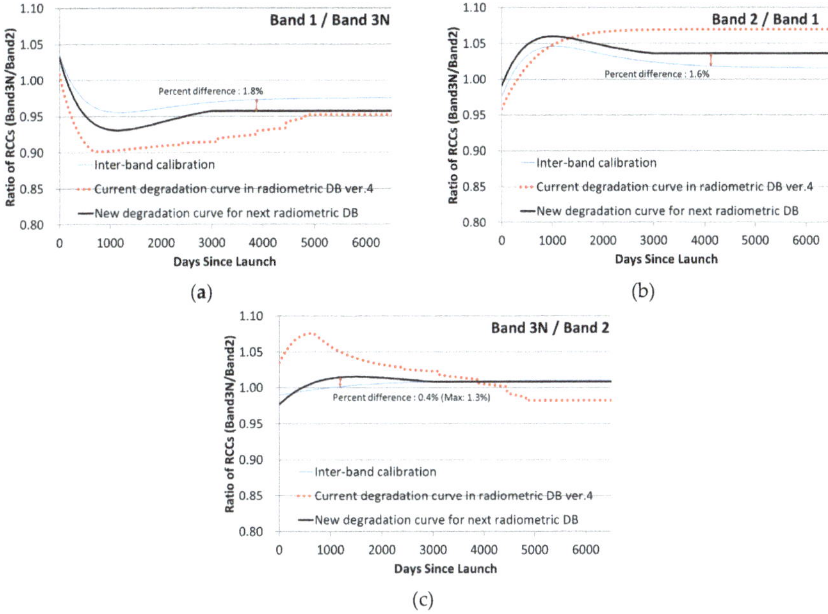

Figure A2. The RCC ratio between bands for each degradation curve for comparison of inter-band consistency. (**a**) Band 1/Band 3N, (**b**) Band 2/Band 1, and (**c**) Band 3N/Band 2.

Appendix C

The cross calibration is a good method that can be carried out frequently and has lower uncertainty than vicarious calibration. However, we have encountered the two problems with applying it to the ASTER VNIR radiometer. One is the large spread (Band 1: almost 10%, Band 2: 5–10%) in the RCC values, which is calculated from the cross-calibration between ASTER and MODIS, among the sites [7–9]. The other is the large spread (Band 1: 3–7%, Band 2: 7–11%) in the RCC values, which is calculated from the cross-calibration between ASTER and MODIS, between the normal and high gains [7,29,30]. The Band 2 High/Normal gain problem was corrected from the analysis of the onboard electrical calibration [31] at the current radiometric DB, but the other problem is currently under investigation. Moreover, the onboard electrical calibration showed many kinds of problems [31]. Recent studies of the cross calibration have been conducted at only one site (Railroad Valley), but the results show still the 1–4% difference between the studies [10,11]. We have identified and solved the problems inherent in the ASTER VNIR radiometer one by one using the results from the cross calibration. The studies of the cross calibration for the ASTSER will be continued and eventually reflected in the sensor trends.

The pseudo-invariant calibration [36] is a good method that can be carried out frequently and has high accuracy for monitoring of the relative degradation. However, we could not obtain a sufficient number of ASTER images for the study of the pseudo-invariant calibration. The swath width of ASTER image is 60 km, then the ASTER sensor pointing system in cross-track direction has covered 232 km width. It means that the ASTER does not observe the pseudo-invariant calibration sites (PICSs) at every overpass-time, and its observation has been planned according to the operational mission and user request. For example, in Libya-4 desert site, the number of ASTER-VNIR images for 20 years after the launch is almost 40 scenes in the daytime, and the number of the cloud-free nadir images (i.e., the cloud-free images at the overpass-time) is almost 10. Moreover, 10 cloud-free and nadir observed scenes include different gain modes. The ASTER-VNIR images observed at different targets and/or by the different gain modes sometimes cause the different results as mentioned above in the cross calibration. We hope that these problems will be solved, then pseudo-invariant calibration will finally be applied to the ASTER VNIR.

References

1. Yamaguchi, Y.; Kahle, A.; Tsu, H.; Kawakami, T.; Pniel, M. Overview of Advanced Spaceborne Thermal Emission and Reflection Radiometer (ASTER). *IEEE Trans. Geosci. Remote Sens.* **1998**, *36*, 1062–1071. [CrossRef]
2. ASTER Science Office ASTER SWIR Data Status Report. Available online: http://www.aster.jspacesystems.or.jp/en/about_aster/swir_en.pdf (accessed on 9 December 2019).
3. Ono, A.; Sakuma, F.; Arai, K.; Yamaguchi, Y.; Fujisada, H.; Slater, P.N.; Thome, K.J.; Palluconi, F.; Kieffer, H. Preflight and in-flight calibration plan for ASTER. *J. Atmos. Ocean. Technol.* **1996**, *13*, 321–335. [CrossRef]
4. Sakuma, F.; Ono, A.; Tsuchida, S.; Ohgi, N.; Inada, H.; Akagi, S.; Ono, H. Onboard calibration of the ASTER instrument. *IEEE Trans. Geosci. Remote Sens.* **2005**, *43*, 2715–2724. [CrossRef]
5. Thome, K.; Arai, K.; Tsuchida, S.; Biggar, S. Vicarious calibration of ASTER via the reflectance-based approach. *IEEE Trans. Geosci. Remote Sens.* **2008**, *46*, 3285–3295. [CrossRef]
6. Kouyama, T.; Kato, S.; Kikuchi, M.; Sakuma, F.; Miura, A.; Tachikawa, T.; Tsuchida, S.; Obata, K.; Nakamura, R. Lunar Calibration for ASTER VNIR and TIR with Observations of the Moon in 2003 and 2017. *Remote Sens.* **2019**, *11*, 2712. [CrossRef]
7. Yamamoto, H.; Kamei, A.; Nakamura, R.; Tsuchida, S. Long-term cross-calibration of the Terra ASTER and MODIS over the CEOS calibration sites. *Proc. SPIE* **2011**, *8153*, 815318.
8. McCorkel, J. Cross-calibration of Earth Observing System Terra satellite sensors MODIS and ASTER. *Proc. SPIE* **2014**, *9218*, 92180X.
9. Thome, K.; McCorkel, J.; Czapla-Myers, J. In-situ transfer standard and coincident-view intercomparisons for sensor cross-calibration. *IEEE Trans. Geosci. Remote Sens.* **2013**, *51*, 1088–1097. [CrossRef]
10. Yuan, K.; Thome, K.; Mccorkel, J. Radiometric cross-calibration of Terra ASTER and MODIS. *Proc. SPIE* **2015**, *9607*, 1–9.
11. Obata, K.; Tsuchida, S.; Yamamoto, H.; Thome, K. Cross-Calibration between ASTER and MODIS Visible to Near-Infrared Bands for Improvement of ASTER Radiometric Calibration. *Sensors* **2017**, *17*, 1793. [CrossRef]
12. Obata, K.; Tsuchida, S.; Iwao, K. Inter-Band Radiometric Comparison and Calibration of ASTER Visible and Near-Infrared Bands. *Remote Sens.* **2015**, *7*, 15140–15160. [CrossRef]
13. ASTER Science Team. *Algorithm Theoretical Basis Document for ASTER Level-1 Data Processing (Ver. 3.0)*; ASTER Science Team, Japan: Tokyo, Japan, 1996; Available online: http://www.aster.jspacesystems.or.jp/en/documnts/pdf/1a1b.pdf (accessed on 9 December 2019).
14. Sakuma, F.; Kikuchi, M.; Inada, H.; Akagi, S.; Ono, H. Onboard calibration of the ASTER instrument over twelve years. *Proc. SPIE* **2012**, *8533*, 853305.
15. ASTER User's Guide Part I General (Ver.4.0). Earth Remote Sensing Data Analysis Center. 2005. Available online: http://www.aster.jspacesystems.or.jp/en/documnts/users_guide/part1/pdf/Part1_4E.pdf (accessed on 9 December 2019).
16. Tsuchida, S.; Yamamoto, H.; Kamei, A. Long-term vicarious calibration of ASTER VNIR bands. In Proceedings of the 52nd Conference of the Remote Sensing Society of Japan, Tokyo, Japan, 23–24 May 2012; pp. 85–86, (In Japanese with English abstract).
17. Tachikawa, T. ASTER Science Team Meeting. *Earth Obs.* **2014**, *26*, 18–21.
18. Slater, P.; Biggar, S.; Holm, R.; Jackson, R.; Mao, Y.; Moran, M.; Palmer, J.; Yuan, B. Reflectance- and radiance-based methods for the in-flight absolute calibration of multispectral sensors. *Remote Sens. Environ.* **1987**, *22*, 11–37. [CrossRef]
19. Slater, P.; Biggar, S.; Thome, K.; Gelman, D.; Spyak, P. Vicarious Radiometric Calibration of EOS Sensors. *J. Atmos. Oceanic Technol.* **1996**, *13*, 349–359. [CrossRef]
20. Tsuchida, S.; Sato, I.; Yamaguchi, Y.; Arai, K.; Takashima, T. Vicarious calibration in the visible to infrared region based on reflectance-based method at the snow fields in Hokkaido. *J. Remote Sens. Soc. Jpn.* **1998**, *18*, 12–31, (In Japanese with English abstract).

21. Liu, C.-C.; Kamei, A.; Hsu, K.-H.; Tsuchida, S.; Huang, H.-M.; Kato, S.; Nakamura, R.; Wu, A.-M. Vicarious calibration of the Formosat-2 remote sensing instrument. *IEEE Trans. Geosci. Remote Sens.* **2010**, *48*, 2162–2169.

22. Kamei, A.; Nakamura, K.; Yamamoto, H.; Nakamura, R.; Tsuchida, S.; Yamamoto, N.; Sekiguchi, S.; Kato, S.; Liu, C.C.; Hsu, K.H.; et al. Cross calibration of formosat-2 Remote Sensing Instrument (RSI) using Terra Advanced Spaceborne Thermal Emission and Reflection Radiometer (ASTER). *IEEE Trans. Geosci. Remote Sens.* **2012**, *50*, 4821–4831. [CrossRef]

23. Working with ASTER L1T Visible and Near Infrared (VNIR) Data in R. Available online: https://lpdaac.usgs.gov/resources/e-learning/working-aster-l1t-visible-and-near-infrared-vnir-data-r/ (accessed on 9 December 2019).

24. Thome, K.J.; Biggar, S.F.; Slater, P.N. Effects of assumed solar spectral irradiance on intercomparisons of Earth-observing sensors. In *Sensors, Systems, and Next-Generation Satellites V*; International Society for Optics and Photonics: Bellingham, WA, USA, 2001; Volume 4540, pp. 260–269. [CrossRef]

25. Kieffer, H.; Stone, T. The spectral irradiance of the Moon. *Astron. J.* **2005**, *129*, 2887–2901. [CrossRef]

26. Yokota, Y.; Matsunaga, T.; Ohtake, M.; Haruyama, J.; Nakamura, R.; Yamamoto, S.; Ogawa, Y.; Morota, T.; Honda, C.; Saiki, K.; et al. Lunar photometric properties at wavelengths 0.5-1.6 µm acquired by SELENE Spectral Profiler and their dependency on local albedo and latitudinal zones. *Icarus* **2011**, *215*, 639–660. [CrossRef]

27. Kouyama, T.; Yokota, Y.; Ishihara, Y.; Nakamura, R.; Yamamoto, S.; Matsunaga, T. Development of an application scheme for the SELENE/SP lunar reflectance model for radiometric calibration of hyperspectral and multispectral Sensors. *Planet. Space Sci.* **2016**, *214*, 76–83. [CrossRef]

28. Tsuchida, S.; Sakuma, F.; Iwasaki, A.; Ogi, N.; Inada, H. Degradation models and functions for ASTER/VNIR sensor. In Proceedings of the 38th Conference of the Remote Sensing Society of Japan, Chiba, Japan, 20 May 2005; pp. 99–100, (In Japanese with English abstract).

29. Yamamoto, H.; Kamai, A.; Nakamura, R.; Tsuchida, S. Radiometric evaluation of ASTER VNIR/SWIR bands be long-term Terra ASTER/MODIS cross-calibration over CEOS Reference Test Sites. In Proceedings of the 52nd Conference of the Remote Sensing Society of Japan, Tokyo, Japan, 23–24 May 2012; pp. 87–88, (In Japanese with English abstract).

30. Yamamoto, H.; Koyama, T.; Nakamura, R.; Tsuchida, S. Status of ASTER/HISUI radiometric calibration—Vicarious calibration and cross-calibration. In *CEOS IVOS 25th Meeting*; ESA ESRIN: Frascati, Italy, 2013; Available online: http://calvalportal.ceos.org/ceos-wgcv/ivos/ivos25 (accessed on 5 December 2019).

31. Sakuma, F.; Kikuchi, M.; Inada, H. Onboard electrical calibration of the ASTER VNIR. *Proc. SPIE* **2013**, *8889*, 888903.

32. Abrams, M.; Yamaguchi, Y. ASTER Science Team Meeting. *Earth Obs.* **2019**, *31*, 21–27.

33. Slater, P.; Biggar, S.; Palmer, J.; Thome, K. Unified approach to absolute radiometric calibration in the solar-reflective range. *Remote Sens. Environ.* **2001**, *77*, 293–303. [CrossRef]

34. Biggar, S.; Slater, P.; Gellman, D. Uncertainties in the in-flight calibration of sensors with reference to measured ground sites in the 0.4–1.1 m range. *Remote Sens. Environ.* **1994**, *48*, 245–252. [CrossRef]

35. Ono, A.; Arai, K. Radiometric calibration of spaceborne optical sensors, lecture series 5: Integrating analysis of calibration data. *J. Remote Sens. Soc. Jpn.* **2017**, *37*, 375–384. (In Japanese)

36. Helder, D.; Thome, K.J.; Mishra, N.; Chander, G.; Xiong, X.; Angal, A.; Choi, T. Absolute Radiometric Calibration of Landsat Using a Pseudo Invariant Calibration Site. *IEEE Trans. Geosci. Remote Sens.* **2013**, *51*, 1360–1369. [CrossRef]

Article

Toward the Detection of Permafrost Using Land-Surface Temperature Mapping

Jigjidsurengiin Batbaatar [1,*], Alan R. Gillespie [1], Ronald S. Sletten [1], Amit Mushkin [1,2], Rivka Amit [2], Darío Trombotto Liaudat [3], Lu Liu [1] and Gregg Petrie [1]

[1] University of Washington, Department of Earth and Space Science and Quaternary Research Center, Box 351310, Seattle, WA 98195-1310, USA; arg3@uw.edu (A.R.G.); sletten@uw.edu (R.S.S.); mushkin@uw.edu (A.M.); liul99@uw.edu (L.L.); gregg@hiker.org (G.P.)

[2] Geological Survey of Israel, 32 Yishayahu Leibovitz St., Jerusalem 9692100, Israel; rivka@gsi.gov.il

[3] Geocryology, Instituto Argentino de Nivología, Glaciología y Ciencias Ambientales (IANIGLA), CCT CONICET Mendoza, Bajada del Cerro s/n, P.O. Box 330, Mendoza 5500, Argentina; dtrombot@mendoza-conicet.gob.ar

* Correspondence: bataa@uw.edu

Received: 14 January 2020; Accepted: 17 February 2020; Published: 20 February 2020

Abstract: Permafrost is degrading under current warming conditions, disrupting infrastructure, releasing carbon from soils, and altering seasonal water availability. Therefore, it is important to quantitatively map the change in the extent and depth of permafrost. We used satellite images of land-surface temperature to recognize and map the zero curtain, i.e., the isothermal period of ground temperature during seasonal freeze and thaw, as a precursor for delineating permafrost boundaries from remotely sensed thermal-infrared data. The phase transition of moisture in the ground allows the zero curtain to occur when near-surface soil moisture thaws or freezes, and also when ice-rich permafrost thaws or freezes. We propose that mapping the zero curtain is a precursor to mapping permafrost at shallow depths. We used ASTER and a MODIS-Aqua daily afternoon land-surface temperature (LST) timeseries to recognize the zero curtain at the 1-km scale as a "proof of concept." Our regional mapping of the zero curtain over an area around the 7000 m high volcano Ojos del Salado in Chile suggests that the zero curtain can be mapped over arid regions of the world. It also indicates that surface heterogeneity, snow cover, and cloud cover can hinder the effectiveness of our approach. To be of practical use in many areas, it may be helpful to reduce the topographic and compositional heterogeneity in order to increase the LST accuracy. The necessary finer spatial resolution to reduce these problems is provided by ASTER (90 m).

Keywords: land-surface temperature; zero curtain effect; MODIS; ASTER; permafrost; phase change

---

1. Introduction

Permafrost, defined as ground that stays below freezing for more than two years, is an integral part of the cryosphere that is predicted to rapidly degrade under current warming climatic conditions (e.g., [1,2]). A quarter of the land surface of the Earth is subject to permafrost-related processes [3] and 23–26% of the land area in the northern hemisphere is permafrost [4–6]. The loss of permafrost affects the regional water balance and changes landscapes and ecosystems in cold regions. Erosion of soil due to melting permafrost not only poses a significant uncertainty in infrastructure stability but may also accelerate the release of carbon from the permafrost into the atmosphere [1,7].

Traditional mapping send monitoring techniques of permafrost are accurate but are presently labor-intensive and rely on interpolating small numbers of locally measured data across large areas in order to study regional-scale processes. Remote sensing addresses this difficulty by making spatially dense measurements over vast regions. Previous remote sensing studies have inferred permafrost extent from the instantaneous zero isotherm at the land surface or from changes in the dielectric marking the phase transition of liquid water to ice [8–10].

We propose to use land-surface temperature (LST) measured from space-borne thermal-infrared (TIR) sensors to map patterns in the annual cycle of LST in order to ultimately detect freeze/thaw surface dynamics. Previous studies demonstrated that in autumn or spring, the LST of moist/frozen soils remains at ~0 °C while the phase transition between water and ice occurs. This period is a well-known phenomenon called the "zero curtain" in permafrost studies (e.g., [11]) and its presence can be used to distinguish seasonally freezing ground from freezing dry ground, in which there is no zero curtain. The premise of our approach is that if the zero curtain lasts for several days, its presence can potentially be recognized and mapped with time series of daily LST images.

Repetition of the zero curtain for two or more annual LST cycles can be used to identify candidate permafrost pixels, and seasonal asymmetry in the duration of the zero curtain can likely be used to distinguish seasonally frozen soil from permafrost. Putkonen [12] suggested from modeling, substantially confirmed by Yi et al. [13], that the zero curtain should be present over near-surface permafrost during freeze-up, but not during the thaw, and vice versa in seasonally frozen soil. Over deep permafrost (~130 cm), de Pablo et al. [14] reported zero curtains in both seasons but shorter at the end of the freezing season. Although the zero-curtain duration is affected by seasonal changes in soil moisture and by the onset date of snow cover [13], this asymmetry should allow further winnowing of candidate pixels.

This paper does not reach so far as to identify permafrost, but simply establishes that precursor maps of the zero curtain are feasible to make with satellite-borne TIR imagers. We tested our approach in an arid region of the Atacama Andes near the Ojos del Salado volcano, close to the border between Chile and Argentina, where previous studies had documented the presence of frozen ground and near-surface permafrost [15–18].

1.1. Zero Curtain in the Surface and Soil Temperatures

The occurrence of a zero curtain in the annual ground-temperature cycle is attributed to the phase change of moisture in the soil due to the release of excess latent heat during the freezing and thawing seasons (Figure 1). Because keeping soil near 0 °C for several days generally requires moisture in the soil, the detection of the zero curtain implies the presence of soil moisture. Soil moisture may be ephemeral or seasonal, but it may further imply the presence of buried ice-rich permafrost. Putkonen [12] proposed that, in addition to the presence of moisture in the soil, the thermal gradient in the soil must become 0 °C m^{-1} above the base of the active layer (depth of the thaw) for the zero curtain to form. It follows that the timing of the zero curtain indicates the presence of permafrost. During the thaw season, a high thermal gradient between the warming surface and the cold permafrost effectively conducts heat through the soil without forming a zero curtain; during freeze-up, the cold ground surface and the cold permafrost create a low thermal gradient, ensuring a condition at some depth in the soil where the temperature is at or near 0 °C for a prolonged period (Figure 1B). In other words, seasonally frozen ground not overlying permafrost may exhibit a zero curtain during the thaw, whereas soil underlain by permafrost may exhibit a zero curtain only during freeze-up.

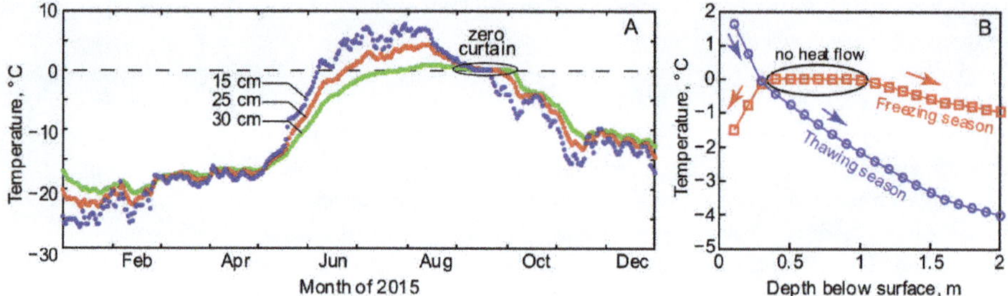

Figure 1. The zero curtain. (**A**) Daily ground temperatures (2 PM local) measured at various depths in the active layer above permafrost (location at 76.55414° N, 68.561680° W, 190 m asl) near Qaanaaq, Greenland. During 2015, there was a prominent zero-curtain effect due to the phase transition of water in the soil. During this time, the ground temperature was maintained at or near 0 °C at depths of 15 and 25 cm. Deeper than 30 cm the temperature never exceeded freezing, and near the immediate surface (not shown) the ground temperatures were commonly variable due to solar heating of the surface, complicating the effects of wind and cloudiness. This zero curtain was only observed from September to early October, during freeze-up, but not during the spring thaw. Unpublished data from R.S. Sletten. (**B**) Schematic from Putkonen [12] showing the modeled heat flow during freezing and thawing seasons. During freeze-up (red squares), the soil temperatures at the 0.5–1 m depth quickly reaches the freezing point, such that no heat diffusion upward (driven by the temperature gradient) can occur. The zero curtain will be maintained until all water in the active layer freezes. During the thaw (blue circles), the high thermal gradient in the soil ensures efficient heat flow and the soil will consistently warm with no zero curtain.

Yi et al. [13] observed that the duration of the zero curtain revealed information about the permafrost, specifically the depth of the active layer. The heat-flow plateau at ~0 °C shown schematically in Figure 1B (from Putkonen [12]) would persist longer over thicker active layers but would break down more rapidly over thin layers. Measuring and mapping the duration of the zero curtain, therefore, would supply important information about the weather conditions, their change with altitude and geographic position, and changes over time relevant to climate change.

Significant characteristics of the zero curtain may be measured by three parameters: (1) The starting date of the zero curtain; (2) the duration of the zero curtain; and (3) seasonal asymmetry in the duration. While the zero curtain itself indicates the presence of soil moisture, its starting date reveals the beginning of the annual thaw and/or freeze-up and its change over the years may indicate the changing nature of the climate and seasonal fluctuations of soil moisture. The duration of the zero curtain may respond to the thermal regime of the soil and the presence of underlying permafrost and the depth of the active layer.

The analysis of daily measurements of LST is less sensitive to measurement precision and cloud cover than a single measurement of LST. The occurrence of the zero curtain in LST, regardless of the seasons, strongly indicates the presence of moisture in the soil and provides an opportunity to distinguish between dry and wet soil and map at a regional scale. Therefore, such a remote-sensing approach may be appropriate for detecting and mapping ground ice at moderate (10^2–10^3 m) resolutions.

1.2. Remote Sensing and Numerical Modeling Studies of Permafrost

Most of the traditional permafrost studies rely on the recognition of landscape changes in the field and extrapolating over large regions (e.g., [19]). Such field expeditions can be expensive and prohibitive in remote regions. If the changes in the landscape are large enough to be seen with remotely sensed data, the permafrost-related landscapes and their changes can be mapped at a regional scale. For example, Nitze et al. [20] used visible/near-infrared Landsat imagery (30 m resolution) from 1999 to 2014 to quantify the expansion of thermokarst lakes and the increased slumping of frozen coastal areas due to melting of permafrost. Such studies are invaluable to document landscape changes in permafrost areas but are not sufficient to explain the changes in the thermal regime of the soil.

Microwave sensors are sensitive to soil moisture and temperature (e.g., [21]), and exploiting the fact that freezing changes the dielectric constant of moist soils, numerous studies of radar measurements have been used to map the state of freeze and thaw in the soils (e.g., [22,23]). Such regional-scale mapping approaches based on microwave sensors are useful to infer the changes in seasonally frozen grounds. However, the existence of permafrost underlying the seasonally frozen ground has not been inferred from just the dielectric constant or the state of freeze/thaw, and the resolution of the microwave sensors (~6–60 km) can be a limiting factor to detect small changes in lateral extent of permafrost areas. With the goal of mapping the depth of the active layer, Liu et al. [24] used interferometric synthetic aperture radar (InSAR) to monitor surface deformation over permafrost. Because ice and water have different densities, the ground surface settles during the thaw and the amount of surface subsidence can be used to estimate the thickness of the active layer if the vertical distribution of pore water within the soil is known. Wang et al. [25] used multi-temporal TerraSAR-X backscatter intensity and interferometric coherence along with the relationship between vegetation cover and the permafrost as the basis for classifying and mapping permafrost landscape features. Chen et al. [26] proposed a method to estimate the active layer thickness using time-series P-band polarimetric synthetic aperture radar (SAR) observations, achieving retrieval with errors <0.1 m in some cases. Yi et al. [13] combined field observations, active layer thickness, and soil moisture maps derived from low-frequency (L + P-band) airborne radar measurements, and global satellite environmental observations to investigate the sensitivity of the active layer to recent climate trends.

Thermal infrared (TIR) sensors provide direct measurements of skin temperatures of land surfaces. Widely used LST products are derived from the Advanced Spaceborne Thermal Emission and Reflection Radiometer (ASTER) onboard the Terra spacecraft, the Thematic Mapper (onboard Landsat), and from the Moderate Resolution Imaging Spectroradiometer (MODIS) onboard two satellites: Terra (with a morning overpass) and Aqua (with an afternoon overpass). Terra, Landsat, and Aqua share a common sun-synchronous orbit with different overpass times. Among the TIR imagers on these satellites, only MODIS provides daily measurements of LST. Numerical modeling studies based on the empirical relationship between the ground temperature and air temperature have taken advantage of the MODIS LST products (e.g., [27–29]) to derive static maps of the spatial distribution of permafrost. Recently, a numerical modeling study has combined the MODIS LST products with gridded air temperature data to derive time-series maps of permafrost extent [5]. These dynamic maps of changes in permafrost are important to assess the impact of a warming climate to the state of the permafrost. The approach proposed in this article is a contribution toward creating temporally and spatially dynamic maps of the quantified state of permafrost and seasonally frozen ground, complementary to the previous regional-scale studies.

2. Materials and Methods

2.1. Remotely Sensed and Climate Data

MODIS provides daily LST products (MOD11A1 and MYD11A1) at 1-km spatial resolution using TIR data measured from the Aqua and Terra platforms, respectively [30,31]. We used MYD11A1 (Aqua) daytime LST products from 2003 to 2017. The MODIS LST is estimated using the split-window algorithm [32] with an estimated uncertainty of ~4.5 °C [33]. We compiled the daily MODIS LST over the study area and converted the data for processing with MATLAB. Daily coverage is possible using oblique views from adjacent overpasses. The local time for MODIS measurements in the LST compilations, therefore, ranged from 1:10 to 2:50 PM.

We used an ASTER Surface Kinetic Temperature (AST_08; 90 m pixel^{-1}) and visible/near-infrared (VNIR) ASTER products (AST_L1B; 15 m pixel^{-1}) to assess the variability of the land surface and the related temperatures within a single MODIS pixel, and to check the distribution of snow. The ASTER products were derived from a scene acquired on 8 April, 21 August, 22 September, and 8 October of 2017 at 11:43 AM local time [34].

Air temperature variations over different altitudes were evaluated using the long-term (1981–2010) monthly mean air temperature data from the Global Historical Climatology Network and Climate Anomaly Monitoring System (GHCN CAMS; 0.5° pixel^{-1} [35]). These low-resolution air temperature data were resampled at a 30 m pixel^{-1} resolution using the SRTM 1 arc-second DEM [36] and the long-term (1981–2010) monthly mean lapse rate derived from the National Centers for Environmental Prediction and the National Center for Atmospheric Research reanalysis data (NCEP/NCAR; 2.5° pixel^{-1} [37]). We followed the procedures described in [38] to resample the air temperature data to a higher spatial resolution to see the fine spatial variations of air temperature.

2.2. "Threshold Window" Filtering Algorithm

We developed an algorithm in the MATLAB environment that analyzes the daily MODIS LST during the first and second six months of the year at each pixel, and tests whether the daily LST satisfies the following rules indicating the occurrence of a zero curtain:

(1)   LST must be between −3.5 and 3.5 °C (defined below as the LST during a zero-curtain event, or "zero-curtain LST");

(2)   The number of consecutive zero-curtain LSTs must be >3;

(3)   The number of days with missing data between two identified zero-curtain LSTs must be <3; and

(4)   The total number of zero-curtain LSTs must be >5.

In other words, we split the annual daily LST data into two halves, each containing the thaw or freeze-up period. In each half, we counted the number days during which the −3.5 < LST < 3.5 °C. We identified a zero curtain if the number of consecutive days with −3.5 < LST < 3.5 °C exceeded five. At the end of the counting iteration for a pixel, the starting and end date for the zero curtain were recorded, from which we calculated the difference, which we defined as the duration of the zero curtain. Currently, no correction is made for snow cover. Our algorithm returns as output for every pixel matrix for the zero-curtain duration and the starting dates for thawing and freezing seasons for each year. The output matrix was converted back to georeferenced raster files using the same gridding scheme and spatial resolution of MYD11A1 products. All maps and raster data produced from this study are available in georeferenced TIFF files, as well as the detailed descriptions of the data, in the online supplement to this article (Table S1).

2.3. Validation Sites

We chose three validation sites near the Ojos del Salado volcano in the Andes east of the Atacama Desert in Chile (Figure 2). At least one 1-km^2 area surrounding each study site is homogeneous in terms of composition and small-scale roughness, topographically flat, and sparsely vegetated. Climate data are given in Table 1. Only ~1% of the total surface area over the whole region including the validation sites is covered with shrubs, grassland, water bodies, and permanent snowfields or glaciers (supplementary Table S2). Two of the sites were chosen in areas that were previously found to have periglacial features [39] and possibly permafrost [16,40]. However, intermittent or thin snow cover on Ojos del Salado implies complicating factors due to increased soil moisture during the melt season and the possibility of near-surface ground ice. The sites were chosen at different altitudes, ranging from 3815 to 4910 m asl, to assess the variability in the zero curtain at various altitudes. The lowest site lacked periglacial features and was intended to act as control for the higher ones.

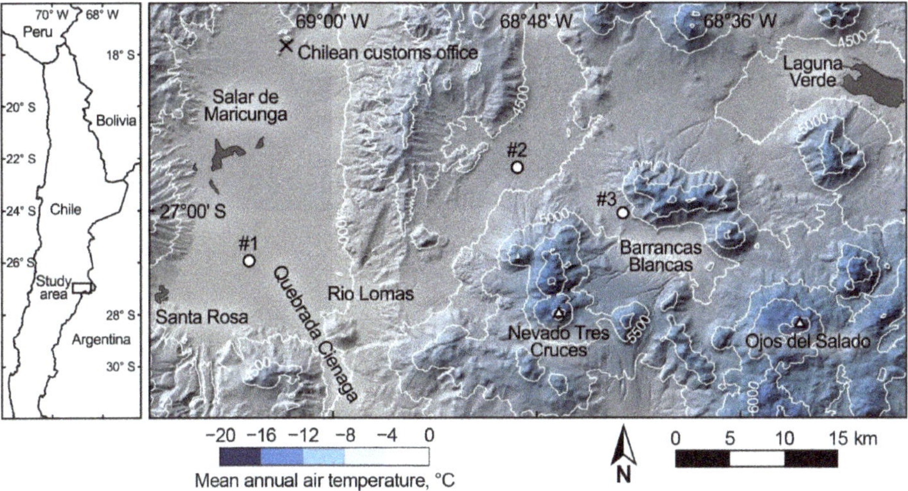

Figure 2. Maps showing the study area and the mean annual air temperature for 2018. The numbered circles indicate the locations of validation sites where we measured subsurface ground temperatures. The Chilean customs office is marked with an ✕ symbol. The air temperatures were calculated from the GHCN CAMPS long-term (1981–2010) monthly mean dataset of air temperature (0.5° pixel^{-1} [35]) and were scaled to the resolution of SRTM DEM (30 m pixel^{-1}) using the long-term (1981–2010) monthly mean NCEP/NCAR reanalysis dataset of lapse rate (2.5° pixel^{-1} [37]). The background hillshade image was constructed from 1 arcsec SRTM DEM (SRTMGL1).

Site #1 is located on one of the low-gradient (1°) alluvial fans (~140 km^2) emanating from the rivers Quebrada Cienaga and Rio Lomas (Figure 2). These rivers form braided channels that lead to Salar de Maricunga, but the channels remain dry most of the time. The surface of the fan is paved with rhyodacitic/andesitic (cf. Baker et al. [41]) gravel 2–3 cm in diameter and 1 cm thick, overlying silty sand. A few of the clasts are 3–5 cm in diameter and <2 cm thick. Site #2 is located on an alluvial fan (~13 km^2) below the volcanoes Nevado Tres Cruces. The surface of this fan is a low-gradient (~2°) rhyodacitic/andesitic/dacitic gravel pavement similar to Site #1, with gravel 2 cm in diameter and 1 cm thick. Site #3 (~1 km^2) is on a steeper alluvial fan (~10%) in the Valle de Barrancas Blancas and has the most complex surface in terms of composition (pyroclastic, rhyolitic, dacitic) and topography. It is covered with gravel 3–5 cm in diameter and <3 cm thick. Some large wind-polished boulders are scattered on the surface. Almost-parallel, shallow, dry channels ~20 cm deep dissect the surface. Small grasses and flowers may cover the site in summer. At all three sites, the percent of the surface covered by gravel was >90%. Photos showing the surface features at the validation sites are provided in supplementary Figures S1–S7.

Table 1. Validation sites, their locations, and local climate. Precipitation data are too low resolution to show differences among the sites.

Site	Location	Altitude, m asl	Mean Annual Climate Parameters		
			Precipitation [1], mm w.e.	Air Temperature [2], °C	Lapse Rate [3], °C km^{-1}
#1	27°02′54″ S, 69°04′52″ W	3815	58	−0.8	5.5
#2	26°57′31″ S, 68°49′09″ W	4415	58	−3.1	5.5
#3	27°00′09″ S, 68°42′55″ W	4910	58	−5.0	6.3

[1] All sites show the same precipitation due to 0.5° resolution of GPCC V7 data [42]; [2] [35]; [3] [37].

Surface features in the study region indicate periglacial processes and seasonal freeze and thaw. For example, in the Valle de Barrancas Blancas (Figure 2), sorted and unsorted patterned ground, surface extrusion due to cryoturbation, rock glaciers, cryoplanation surfaces, and slopes affected by gelifluction have been observed by Buchroithner and Trombotto Liaudat [43]. Nagy et al. [16] measured ground temperatures at six sites at different altitudes, ranging from 4200 to 6890 m asl, around Ojos del Salado during 2012–2016 to show that the seasonal freezing front reaches depths of at least 35–60 cm at these altitudes. Zero curtains were observed in both spring and autumn for two more weeks at all six sites, except for the two highest sites, at 6750 and 6890 m asl, where the ground temperatures remained below 0 °C for most of the year (see Section 2.4 below). On the basis of these measured ground temperatures, Nagy et al. [16] concluded that permafrost likely exists above 5250 m asl. The likelihood of permafrost at these altitudes is consistent with suggestions based on the mapping of permafrost and periglacial landforms [39,44] and numerical modeling of air temperatures [40]. Nagy et al. [16] concluded that it is unlikely for permafrost to exist below 4550 m asl.

2.4. In-Situ Measurements at Validation Sites

In November 2016, we installed TMC6-HC temperature sensors with an accuracy of ±0.25 °C at each site to measure subsurface ground temperatures every hour at depths of 2, 10, 20, and 40 cm (Figure 2; Table 1). The data from the thermometers were collected using HOBO® H-8 4-channel data loggers. The near-surface sensor was for comparison to the MODIS LST data, averaged over 1 km². The deeper subsurface sensors were to evaluate the 2-cm temperature in terms of the diurnal cycle. When we removed the probes (March 2018), we examined the soils. See soil descriptions at Site #1 (Table S3), at Site #2 (Table S4), and at Site #3 (Table S5) in the online Supplementary Materials.

3. Results

3.1. The Zero Curtain in the LST Data

The LST for the ~2 PM MODIS overflight of Ojos del Salado during the thawing season of 2017 is shown for Site #2 in Figure 3. The plateau near 0 °C between June 20 and July 14 is the zero curtain. 2 PM is near the peak of the diurnal temperature cycle, and thin ice films forming at night will likely not register then. The lack of LST records reflects the cloudy days.

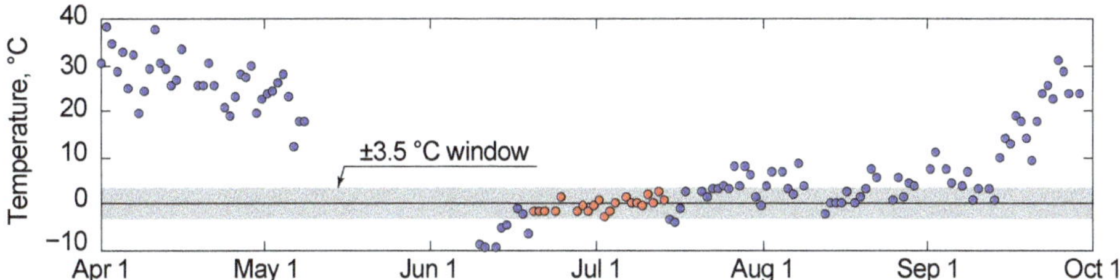

Figure 3. The LST time series for the thawing season, 2017, at Site #2. The zero-curtain (red data points) followed a protracted period of cloudiness during which there were no LST measurements of the ground surface.

MODIS LST images were consistently recorded for the three validation sites between 2003 and 2017. The seasonal temperature cycle is clearly visible in Figure 4, although the short zero curtains are not evident at this compressed temporal scale. The persistence of low winter temperatures was less at all three sites than the persistence of high summer temperatures. Gaps in the record are times when clouds obscured the ground. The maximum temperatures ranged among sites from ~50 °C at sites #1 and #2 to ~35 °C at Site #3. The minimum temperatures ranged much less, only rarely dipping below 0°C during the afternoon even at Site #3 (4910 m asl), 460 m above the altitude at permafrost was thought possible by Nagy et al. [16]. The LST at the lowest site, Site #1, consistently stayed above freezing for almost all years, consistent with the low probability of permafrost existence based on the surface mapping and numerical modeling of air temperatures [39,40,44]. The highest site, Site #3, near Barrancas Blancas shows consistent zero curtains during both freezing and thawing seasons.

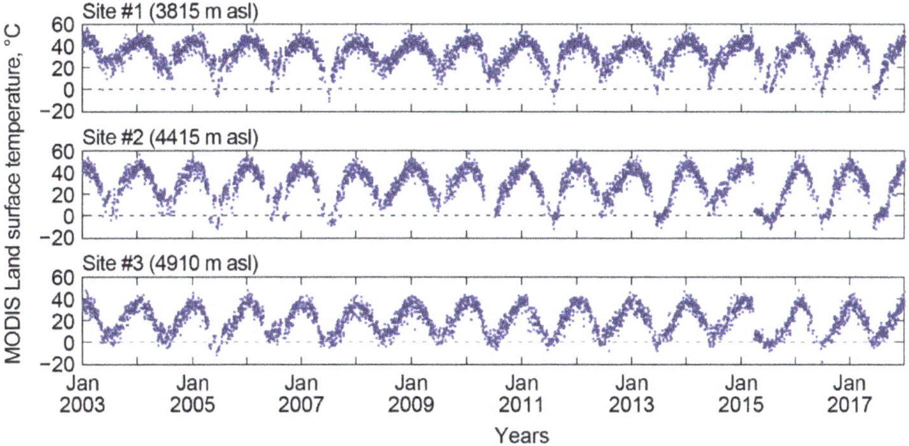

Figure 4. MODIS land surface temperatures at 2 PM (local) from 2003 to 2017 for the three validation sites.

3.2. Occurrence of Zero Curtains in the Study Area

We analyzed 3393 pixels (87 × 39 km grid) with a 1-km spatial resolution in the study area for each of the 15 years (2003–2017). Approximately 60% of the study area is located above 4550 m asl, the altitude above which permafrost likely exists per Nagy et al. [16]. We produced two types of maps for thawing and freezing seasons of the years 2003–2017 using the Threshold Window algorithm: (a) Number of days with zero curtain (duration), and (b) starting date of the zero curtain. The number of pixels with detected zero curtains (Figure 5) shows significant differences between the thawing and freezing seasons. The area with zero curtains during the freezing season remained largely constant throughout the years. However, the area with zero curtain during the thaw has increased over the years. A notable exception was during 2015, where the number of pixels with zero curtain during the freezing season was anomalously high.

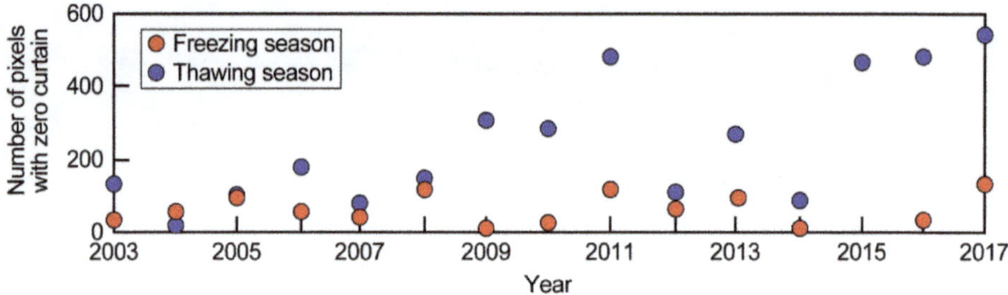

Figure 5. The number of pixels with zero curtain from 2003 to 2017. Out of the 3393 pixels analyzed for the study area (87 × 39 grid) the number of pixels with zero curtain during the thawing seasons have increased over time, while the values for the freezing seasons have remained approximately constant. Note the exception in 2015, for which the number of pixels with zero curtain during freeze-up was not plotted due to an anomalously high number of zero curtain pixels (1260).

Zero curtains were observed in ~5% to 40% of areas above 4550 m asl between 2003 and 2017. In the majority of cases, the zero curtain was observed only during a single season. In each year, except 2015, less than 50 pixels showed zero curtains in both seasons. In other words, there was very little overlap between the areas with zero curtain detected only during freeze-up and the zero curtain detected only in the thaw (supplementary Figure S8). During the thaw, the zero curtain was detected largely in areas at low altitudes (e.g., Figure 6).

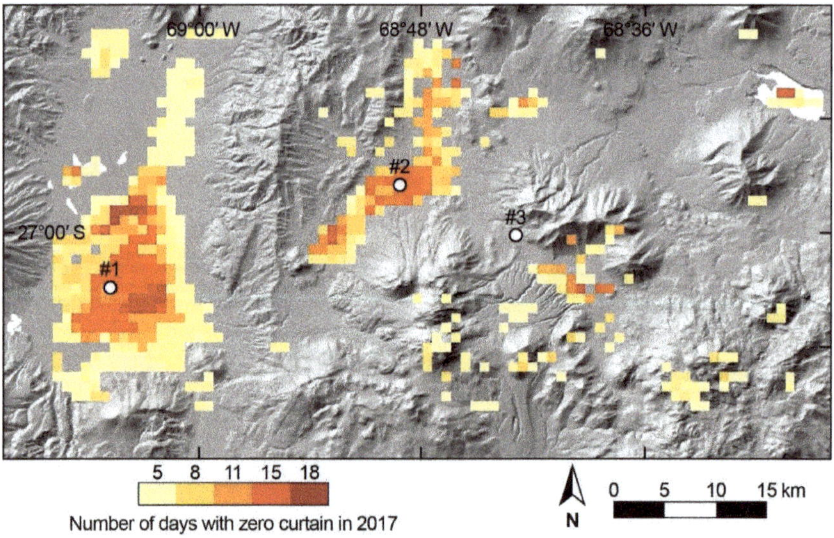

Figure 6. Number of days with zero curtains during the thaw season of 2017.

During freeze-up between 2003 and 2017, zero curtains were detected in late April to late May (supplementary Figure S9). In the higher altitude areas, the zero curtain started earlier than at lower altitudes, suggesting that the MODIS LST measurements reflect the temperature decrease with increasing altitude. During the thaw, zero curtains were detected between late June and early October.

Zero curtains during the thaw of 2003–2010 were detected mostly in August. However, after 2010, zero curtains were detected earlier, during late June to the middle of July (e.g., Figure 7). These earlier starting dates during the thaw seasons may reflect the recent warming trends in the air temperatures over the Andes [45].

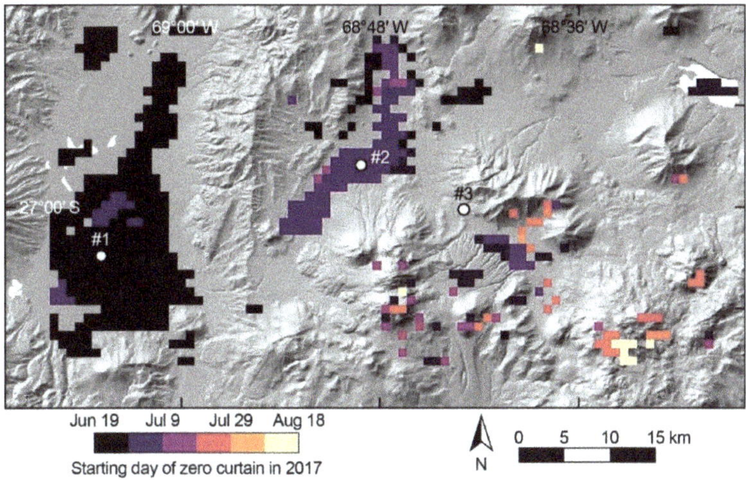

Figure 7. The starting day of the zero curtain, determined using the Threshold Window algorithm, during the thaw season of 2017.

3.3. Comparison Between MODIS LST and In Situ Measured Ground Temperatures

The MODIS daytime LST is a measure of the ~2 PM skin temperature integrated over each 1 km^2 pixel. The skin temperatures are subject to short-term fluctuations caused by gusts of wind and the shadows of high-altitude clouds located over adjacent pixels. The km-scale skin temperature is weighted by the temperatures of the gravel of the pavements on the fans, whereas the subsurface temperature probes record the temperature of the soil beneath the pavement. Despite such different representations of ground temperatures, the measurements of daily MODIS LST and subsurface ground temperatures at the 2 cm depth in the three validation sites during 2017 agree well during the warm seasons (Figure 8), demonstrating the robustness of MODIS LST measurements for monitoring purposes. However, the correlation is less well established during the cold seasons. During 2017, the ground temperatures at the three validations sites were consistent at all depths (2–40 cm) during May–September, suggesting that the surface was covered with snow (see MODIS snow cover data, MYD10A1 [46], in Figure 8). Both MODIS snow cover data and the in situ measurements of ground temperatures at Site #3 show the fewest number of days with snow cover.

The MODIS LST at sites #1 and 2 show 15 days of sustained zero curtain during the thawing season of 2017. However, during this time at both sites, the subsurface temperatures measured at 2 cm were ~5–10 °C lower than the LST values. Only at Site #2 did the ground temperatures at the 40 cm depth show a persistent zero curtain, during which time the MODIS LST appear to agree (Figure 8B). We discuss in Section 4 below the complexities in the TIR remote sensing and their implications for the MODIS LST data.

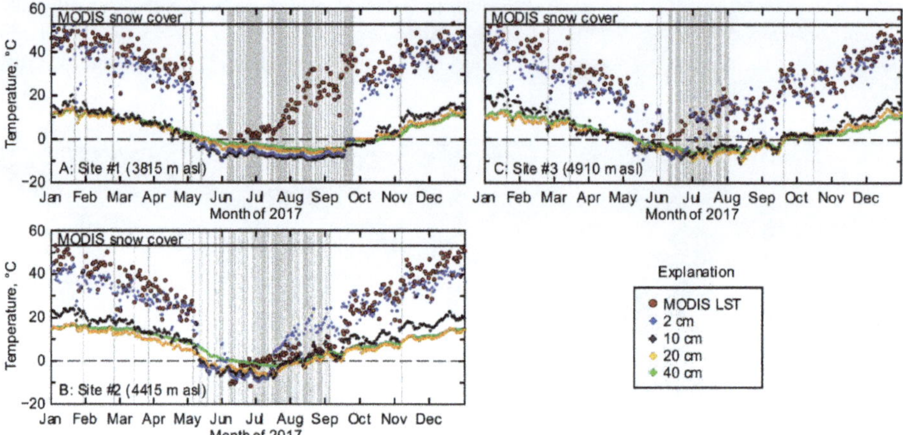

Figure 8. Validation site temperatures at 2 PM. (**A**) Site #1. Between early June and mid-July, LSTs were near freezing and the 2 cm temperatures were about the same as at 10 cm, by late September increasing to 0 °C, later rising to match the LST. (**B**) Site #2. There was a late-June zero curtain in the LST. After the thaw, the LST was lower than the 2 cm temperature. (**C**) Site #3. Subsurface zero curtains, but none in the LST, occurred during both and freeze-up. The MODIS snow cover data is MYD10A1 [46].

4. Discussion

4.1. The MODIS LST Product

The basis for the detection of the zero curtain in this study is the MODIS LST product, of low (1 km) spatial but high (daily) temporal resolution. The MODIS LST is constructed from MODIS bands 31 and 32 (10.78–11.28 and 11.77–12.27 µm) using a split-window algorithm that corrects the data for atmospheric effects. Calculating LST from atmospherically corrected surface-emitted radiance is sensitive to the thermal emissivity: An error of 0.01 at 11 µm (MODIS band 31) corresponds to a temperature error of 0.6 K in the inversion of the Planck equation for the single wavelength. The default surface type (and emissivity in MODIS band 31) is soil (0.97: [47]). Although the MODIS LST algorithm assumes emissivity values for different surface types [48–50], only two are relevant in our validation sites on gravel pavements: Rock and snow (band 31: ε = 0.98). Coincidentally, the dacitic and volcanic pebbles composing the surfaces of the validation sites have the same emissivity at 11 µm used by the MODIS LST algorithm for soil. Therefore, the composition of the geological surface is unlikely to introduce temperature errors in the output images. MODIS snow-cover products (MOD10 and MYD10) are integrated into MODIS LST so that the calculation uses the proper emissivities, whether for the gravel surface or snow. Any discrepancy in LST due solely to the emissivities of the geologic surface and snow cover would be too small to detect readily in the time histories of LST for the three validation sites (Figure 4).

The cloud cover is detected and masked out in making the MODIS LST products [51] so that only the unobscured ground surface is represented. However, the lack of daily LST data during the seasonal thaw and freezing seasons may hinder the opportunities to detect the zero curtain occurrence. The percentage of data loss due to clouds over the study region from 2003 and 2017 ranged from ~20% to 60% in the mountainous areas. The percentage of data loss during March–September from 2003 to 2017 at the validation Site #1 ranged from 16% to 42% whereas at sites #2 and 3 it ranged from 27% to 56% (supplementary Table S6 and Figure S10).

The MODIS LST represents the dynamic balance between heating due to absorption of sunlight, diffusion of heat into the soil, and cooling due to emitted thermal-infrared radiance. The 'true' kinetic surface temperature, on the other hand, is susceptible to additional environmental changes, such as wind, evaporation, or condensation of water. These typically will make the LST more variable than the subsurface measurements, which are buffered by a large mass of adjacent soil. During the zero curtain times in 2017, the difference between the MODIS LST and the in situ ground temperatures at the 2 cm depth ranged between 1 and 6 °C at the validation sites (Figure 8). Considering the accuracy of the MODIS LST product of ~1 K [52] and the in situ ground temperatures, especially at the 2 cm depth, were comparable for much of the year. Exceptions to this rule evident in Figure 8 are discussed below.

4.2. MODIS LST and Subsurface Temperature Profiles

Figure 8 compares the LST to the subsurface kinetic temperatures measured at the same time at the three validation sites. Although the LST and subsurface soil temperatures differ significantly depending on the type of vegetation covers (e.g., [22]), the surface at the validation sites remain barren most of the time (e.g., Table S2). As expected, the temperature at the 2 cm depth generally tracks the LST closely at all three sites, although it is lower by up to ~10 °C in summer. This may be due to the phase lag as the solar heating wave slowly penetrates the ground. In winter, the 2-cm temperature lowers to match the other subsurface values, but the afternoon LST rarely drops below zero.

Snow is highly reflective in the visible part of the spectrum (50–80%), yet 5% of the incoming solar energy is transmitted through cover 10–20 cm thick [53]. In contrast, snow absorbs efficiently in the thermal-infrared spectrum, such that the emitted radiation may originate only in the top 2 mm of the snow surfaces [54,55]. The nearly isothermal character of the subsurface as shown in Figure 8A is likely due to an insulating blanket of snow 10 cm or more in thickness, allowing the shallow subsurface to lose heat to deeper soil even as the snow surface is heated by the sun. In the limit, the LST of the snow cover could exceed the freezing point by at most a few °C, yet temperatures as high as 34 °C were recorded, even as the subsurface temperatures continued to decrease. This suggests that numerous large, sun-heated cobbles protruded from the thinning snowpack or that unresolved patches of snow-free ground had appeared but not over the subsurface sensors. By mid-October, the 2-cm temperature values have risen to match the LST, likely as the snow melted.

At Site #2 (Figure 8B) from mid-July until late September, the LST was lower than the temperature 2 cm down. The MODIS snow product indicates that there was snow cover during this time. This situation is the opposite of what occurred at the same season in Site #1. How could this happen, and what does it imply for the detection by MODIS of a zero curtain?

A likely explanation is that the snow cover was a thin veneer, thick enough so that the thermal infrared signal was radiated from its surface, unaffected by the ground beneath. On the other hand, the veneer was thin enough (i.e., <10–20 cm) that the ground was heated by irradiance from the sun at visible wavelengths. Thus, in mid-May, both the LST and the 2 cm temperature lowered to a few degrees below zero as the first snowfall was recorded.

During the time the ground was snow-covered or obscured by cloud, no meaningful zero curtain could be detected. This period included the likely freeze-up. By early June, the cloudiness cleared, and the surface of the snow rose from −10 °C to zero as the sun heated and thinned the snow cover. From mid-June to mid-July, the LST indicated a zero curtain. Thus, misleading zero curtains that pertain to thawing snow can be returned by the "Threshold Window" filtering algorithm.

At the high-altitude Site #3 (Figure 8C), subsurface zero curtains (10, 20, and 40 cm depth) occurred during both thaw (early May) and freeze-up (early October), but none occurred at the 2 cm depth or at the surface, both of which were 20 °C or warmer. Furthermore, except for the temperatures at 2 cm, the subsurface sites had similar temperatures throughout the winter, between the zero curtains. MODIS reported no snow cover before June or after August, so that snow has no role in explaining these relationships then. During the winter, however, on individual days, temperatures at both 2 cm and the immediate surface dropped ~12 °C to near zero, an occurrence likely associated with ephemeral snow cover. During the summer, temperatures at 20 and 40 cm remained close to each other while temperatures at 10 cm rose to intermediate values between the temperatures at the deeper sensors and the 2 cm depth. These observations likely resulted from solar warming of the surface. At Site #3, there was sufficient soil moisture to allow zero curtains at depth, but the circumstances precluded any at the surface.

4.3. Effects of Scene Roughness

The alluvial fan surfaces at the validation sites were compositionally homogeneous and flat on a coarse km scale. However, we noticed that local fine-scale topography was variable, ranging from dry channels to small periglacial patterned grounds (supplementary Figures S1–S5). The manifestation of topographical variation in the emitted radiance at the thermal infrared spectrum can be seen in Figure 9, where subtle topographic features not discernible in the photograph can be resolved as bright pixels in the FLIR image. Inclusion of these features in low-resolution pixels causes the histogram of temperatures within those pixels to broaden. For example, the standard deviations within pixel aggregations simulating lower-resolution data in the plain in Figure 9 range from about 0.2 K at 2 m pixel^{-1} to 2.5 K at 380 m pixel^{-1}. Topographic effects influencing the effective surface temperatures at the ASTER or MODIS scales are not large enough to affect the analysis in this study significantly. However, during the thaw, the radiance from unresolved patches of snow will mix with that from adjacent bare soil to lower the effective daytime LST.

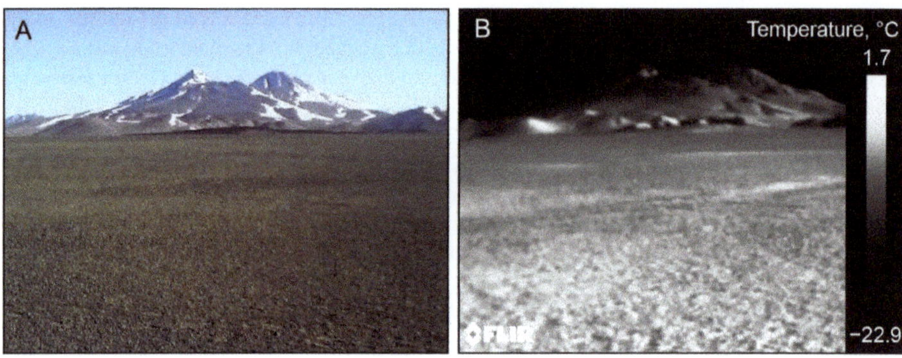

Figure 9. Validation Site #2 (26.9586° S and 68.8190° W, 4415 m asl) looking east over on a flat homogeneous alluviated plain. (**A**) View from the ground; (**B**) FLIR thermal-infrared image taken from the same spot and view angle. The images are about 6 km across at the foot of the mountain. The resolution of the FLIR image ranges from ~2 cm pixel^{-1} in the foreground to ~15 m pixel^{-1} at the foot of the mountain, ~12 km away.

4.4. Spatial Resolution and Radiance Mixing

Although the validation sites appear to be homogeneous from the standpoint of the MODIS LST data, analysis of zero curtains elsewhere will encounter more complex surfaces and in any case there may be seasonal patches of snow unresolved by MODIS. ASTER, at 90 m pixel^{-1}, will produce more unmixed ("pure") pixels in complex terrain, and thus will produce more pixels in which reliable assessments of the zero curtain can be made (Figure 10). Small seasonal snow patches provide one example, important in this study, of temperature mixing that may be reduced with improved resolution.

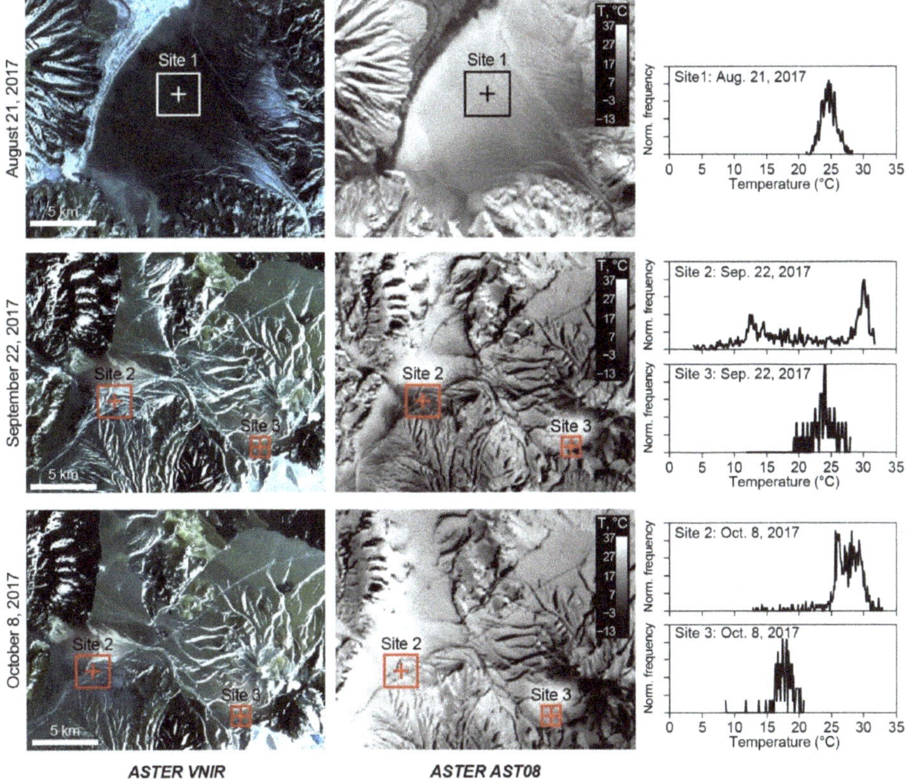

Figure 10. ASTER image pairs (VNIR and Surface Kinetic Temperature, AST08) for three dates in 2017. On 21 August, a 1-km^2 MODIS pixel indicated Site #1 to be snow-covered, which the 15-m pixel^{-1} ASTER VNIR image disagrees with. On 22 September, MODIS indicated both sites #2 and #3 to be snow-free; ASTER shows them to be largely bare. On 8 October, MODIS indicates both sites #2 and #3 to be snow-free, which ASTER confirms. The MODIS snow classifications are binary (snow or no-snow). Modal temperatures (11:43 AM local AST08) for snow-free pixels are 23–30 °C and 12 °C for largely snow-covered pixels (22 September, SE corner Site #2). Even the lowest temperatures for snow cover or partial cover were above zero, implying significant temperature mixing with sun-heated rocks rising above the snow.

For this study, the temperature window in the Threshold Window filtering algorithm was set at ±3.5 °C to account for the MODIS LST accuracy of ~1 °C (1 σ), any variations in emissivity of different types of rocks, and effects of local topographic features that cannot be resolved at the spatial resolution of MODIS LST data. Many but not all mixed pixels of soil and snow will be filtered out with this setting.

Adjusting the size of the Threshold Window decreases or increases the number of data points to be counted as part of a zero curtain, decreasing or increasing false negatives or false positives, but a better approach might be to reduce the pixel size. For example, horizontal changes in the extent of permafrost in Tibet Plateau can be estimated to ~460–920 m in 30 years after accounting for the maximum vertical change of permafrost base during the same period (80 m [56,57]) and the average slope of the interior of the plateau (~5–10° [58]). The threshold window algorithm cannot capture changes at such a 100 m scale due to the relatively low resolution of the MODIS LST data. This limitation might be overcome by using ASTER data, with its higher spatial resolution than MODIS, but repeat acquisitions of ASTER images are too infrequent. A compromise may be possible with data fusion: Surface compositions can be mapped using the 15-m pixel−1 visible and 30-m pixel−1 near-infrared data while variations in the surface emissivity can be resolved from the 90-m pixel−1 ASTER products (Figure 11). With this information and the daily LST, it may be possible to generate the higher 90-m spatial resolution of the ASTER surface temperature product using the STARFM algorithm [59]. Data fusion might also be used with the high spatial but irregular and temporal resolution data of ECOSTRESS (Table 2). However, to monitor subtle changes in permafrost extent in the future, we likely will need data with both high spatial and high temporal resolution.

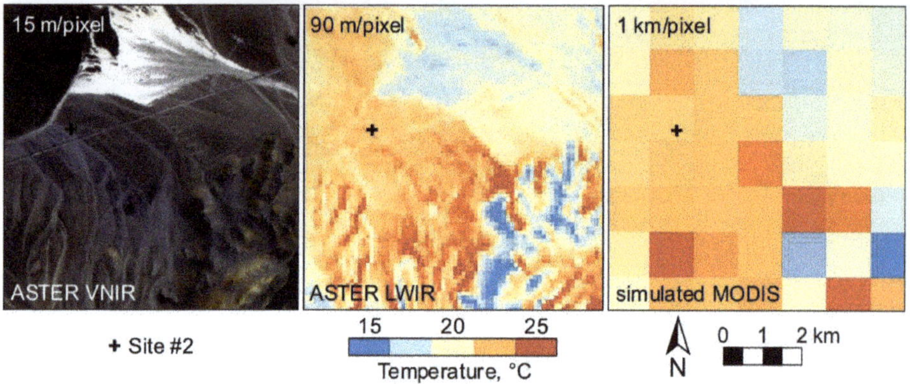

Figure 11. Effect of resolution. The heterogeneity in topographic and geologic features visible from an ASTER 15-m VNIR image (left, AST_L1B_00304082017144319 (11:43 AM local): RGB = band 3 (0.81 μm), band 2 (0.66 μm), band 1 (0.56 μm)) are resolved less well in an ASTER 90-m Surface Kinetic Temperature product (middle, AST_08_00304082017144319) produced from TIR bands (RGB=band 13 (10.6 μm), band 11 (9.1 μm), band 10 (8.3 μm)). Validation Site #2 (marked with +) was chosen on an alluvial plain so that the variation of topography and surface composition under a single MODIS pixel is minimal. An image of the ASTER Surface Kinetic Temperature image resampled to the 1-km resolution of MODIS (left) shows significant mixing in high relief and compositionally varied sites (e.g., lower right part of the scene). Higher resolution is necessary for accurate estimations of LST.

Table 2. Comparison of thermal-infrared sensors.

Sensor	Resolution		Launch Date
	Spatial	Temporal	
GOES	Low (4 km)	High (3 h)	1981
AVHRR (NOAA)	Moderate (1 km)	High (daily)	1979
MODIS	Moderate (1 km)	High (daily)	2000
ASTER	High (90 m)	Low (16 days)	2000
Thematic Mapper	High (60 m)	Low (16 days)	1982
ECOSTRESS [1]	High (38 × 69 m)	1 h of science data day^{-1}	2019

[1] The ECOSTRESS products are in development and data are available only through "Early Adopters" program.

4.5. Seasonal Zero-Curtain Duration

Putkonen [12] stated that there is an asymmetry in the occurrence of the zero curtain during spring and autumn over permafrost, where the zero curtain appears during freeze-up, and over seasonally frozen ground, where the zero curtain appears during thaw. Figure 12 examines this asymmetry both as a correlation between the duration of freeze-up and thaw and as a function of altitude in the validation sites during 2015. Above ~5200 m asl, the zero-curtain duration of both freeze-up and thaw is about the same; at lower altitude, the duration of freeze-up exceeds the duration of thaw~40%.

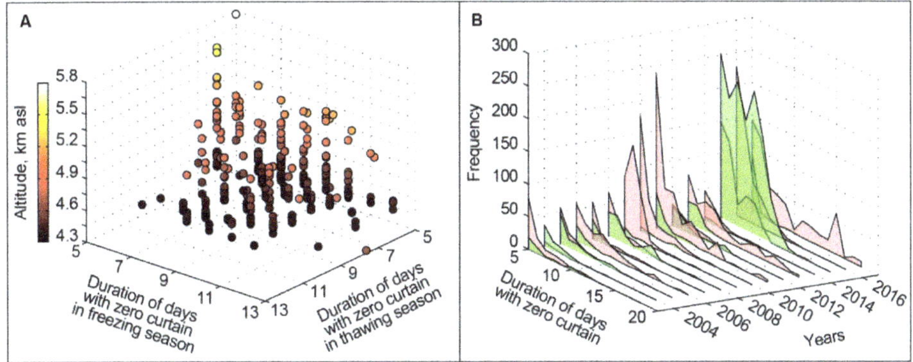

Figure 12. The zero-curtain duration for all pixels in which zero curtains were detected during both freeze-up and thaw. The zero curtain was only recognized for durations of five days or more. (**A**) Versus altitude (2015 only). Two-season zero curtains were not detected below 4300 m asl due to warm temperatures there. Above 5800 m asl, the ground was too cold for persistent thaws. At intermediate altitudes, there was sufficient soil moisture for both near-surface freezing and thawing. (**B**) Zero-curtain duration during freeze-up (red) and thaw (green) plotted against the corresponding years.

Is the weak asymmetry revealed in Figure 12A due to a systematic trend, or is it the result of weather anomalies? In late March of 2015, before the general freeze-up, an extreme heat event occurred in central and northern Chile followed by an anomalously high precipitation event (>20 mm day^{-1} [60]). Compared to other years, this extreme weather event must have supplied enough moisture to the soil to have resulted in the anomalously high number of pixels with zero curtains during both freeze-up and thaw that year (Figure 12A), so the anomaly did not increase any apparent asymmetry. During freeze-up, the zero curtain was detected in early April in high-altitude areas and in early June in low-altitude areas. The zero curtain during this time persisted for about two weeks. Later in the year, during the thaw, the zero curtain was detected only around early July with no contrast between the high- and low-altitude areas. The anomalously high precipitation from the March event may have been preserved in the soil over the winter and the increased soil moisture may have contributed to the sustained zero curtain during the thaw of 2015.

The weather anomaly aside, Figure 12B shows that the in the study area the maximum duration of the zero curtain during freeze-up, but not during thaw, may have increased erratically over the study period. The increase during freeze-up suggests that changes in the zero curtain can be detected on the decadal scale. The more constant and smaller number of zero curtain days during thaw indicates that the results are not random and that there has been little change. Taken together, the data plotted in Figure 12 suggest that there is little evidence at Ojos del Salado for the strong asymmetry predicted by Putkonen [12] for the zero-curtain freeze-up/thaw duration over permafrost. It may be that summer, but not winter, precipitation has been increasing in recent years but is not sufficient to carry over to the thaw. In all, these results are encouraging for the use of the zero curtain in climate-change studies.

4.6. Potential for Quantified Mapping of Seasonally Frozen Ground and Permafrost

The occurrence of the zero curtain requires moisture in the soil and low temperatures. Therefore, the identification of a zero curtain is a useful indicator of ice-rich frozen ground and possibly permafrost. We identify the following four parameters for the zero curtain that can be quantified using the Threshold Window algorithm: (a) Starting date, (b) duration, (c) seasonality, and (d) persistence over more than two seasons. The changes in the starting date of a zero curtain over time may indicate changes in the thermal regime of the soil due to changing environmental and climate conditions. According to Putkonen [12], the thermal regime in the soil would be different depending on the existence of permafrost underneath, and the asymmetric seasonality of the zero curtain may indicate whether it is due to seasonally frozen ground or the possible existence of permafrost. The intensity of freezing or thawing may dictate the duration of zero curtain, the changes of which can be mapped over time. On the basis of these parameters and the criteria for classifications of permafrost conditions (e.g., [61]), it is possible that a quantitative map of permafrost extent and the depth of the active layer can be estimated.

The probabilistic horizontal extent of permafrost based on the empirical relationship between ground temperature and mean annual air temperature has been mapped (e.g., [40]). The detection of permafrost using the Threshold Window algorithm, therefore, can validated using the probabilistic maps of permafrost produced from climate and geomorphological data. Further, the depth of the active layer can be estimated by numerically modeling heat diffusion into the ground using surface temperatures estimated from ASTER or MODIS data as constraints. For example, a heat-transfer model of soils that accounts for different estimates of moisture content [62,63] was run using MODIS LST (8-day mean) as an initial condition and used to numerically construct the thermal profiles of soils in a well-known permafrost region of Tibet (Figure 13; [64]). The ground temperatures calculated at various depths over time allowed the depth to the zero isotherm to be estimated for 2002 and 2009, consistent with the depths to the base of the active layer measured in the field [65]. Figure 13 shows that the depth of the zero isotherm decreased about 20 cm over the 7-year period. The model suggests that there was an extended period when −5 °C < LST <5 °C that may have been a zero curtain (intersection of the dark blue color and the top of the plot) in April–May 2002, and that it was shorter in 2009. Similarly, a zero curtain was predicted for October, and it too was shorter in 2009. Thus, this model suggests that there may have been a zero curtain not just at freeze-up but also during the spring thaw, inconsistent with Putkonen [12] (Figure 4). Further testing of the model will be necessary to resolve this issue.

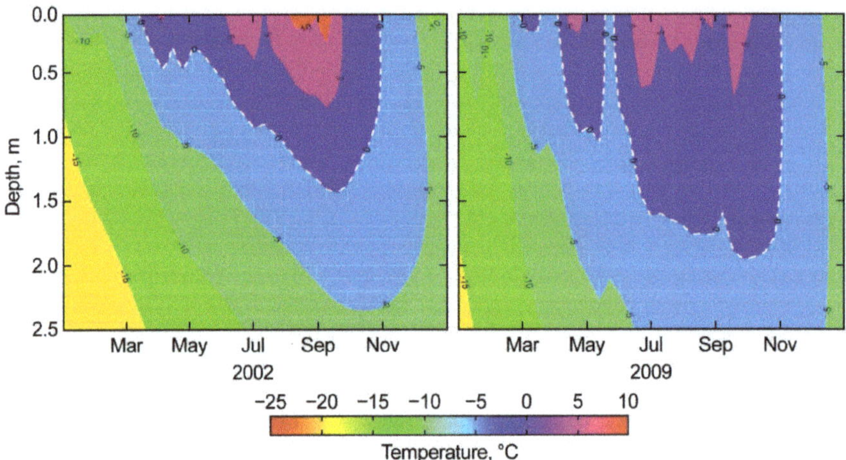

Figure 13. The thermal profiles calculated using a heat-transfer model [62–64] for an ice-rich permafrost region in the Fenghuo Shan site (34.69°N and 92.89°E, 4938 m asl) in Tibet. The white dashed lines are the 0 °C isotherms that are used to determine the active-layer thickness. The modeled values are consistent with in-situ measured changes in active-layer thickness from 1.4 to 1.9 m between 2002 and 2009 [65].

5. Summary and Conclusions

We proposed a new approach to detect the occurrence of the zero curtain in land-surface temperature images, maintenance of the temperature at or near 0 °C due to the release of latent heat from freezing or thawing moisture in the soil. We used a "Threshold Window" filtering algorithm to analyze daily MODIS land surface temperature (LST) data over a cold region of the Atacama Andes in Chile, where periglacial features have been documented. Our results demonstrated that the zero curtain could be consistently identified in the MODIS data. This demonstration opens the path to identifying ice-rich permafrost using time series of LST images.

The duration, seasonality, and the starting time of the zero curtain for each year between 2003 and 2017 were mapped. We measured subsurface ground temperatures at depths of 2–40 cm at three validation sites located at altitudes of 3815, 4415, and 4910 m asl in 2017. The in situ observations of subsurface temperatures complemented the MODIS LST records and showed that, with exceptions during the spring, the LST agreed within a few degrees with temperature at the 2 cm depth. The subsurface temperature records showed clearly the presence of the zero curtain at depth, out of phase with the zero curtain at the surface, and illustrated the effects of thinning snow cover on the recovered zero curtain. Over the period of the study, the duration of the zero curtain during the thaw and freeze-up seasons appeared to fluctuate, consistent with what might be expected from annual changes in weather conditions. However, on Ojos del Salado, the zero curtains at freeze-up tended to be longer than those at thaw, and also may have shown a tendency to lengthen over the study period. We did not find convincing evidence of permafrost in our data or in the field, although there was local evidence of segregation ice.

Our test sites were chosen to minimize vegetation cover that would interfere with satellite measurement of the land-surface temperature, but in practical applications, vegetation cover and snow may hinder the availability of remotely sensed LST for this approach. Small-scale variations in surface composition and topographic conditions create different thermal components that are integrated over large areas for products with moderate spatial resolution, such as MODIS LST. Thermal-infrared sensors with higher spatial resolution, such as ASTER and ECOSTRESS, can be used to resolve the 100-m-scale horizontal changes in the extent of seasonally frozen ground and permafrost. Numerical modeling of the thermal regime of the ground using these surface temperature records may be useful to estimate the depth of thaw in ice-rich frozen grounds and permafrost.

Supplementary Materials: The following are available online at http://www.mdpi.com/2072-4292/12/4/695/s1: Figure S1: Surface features at Site #1, Figure S2: Surface features at Site #2, Figure S3: Platy structure and patterned ground at Site #2, Figure S4: Frost cracks and platy structure in the soil at Site #2, Figure S5: Surface features at Site #3, Figure S6: The 'nieves penitentes' near the Site #3, Figure S7: The nieves penitentes near Laguna Verde, Figure S8: Maps of zero-curtain duration from 2003 to 2017, Figure S9: Maps of zero-curtain starting day from 2003 to 2017, Figure S10: Map of data loss due to clouds in 2017, Table S1: Details of raster data provided as supplementary, 15 geotiff files for annual zero curtain duration, 15 geotiff files for annual zero curtain starting time, Table S2: Annual land-cover classification estimated from MODIS over the study region, Table S3: Soil descriptions at Site #1, Table S4: Soil descriptions at Site #2, Table S5: Soil descriptions at Site #3, Table S6: MODIS LST data loss due to clouds from 2003 to 2017.

Author Contributions: Conceptualization, A.R.G., and J.B.; methods, J.B., A.R.G., and R.S.S.; validation, J.B., A.R.G., R.S.S.; data analysis, J.B.; field investigation, J.B., A.R.G., R.S.S., A.M., R.A., D.T.L.; writing—original draft preparation, J.B.; writing—review and editing, all authors; visualization, J.B., A.M., L.L. All authors have read and agreed to the published version of the manuscript.

Funding: Fieldwork was funded by the Quaternary Research Center, University of Washington. Partial funding for J.B. was from the College of the Environment, University of Washington. Funding for A.R.G. was from the Bear Fight Institute, NASA subcontract No. 1545008.

Acknowledgments: Thanks to Laura Gilson for downloading and the initial processing of the MODIS data. The ASTER images used in this study are courtesy of NASA/METI, LP DAAC. The MODIS images are courtesy of NASA/GSFC/SED/ESD/TISL/MODAPS, LP DAAC. We thank Sebastián Ruiz Pereira, Pontificia Universidad Católica de Chile for the weather data, and two anonymous reviewers for helpful suggestions.

Conflicts of Interest: The authors declare no conflict of interest.

References

1. Schuur, E.A.G.; McGuire, A.D.; Schadel, C.; Grosse, G.; Harden, J.W.; Hayes, D.J.; Hugelius, G.; Koven, C.D.; Kuhry, P.; Lawrence, D.M.; et al. Climate change and the permafrost carbon feedback. *Nature* **2015**, *520*, 171–179. [CrossRef]
2. Biskaborn, B.K.; Smith, S.L.; Noetzli, J.; Matthes, H.; Vieira, G.; Streletskiy, D.A.; Schoeneich, P.; Romanovsky, V.E.; Lewkowicz, A.G.; Abramov, A.; et al. Permafrost is warming at a global scale. *Nat. Commun.* **2019**, *10*, 264. [CrossRef]
3. French, H.M. *The Periglacial Environment*, 3rd ed.; John Wiley & Sons: Chichester, UK, 2007.
4. Smith, S.; Brown, J. *Assessment of the Status of the Development of Standards for the Terrestrial Essential Climate Variables—T7—Permafrost and Seasonally Frozen Ground*; Global Terrestrial Observing System: Rome, Italy, 2009.
5. Obu, J.; Westermann, S.; Bartsch, A.; Berdnikov, N.; Christiansen, H.H.; Dashtseren, A.; Delaloye, R.; Elberling, B.; Etzelmüller, B.; Kholodov, A.; et al. Northern Hemisphere permafrost map based on TTOP modelling for 2000–2016 at 1 km^2 scale. *Earth-Sci. Rev.* **2019**, *193*, 299–316. [CrossRef]
6. Dobiński, W. Northern Hemisphere permafrost extent: Drylands, glaciers and sea floor. Comment to the paper: Obu, J.; et al. 2019. Northern Hemisphere permafrost map based on TTOP modeling for 2000–2016 at 1 km^2 scale. *Earth-Sci. Rev.* **2019**, *193*, 103037. [CrossRef]
7. Turetsky, M.R.; Abbott, B.W.; Jones, M.C.; Anthony, K.W.; Olefeldt, D.; Schuur, E.A.G.; Koven, C.; McGuire, A.D.; Grosse, G.; Kuhry, P.; et al. Permafrost collapse is accelerating carbon release. *Nature* **2019**, *569*, 32–34. [CrossRef]
8. He, H.; Dyck, M. Application of Multiphase Dielectric Mixing Models for Understanding the Effective Dielectric Permitivity of Frozen Soils. *Vadose Zone J.* **2013**, *12*. [CrossRef]
9. Duncan, B.N.; Ott, L.E.; Abshire, J.B.; Brucker, L.; Carroll, M.L.; Carton, J.; Comiso, J.C.; Dinnat, E.P.; Forbes, B.C.; Gonsamo, A.; et al. Space-Based Observations for Understanding Changes in the Arctic-Boreal Zone. *Rev. Geophys.* **2020**, *58*, e2019RG000652. [CrossRef]
10. Du, J.; Watts, J.D.; Jiang, L.; Lu, H.; Cheng, X.; Duguay, C.; Farina, M.; Qiu, Y.; Kim, Y.; Kimball, J.S.; et al. Remote Sensing of Environmental Changes in Cold Regions: Methods, Achievements and Challenges. *Remote Sens.* **2019**, *11*, 1952. [CrossRef]
11. Outcalt, S.I.; Nelson, F.E.; Hinkel, K.M. The zero-curtain effect: Heat and mass transfer across an isothermal region in freezing soil. *Water Resour. Res.* **1990**, *26*, 1509–1516.
12. Putkonen, J. What dictates the occurrence of zero curtain effect? In *Ninth International Conference on Permafrost*; Kane, D.L., Hinkel, K.M., Eds.; Institute of Northern Engineering, University of Alaska Fairbanks: Fairbanks, AK, USA, 2008; pp. 1451–1455.
13. Yi, Y.; Kimball, J.S.; Chen, R.H.; Moghaddam, M.; Reichle, R.H.; Mishra, U.; Zona, D.; Oechel, W.C. Mapping permafrost landscape features using object-based image classification of multi-temporal SAR images. *Cryosphere* **2018**, *12*, 145–161. [CrossRef]
14. De Pablo, M.A.; Ramos, M.; Molina, A. Thermal characterization of the active layer at the Limnopolar Lake CALM-S site on Byers Peninsula (Livingston Island), Antarctica. *Solid Earth* **2014**, *5*, 721–739. [CrossRef]
15. Trombotto Liaudat, D. Geocryology of southern South America. In *Late Cenozoic of Patagonia and Tierra del Fuego*; Developments in Quaternary Sciences; Rabassa, J., Ed.; Elsevier: Amsterdam, The Netherlands, 2008; Volume 11, pp. 255–268.
16. Nagy, B.; Ignéczi, A.; Kovács, J.; Szalai, Z.; Mari, L. Shallow ground temperature measurements on the highest volcano on Earth, Mt. Ojos del Salado, Arid Andes, Chile. *Permafr. Periglac.* **2019**, *30*, 3–18. [CrossRef]
17. Baldis, C.T.; Trombotto Liaudat, D. Rockslides and rock avalanches in the Central Andes of Argentina and their possible association with permafrost degradation. *Permafr. Periglac.* **2019**, *30*, 300–304.
18. Obu, J.; Westermann, S.; Kääb, A.; Bartsch, A. *Ground Temperature Map, 2000–2016, Andes, New Zealand and East African Plateau Permafrost*; PANGAEA: Oslo, Norway, 2019.
19. French, H. Recent Contributions to the Study of Past Permafrost. *Permafr. Periglac.* **2008**, *19*, 179–194. [CrossRef]
20. Nitze, I.; Grosse, G.; Jones, B.M.; Romanovsky, V.E.; Boike, J. Remote sensing quantifies widespread abundance of permafrost region disturbances across the Arctic and Subarctic. *Nat. Commun.* **2018**, *9*, 5423. [CrossRef]
21. Kumar, S.V.; Dirmeyer, P.A.; Peters-Lidard, C.D.; Bindlish, R.; Bolten, J. Information theoretic evaluation of satellite soil moisture retrievals. *Remote Sens. Environ.* **2018**, *204*, 392–400. [CrossRef]

22. Prakash, S.; Norouzi, H.; Azarderakhsh, M.; Blake, R.; Khanbilvardi, R. Potential of satellite-based land emissivity estimates for the detection of high-latitude freeze and thaw states. *Geophys. Res. Lett.* **2017**, *44*, 2336–2342. [CrossRef]

23. Kim, Y.; Kimball, J.; Glassy, J.; McDonald, K. *MEaSUREs Northern Hemisphere Polar EASE-Grid 2.0 Daily 6 km Land Freeze/Thaw Status from AMSR-E and AMSR2, Version 1*; National Snow and Ice Data Center: Boulder, CO, USA, 2018.

24. Liu, L.; Schaefer, K.; Zhang, T.; Wahr, J. Estimating 1992–2000 average active layer thickness on the Alaskan North Slope from remotely sensed surface subsidence. *J. Geophys. Res.* **2012**, *117*, F01005. [CrossRef]

25. Wang, L.; Marzahn, P.; Bernier, M.; Ludwig, R. Mapping permafrost landscape features using object-based image classification of multi-temporal SAR images. *ISPRS J. Photogramm.* **2018**, *141*, 10–29. [CrossRef]

26. Chen, R.H.; Tabatabaeenejad, A.; Moghaddam, M. Retrieval of permafrost active layer properties using time-series P-band radar observations. *IEEE T. Geosci. Remote* **2019**, *57*, 6037–6054. [CrossRef]

27. Westermann, S.; Langer, M.; Boike, J.; Heikenfeld, M.; Peter, M.; Etzelmüller, B.; Krinner, G. Simulating the thermal regime and thaw processes of ice-rich permafrost ground with the land-surface model CryoGrid 3. *Geosci. Model Dev.* **2016**, *9*, 523–546. [CrossRef]

28. Aalto, J.; Karjalainen, O.; Hjort, J.; Luoto, M. Statistical Forecasting of Current and Future Circum-Arctic Ground Temperatures and Active Layer Thickness. *Geophys. Res. Lett.* **2018**, *45*, 4889–4898. [CrossRef]

29. Tao, J.; Koster, R.D.; Reichle, R.H.; Forman, B.A.; Xue, Y.; Chen, R.H.; Moghaddam, M. Permafrost variability over the Northern Hemisphere based on the MERRA-2 reanalysis. *Cryosphere* **2019**, *13*, 2087–2110. [CrossRef]

30. Wan, Z.; Hook, S.; Hulley, G. MOD11A1 MODIS/Terra Land Surface Temperature/Emissivity Daily L3 Global 1km SIN Grid V006, Distributed by NASA EOSDIS Land Processes DAAC. 2015. Available online: https://doi.org/10.5067/MODIS/MOD11A1.006 (accessed on 15 April 2018).

31. Wan, Z.; Hook, S.; Hulley, G. MYD11A1 MODIS/Aqua Land Surface Temperature/Emissivity Daily L3 Global 1km SIN Grid V006, Distributed by NASA EOSDIS Land Processes DAAC. 2015. Available online: https://doi.org/10.5067/MODIS/MYD11A1.006 (accessed on 15 April 2018).

32. Brown, O.B.; Minnett, P.J. *MODIS Infrared Sea Surface Temperature Algorithm*; Algorithm Theoretical Basis Document, NASA Contract Number NAS5-31361; University of Florida: Miami, FL, USA, 1994.

33. Hulley, G.C.; Hughes, T.; Hook, S.J. Quantifying uncertainties in land surface temperature (LST) and emissivity retrievals from ASTER and MODIS thermal infrared data. *J. Geophys. Res.* **2012**, *117*, D23113. [CrossRef]

34. NASA/METI/AIST/Japan Spacesystems, and U.S./Japan ASTER Science Team. ASTER Level 2 Surface Temperature Product, ASTER Kinetic Surface Temperature (AST_08), and Registered Radiance (AST_L1B version 3) Images from April 8, 2017 (Granule UR AST_L1B.003:2248631617), NASA EOSDIS Land Processes DAAC at the USGS Earth Resources Observation and Science (EROS) Center, Sioux Falls, South Dakota. Available online: https://doi.org/10.5067/ASTER/AST_08.003 (accessed on 1 October 2019).

35. Fan, Y.; Van Den Dool, H. A global monthly land surface air temperature analysis for 1948–present. *J. Geophys. Res.* **2008**, *113*, D01103. [CrossRef]

36. NASA JPL. NASA Shuttle Radar Topography Mission Global 1 arc Second, Distributed by NASA EOSDIS Land Processes DAAC at the USGS Earth Resources Observation and Science (EROS) Center, Sioux Falls, South Dakota. 2013. Available online: https://doi.org/10.5067/MEaSUREs/SRTM/SRTMGL1N.003 (accessed on 1 October 2019).

37. Kalnay, E.; Kanamitsu, M.; Kistler, R.; Collins, W.; Deaven, D.; Derber, J.; Gandin, L.; Saha, S.; White, G.; Woollen, J.; et al. The NCEP/NCAR 40-year re-analysis project. *Bull. Am. Meteorol. Soc.* **1995**, *77*, 437–471. [CrossRef]

38. Vargo, L.J.; Galewsky, J.; Rupper, S.; Ward, D.J. Sensitivity of glaciation in the arid subtropical Andes to changes in temperature, precipitation, and solar radiation. *Glob. Planet. Chang.* **2018**, *163*, 86–96. [CrossRef]

39. Trombotto, D.T. *Survey of Cryogenic Processes, Periglacial Forms and Permafrost Conditions in South America*; Revista do Instituto Geológico: São Paulo, Brazil, 2000; pp. 33–55.

40. Gruber, S. Derivation and analysis of a high-resolution estimate of global permafrost zonation. *Cryosphere* **2012**, *6*, 221–233. [CrossRef]

41. Baker, P.E.; Gonzalez-Ferran, O.; Rex, D.C. Geology and geochemistry of the Ojos del Salado volcanic region, Chile. *J. Geol. Soc.* **1987**, *144*, 85–96. [CrossRef]

42. Schneider, U.; Becker, A.; Finger, P.; Meyer-Christoffer, A.; Rudolf, B.; Ziese, M. GPCC Full Data Reanalysis Version 7.0: Monthly Land-surface Precipitation from Rain Gauges Built on GTS Based and Historic Data; Research Data Archive at the National Center for Atmospheric Research, Computational and Information Systems Laboratory. Available online: https://doi.org/10.5065/D6000072 (accessed on 10 January 2016).
43. Buchroithner, D.; Trombotto, D. Cryophenomena in the Cold Desert of Atacama. In *EGU General Assembly Conference Abstracts*; EGU: Vienna, Austria, 2012; Volume 14, p. 654.
44. García, A.; Ulloa, C.; Amigo, G.; Milana, J.P.; Medina, C. An inventory of cryospheric landforms in the arid diagonal of South America (high Central Andes, Atacama region, Chile). *Quatern. Int.* **2017**, *438*, 4–19. [CrossRef]
45. Falvey, M.; Garreaud, R.D. Regional cooling in a warming world: Recent temperature trends in the southeast Pacific and along the west coast of subtropical South America (1979–2006). *J. Geophys. Res.* **2009**, *114*, D04102. [CrossRef]
46. Hall, D.K.; Riggs, G.A.; Salomonson, V.V. *MODIS/Terra Snow Cover 5-Min L2 Swath 500m. Version 5*; NASA National Snow and Ice Data Center Distributed Active Archive Center: Boulder, CO, USA, 2006.
47. Baldridge, A.M.; Hook, S.J.; Grove, C.J.; Rivera, G. The ASTER spectral library version 2.0. *Remote Sens. Environ.* **2009**, *113*, 711–715. [CrossRef]
48. Snyder, W.C.; Wan, Z.; Zhang, Y.; Feng, Y.Z. Classification-based emissivity for land surface temperature measurement from space. *Int. J. Remote Sens.* **1998**, *19*, 2753–2774. [CrossRef]
49. Wan, Z. New refinements and validation of the MODIS land-surface temperature/emissivity products. *Remote Sens. Environ.* **2008**, *112*, 59–74. [CrossRef]
50. Wan, Z. New refinements and validation of the Collection-6 MODIS land-surface temperature/emissivity products. *Remote Sens. Environ.* **2014**, *140*, 36–45. [CrossRef]
51. Ackerman, S.A.; Frey, R. MODIS Atmosphere L2 Cloud Mask Product (35_L2). NASA MODIS Adaptive Processing System at the Goddard Space Flight Center, Greenbelt, MD, USA. 2015. Available online: http://dx.doi.org/10.5067/MODIS/MYD35_L2.006 (accessed on 10 December 2019).
52. Wan, Z.; Zhang, L.; Zhan, Q.; Li, Z.L. Quality assessment and validation of the MODIS global land surface temperature. *Int. J. Remote Sens.* **2004**, *25*, 261–274. [CrossRef]
53. Warren, S.G. Optical properties of ice and snow. *Philos. T. Roy. Soc. A* **2019**, *377*, 20180161. [CrossRef]
54. Brandt, R.E.; Warren, S.G. Solar-heating rates and temperature profiles in Antarctic snow and ice. *J. Glaciol.* **1993**, *39*, 99–110. [CrossRef]
55. Salisbury, J.W.; D'Aria, D.M.; Wald, A. Measurements of thermal infrared spectral reflectance of frost, snow, and ice. *J. Geophys. Res.* **1994**, *99*, 24235–24240. [CrossRef]
56. Cheng, G.; Wu, T. Responses of permafrost to climate change and their environmental significance, Qinghai-Tibet Plateau. *J. Geophys. Res. Earth* **2007**, *112*, F02S03. [CrossRef]
57. Changwei, X.; Gough, W.A.; Lin, Z.; Tonghua, W.; Wenhui, L. Temperature-dependent adjustments of the permafrost thermal profiles on the Qinghai-Tibet Plateau, China. *Arct. Antarct. Alp. Res.* **2015**, *47*, 719–728. [CrossRef]
58. Liu-Zeng, J.; Tapponnier, P.; Gaudemer, Y.; Ding, L. Quantifying landscape differences across the Tibetan plateau: Implications for topographic relief evolution. *J. Geophys. Res-Earth* **2008**, *113*, F04018. [CrossRef]
59. Gao, F.; Masek, J.; Schwaller, M.; Hall, F. On the blending of the Landsat and MODIS surface reflectance: Predict daily Landsat surface reflectance. *IEEE Trans. Geosci. Remote. Sens.* **2006**, *44*, 2207–2218.
60. Barrett, B.S.; Campos, D.A.; Vicencio Veloso, J.; Rondanelli, R. Extreme temperature and precipitation events in March 2015 in central and northern Chile. *J. Geophys. Res. Atmos.* **2016**, *121*, 4563–4580. [CrossRef]
61. Moore, J.P.; Ping, C.L. Classification of Permafrost Soils. *Soil Horiz.* **1989**, *30*, 98–104. [CrossRef]
62. Liu, L.; Sletten, R.S.; Hagedorn, B.; Hallet, B.; McKay, C.P.; Stone, J.O. An enhanced model of the contemporary and long-term (200 ka) sublimation of the massive subsurface ice in Beacon Valley, Antarctica. *J. Geophys. Res. Earth Surf.* **2015**, *120*, 1596–1610. [CrossRef]
63. Liu, L.; Sletten, R.S.; Hallet, B.; Waddington, E.D. Thermal regime and properties of soils and ice-rich permafrost in Beacon Valley, Antarctica. *J. Geophys. Res. Earth* **2018**, *123*, 1797–1810. [CrossRef]

64. Gillespie, A.R.; Batbaatar, J.; Sletten, R.S.; Trombotto, D.; O'Neal, M.; Hanson, B.; Mushkin, A. Monitoring and mapping soil ice/water phase transitions in arid regions. In *Geological Society of America Abstracts with Programs*; Geological Society of America: Boulder, CO, USA, 2017.
65. Zhao, L.; Wu, Q.; Marchenko, S.; Sharkhuu, N. Thermal state of permafrost and active layer in Central Asia during the International Polar Year. *Permafr. Periglac.* **2010**, *21*, 198–207. [CrossRef]

Article

Validation of ASTER Emissivity Retrieval Using the Mako Airborne TIR Imaging Spectrometer at the Algodones Dune Field in Southern California, USA

Amit Mushkin [1,2,*], Alan R. Gillespie [1], Elsa A. Abbott [3], Jigjidsurengiin Batbaatar [1], Glynn Hulley [3], Howard Tan [3], David M. Tratt [4] and Kerry N. Buckland [4]

[1] Department of Earth and Space Sciences, University of Washington, Box 351310, Seattle, WA 98195-1310, USA; arg3@uw.edu (A.R.G.); bataa@uw.edu (J.B.)

[2] Geological Survey of Israel, 32 Yishayahu Leibovitz St., Jerusalem 9692100, Israel

[3] Jet Propulsion Laboratory, Pasadena, CA 91109, USA; elsa.a.abbott@jpl.nasa.gov (E.A.A.); glynn.hulley@jpl.nasa.gov (G.H.); Howard@t-a-n.org (H.T.)

[4] The Aerospace Corporation, El Segundo, CA 90245-4609, USA; david.m.tratt@aero.org (D.M.T.); Kerry.N.Buckland@aero.org (K.N.B.)

* Correspondence: mushkin@uw.edu

Received: 29 January 2020; Accepted: 26 February 2020; Published: 3 March 2020

Abstract: Validation of emissivity (ε) retrievals from spaceborne thermal infrared (TIR) sensors typically requires spatial extrapolations over several orders of magnitude for a comparison between centimeter-scale laboratory ε measurements and the common decameter and lower resolution of spaceborne TIR data. In the case of NASA's Advanced Spaceborne Thermal Emission and Reflection Radiometer (ASTER) temperature and ε separation algorithm (TES), this extrapolation becomes especially challenging because TES was originally designed for the geologic surface of Earth, which is typically heterogeneous even at centimeter and decameter scales. Here, we used the airborne TIR hyperspectral Mako sensor with its 2.2 m/pixel resolution, to bridge this scaling issue and robustly link between ASTER TES 90 m/pixel emissivity retrievals and laboratory ε measurements from the Algodones dune field in southern California, USA. The experimental setup included: (i) Laboratory XRD, grain size, and TIR spectral measurements; (ii) radiosonde launches at the time of the two Mako overpasses for atmospheric corrections; (iii) ground-based thermal measurements for calibration, and (iv) analyses of ASTER day and night ε retrievals from 21 different acquisitions. We show that while cavity radiation leads to a 2% to 4% decrease in the effective emissivity contrast of fully resolved scene elements (e.g., slipface slopes and interdune flats), spectral variability of the site when imaged at 90 m/pixel is below 1%, because at this scale the dune field becomes an effectively homogeneous mixture of the different dune elements. We also found that adsorption of atmospheric moisture to grain surfaces during the predawn hours increased the effective ε of the dune surface by up to 0.04. The accuracy of ASTER's daytime emissivity retrievals using each of the three available atmospheric correction protocols was better than 0.01 and within the target performance of ASTER's standard emissivity product. Nighttime emissivity retrievals had lower precision (<0.03) likely due to residual atmospheric effects. The water vapor scaling (WVS) atmospheric correction protocol was required to obtain accurate (<0.01) nighttime ASTER emissivity retrievals.

Keywords: ASTER; Mako; TES algorithm; temperature; emissivity; validation; Algodones

---

1. Introduction

The Advanced Spaceborne Thermal Emission and Reflection Radiometer (ASTER, [1–4]) was launched into a sun-synchronous orbit on 19 December 1999 onboard NASA's Terra satellite. It includes a five band (8 to 12 μm) thermal infrared (TIR) scanner delivering 60 × 60 km radiance images with 90 m/pixel spatial resolution and with instrumental capability of 1 K accuracy and <0.3 K precision (NEΔT$_{300K}$). The temperature and emissivity (T/ε) separation algorithm "TES" developed for ASTER TIR data [5] was designed to generate the standard land-surface products AST08 (T) and AST05 (ε) for the geologic surface of Earth. TES builds on the relatively high contrast in ε spectra of geologic surfaces for its T/ε separation approach and was not designed to recover the T or relatively low contrast ε spectra of water or vegetation surfaces, for which ε is largely well known a priori (e.g., [6], https://speclib.jpl.nasa.gov/). Before launch, AST08 and AST05 were estimated to have a nominal predicted accuracy of ±1.5 K and precision of 0.015, respectively [5].

Validation tests of ASTER radiance data and its standard T and ε products have been conducted over the years since launch, resulting in adjustments to the TES algorithm itself [7,8], as well as improvements of the atmospheric correction procedures applied to the standard land-leaving radiance product (AST09T) that is input into the TES algorithm [9]. Many validation studies focused on water bodies [8,10] and vegetated or low contrast soil-mantled landscapes [11–13] that provide large homogenous surfaces resolvable by ASTER. These validation studies have shown that the ASTER radiance products are capable of recovering T for water bodies, for example, accurately and precisely. The TES algorithm, however, performs less reliably over these surfaces, due in large measure to the empirical regression between spectral contrast (measured as the $\varepsilon_{maximim} - \varepsilon_{minimum}$ difference, or "MMD") and minimum ε which is at the heart of the algorithm originally designed for bare geologic surfaces [5].

ASTER emissivity retrievals (i.e., AST05) were tested over bare geologic surfaces (e.g., [14,15]). Hulley et al. [16] focused on several sand dune sites in North America to validate the North American ASTER Land Surface Emissivity Database (NAALSED) [17], in addition to a set of global sand dune sites to validate the ASTER Global Emissivity Dataset (GED) v3 [18]. They found an absolute mean difference of 0.016 between ASTER ε retrievals (using TES + water vapor scaling) and representative laboratory spectra from these sites. Sabol et al. [19] focused on bare basalt fields in Hawaii and on the Railroad Valley playa surface in Nevada, USA, and found that the AST05 ε for these sites again was generally within the predicted accuracy limitations of TES. However, their observations of temporal changes in soil moisture and texture across the playa surface, as well as the natural geologic variability in the composition and texture of the Hawaii basalt surfaces, highlighted the inherent complications in the conventional validation approach of comparing "snapshot" ASTER TIR measurements acquired at 90 m/pixel against centimeter-scale laboratory ε spectra of "representative" field samples. This conventional validation approach is not designed to effectively account for sub-pixel heterogeneity in the TIR properties of natural geologic surfaces. This common TIR heterogeneity of geologic surfaces can arise from compositional variability, differential solar heating, or multiple reflections between unresolved landscape elements [20].

Here, we address the fundamental scaling assumption in field-based validation experiments of satellite TIR data, which in the case of testing TES ε retrievals requires a nearly four orders of magnitude extrapolation to link between ASTER's 90 m/pixel measurements and centimeter-scale laboratory TIR spectra. We focus on the Algodones dune field in southern California (Figure 1) and use 2.2 m/pixel hyperspectral thermal data (128 channels between 7.8 and 13.4 μm) acquired by the airborne Mako imaging spectrometer [21,22] to bridge the spatial scale gap between laboratory spectra and the 90 m scale of the AST05 standard ε product. Throughout this paper we use conventional abbreviations (e.g., west-southwest "WSW") to indicate directions on the compass rose. Times are local Pacific Standard Time (PST).

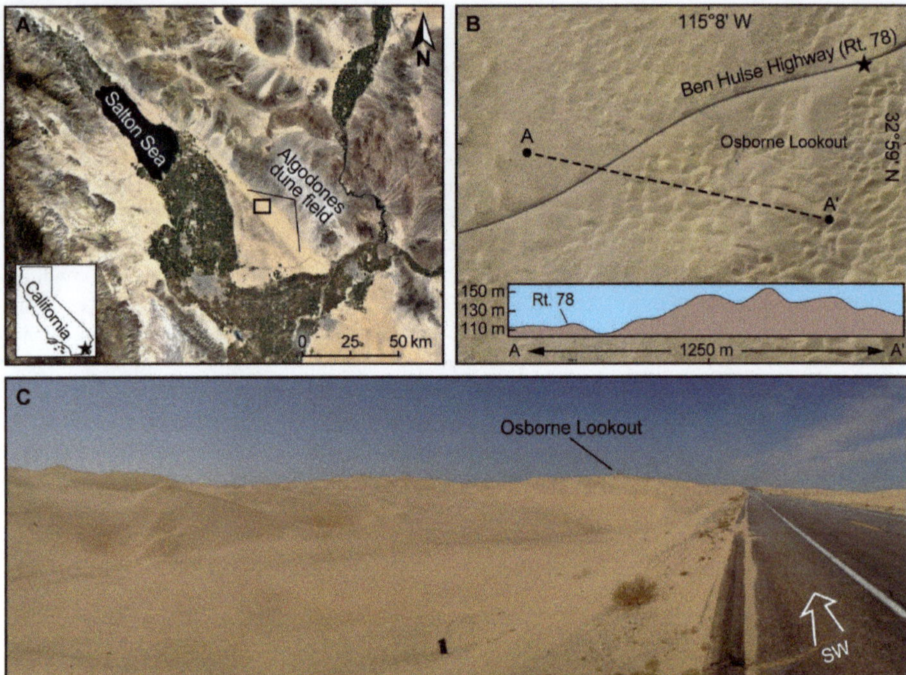

Figure 1. Study site. (**A**) Google Earth image showing the study site (black box) located approximately 50 km SE of the Salton Sea within the Algodones dune field; (**B**) The Algodones study site near the Osborne Lookout south of the Ben Hulse highway. Black star, location of photo in (C), elevation data from the USGS National Elevation Dataset (NED); (**C**) SW view onto the Algodones study site south of the Ben Hulse highway on the right.

2. Background

2.1. Remote-Sensing Instrumentation

The spaceborne ASTER scanner collects thermal-infrared images in five spectral channels (#10, 8.125–8.475 µm; #11, 8.475–8.825 µm; #12, 8.925–9.275 µm; #13, 10.25–10.95 µm; and #14, 10.95–11.65 µm). The Terra satellite, which hosts ASTER, is in a sun-synchronous polar orbit 705 km above the Earth's surface, crossing the equator at about 10:30 and 22:30 local time. Images are acquired on request during the descending daytime overpass and also during the ascending nighttime overpass. Terra passes overhead every 16 days. The image data are calibrated to radiance at the sensor and, then, corrected for the effects of the atmosphere to produce a land-leaving radiance standard product (AST09T), which also includes an image plane containing the calculated downwelling sky irradiance (e.g., [13]). The default atmospheric correction is based on an interpolated National Center for Atmospheric Research/National Centers for Environmental Prediction (NCAR/NCEP) atmosphere characterization adjusted for surface elevation with a km-scale DEM. Optionally, a MODIS (Moderate Resolution Imaging Spectroradiometer, also on Terra) water-vapor special product (MOD07) can be employed [23], or an ASTER-derived water vapor scaling (WVS) can be used to adjust the nominal corrections [9]. WVS can improve the accuracy of AST05 ε data by up to 0.03 under warm, humid atmospheric conditions [24].

Mako is an airborne hyperspectral thermal infrared whiskbroom imaging spectrometer with a 128 × 128 element sensor, designed and built by The Aerospace Corporation (El Segundo, California) and operated from a Twin Otter aircraft [21,22]. It operates in the 7.8 to 13.4 µm spectral region employing a cryogenically cooled grating-based spectrometer with a native spectral resolution of

~4 cm^{-1} at 10 μm (0.04 μm per channel). The instantaneous field of view (IFOV) is 0.55 mrad, which results in ~2 m GSD at a flight elevation of 3.8 km above ground. The nominal NEΔT$_{300K}$ of the Mako data is 0.03 K. Mako whisks are 128 pixels wide, the number of whisks comprising an imaging session being determined by the duration of the data-collection overpass. Imaging sessions with 160 whisks of length 4.9 km were designed to incorporate both the Algodones dune field area-of-interest and the adjacent environs. Visible-wavelength context images are co-collected with the thermal data.

2.2. The TES Algorithm

Surface T and the ε in each of the five ASTER TIR channels for each image pixel are the products of the TES algorithm [5]. Inputs to TES are "land-leaving radiance" [25] and downwelling sky irradiance, and both are in ASTER standard product AST09T. To convert land-leaving radiance to the land-emitted component, the reflected sky irradiance must first be subtracted. To do this it is necessary to know the reflectivity (ρ) of the land surface, but since $\rho = 1 - \varepsilon$ (Kirchhoff's law, [26]), this requires that the ε be first determined. However, this requirement is not met for most geologic surfaces, and therefore has to also be dealt with in TES.

A fundamental hurdle in finding T and ε from remotely sensed land-emitted TIR data is that both are unknowns in the Planck equation, which describes measured radiance R as:

$$R_\lambda = \varepsilon_\lambda \frac{c_1}{\pi\lambda^5}\left(e^{c_2(\lambda T)^{-1}} - 1\right)^{-1} \text{ W m}^{-1}\text{ sr}^{-1}\text{ μm}^{-1} \tag{1}$$

where λ is wavelength, c_1 and c_2 are characteristic constants, and T is in K. TES makes use of two observations to break this indeterminacy as follows: First, for most rocks, at $\lambda > 10$ μm $\varepsilon \approx 0.965$; and second, the minimum value of ε in the spectrum ($\varepsilon_{minimum}$), is related to the spectral contrast as measured in the laboratory as the difference between $\varepsilon_{minimum}$ and the maximum value of ε in the spectrum ($\varepsilon_{maximum}$). This first observation can be used in the "normalized emissivity method" (NEM) to approximate T by inversion of Planck's law, and then the normalized values of ε at other wavelengths can be found since T is now estimated [27,28]. TES makes use of the normalized ε to calculate and, then, subtract the downwelling sky irradiance, given the land-leaving radiance. Then, it essentially uses the new spectral contrast to estimate $\varepsilon_{minimum}$, and hence a refined approximation of T (AST08), and thus the ε data (AST05). Gillespie et al. [5] used 86 laboratory ε spectra to relate $\varepsilon_{minimum}$ for each to their spectral contrast. They gave the relation as a power law, later simplified by Sabol et al. [19] for practical considerations to:

$$\varepsilon_{minimum} = 0.955 - 0.8625 * MMD. \tag{2}$$

The simplification was made to reduce the impact of measurement "noise" in the calculated emissivity images for grey-body targets such as water and vegetation. For geological surfaces with $0.05 < MMD < 0.3$ the difference in ε_{min} between the linear and power-law versions of the equation was less than 0.004.

2.3. Sources for Error in AST05

There are multiple sources of error in calculating AST05. Calibration coefficients are used to convert data collected by ASTER to radiance change as the sensor sensitivities change and do not account for electronic striping (e.g., [8]). Atmospheric correction is prone to error from several sources [25]. The atmosphere is characterized from distant radiosonde launches and must be interpolated to the desired site and, then, adjusted for site elevation using a DEM (with 1 km resolution). The TES algorithm itself can introduce error in its assumed $\varepsilon_{maximum}$ value and, especially, in its empirical regression relating $\varepsilon_{minimum}$ to ε contrast. Before launch, Gillespie et al. [5] considered that uncertainties from these three main sources (radiance measurement, atmospheric characterization, and TES regression) were all about the same magnitude, although more extreme examples were encountered later.

In addition to these three sources of uncertainty in AST05, there are also possible "geological" sources of uncertainty that are the focus of the present study. These "geological" sources relate to radiance mixing in the 90 m ASTER pixels, and to multiple reflections among surface elements, which would affect the recovered effective T and ε regardless of imaging resolution. It is worth recalling that when Terra and its instruments were designed, 1 km resolution in the thermal infrared was regarded as "moderate" resolution, and ASTER with its 90 m data was regarded as the high-resolution "zoom lens" for MODIS. However, the imaged surface is potentially heterogeneous at much finer scales, and radiance mixing from diverse surface materials affects nearly all remote-sensing measurements (e.g., [29]). Geological uncertainties include unresolved topographic facets that are differentially heated or have different compositions and multiple reflections between facets, reducing the spectral contrast and changing the surface temperature (e.g., [30,31]). In the presence of significant topographic structure, these effects can be several percent and can present a challenge for robust validation of 90 m/pixel data against centimeter-scale laboratory emissivity spectra.

3. Approach and Methods

We tested the performance of the TES algorithm and the AST05 standard product over bare geologic terrain having high ε contrast, which is the type of surface TES was originally designed for. We focused on a presumably compositionally homogenous sand dune site and used two Mako datasets to quantitatively map and characterize this assumed homogeneity during daytime and nighttime with 2.2 m/pixel data, which allowed us to effectively bridge the spatial scale gap between centimeter-scale laboratory ε spectra of surface samples and the native 90 m/pixel scale of AST05. To retrieve and validate surface emissivities from the Mako data we employed: (i) laboratory analyses to characterize the primary surface materials within the imaged scene, (ii) radiosonde launches at time of Mako overpasses to drive atmospheric corrections; and (iii) ground-based thermal measurements at time of Mako overpasses to validate the Mako T retrievals. The methodology for these experiments and the validation of AST05 using Mako are described below.

The study site is a 1 km^2 area within the 10 × 70 km Algodones dune field in southern CA (Figure 1). The dune field, which is located ~30 km southeast of the Salton Sea, strikes NW-SE and was previously used for validating NASA's North American Land Surface Emissivity Database (NAALSED) [16,32]. Within the 1 km^2 study site we focused on a 270 × 270 m test area at an elevation above mean sea level of 120 to 150 m just south of Rt. 78, the "Ben Hulse Highway" (Figure 1B,C). The primary physical scene elements within the test area, which appears to be visually homogeneous, are ~10 m high dunes and the interdune flats that occur between them. Herein, we further classify the dune elements into their windward, crest, and slipface sections, as these experience different diurnal surface-temperature cycles due to their different topographic slopes and azimuths. The windward and slipface slopes typically dip ~NW and SE, respectively. The strongest winds in this region are northerly and westerly, and the average annual temperatures ranges from 16.5 to 31 °C. Annual precipitation averages 83 mm and the annual relative humidity averages 32%. Vegetation cover across the study site is <1%, mostly in the low-lying interdune areas. The low vegetation cover is likely due to the instability of the dunes in the frequent wind events, rather than low precipitation, as the dune field is surrounded by a shrub steppe.

3.1. Laboratory Analyses

The sand mineralogy was characterized at the Geological Survey of Israel with X-ray diffraction (XRD) using a PANalytical X'Pert diffractometer. The XRD samples were ground using a porcelain mortar and were measured by X-ray diffraction using CuKα radiation. Mineral phase identification was performed using HighScore Plus® software based on the ICSD database. Mineral phase abundances were based on the reference intensity ratio (RIR) method using in-house RIR values. Particle-size distributions (PSDs) for the sand samples (Figure 2A) were measured at the Geological Survey of Israel through laser diffraction using a Malvem Mastersizer MS-2000 (see [33] for detailed procedure).

Laboratory reflectivity (ρ) spectra for surface materials (Figure 2B–F) were determined at the Jet Propulsion Laboratory using a 520FT-IR Nicolet Fourier transform spectrometer with a Labsphere RSA-N1-700D integrating sphere [6]. The Nicolet reflectivity measurements were converted to emissivity according to $\varepsilon = 1 - \rho$, with a reported accuracy of ±0.002 (0.2%) [34].

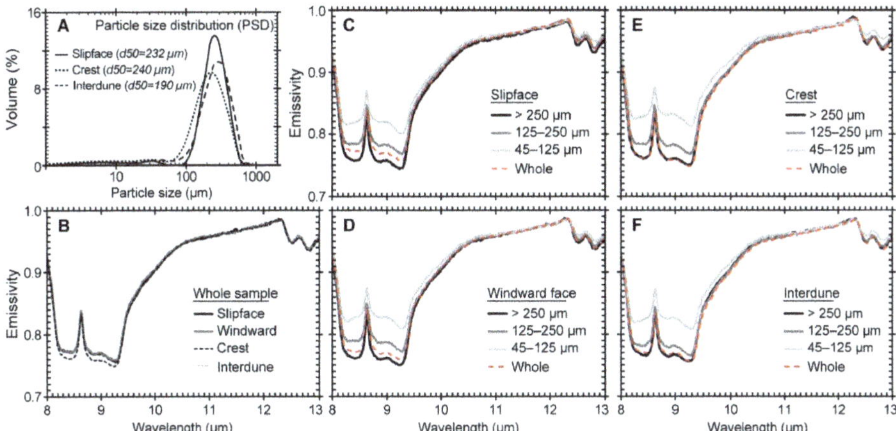

Figure 2. Laboratory characterization of the Algodones site. (**A**) Particle-size distribution (PSD) of the main scene elements. "d50" indicates the median value; (**B**) Nicolet emissivity spectra for the dune elements; (**C–F**) Nicolet spectra for grain-size splits (in μm) of the same scene elements.

3.2. Field-Based Temperature Measurements

Time-series of kinetic surface-temperatures for the dune scene elements at the test site during the Mako overpasses were acquired using iButton DS1921G-F5# sensors (www.ibuttonlink.com) (±1 K accuracy) at 5 min intervals (Figure 3). The sensors were emplaced 3 to 5 mm below the sand surface approximately 12 h before the first overpass with the implied assumption that measured temperatures at these depths can be used to approximate skin temperatures. In addition, ground-based thermal images were acquired using a handheld FLIR T300 camera (www.flir.com) with an NEΔT of 0.05 K. Processing and analysis of the FLIR images were carried out using the FLIR ResearcherIR software package. The FLIR images were used to map temperature homogeneity of the landscape elements at scales from centimeters to tens of meters, bridging the gap between the point measurements of kinetic surface-temperatures with the iButton sensors and the 2.2 m/pixel scale of the Mako data.

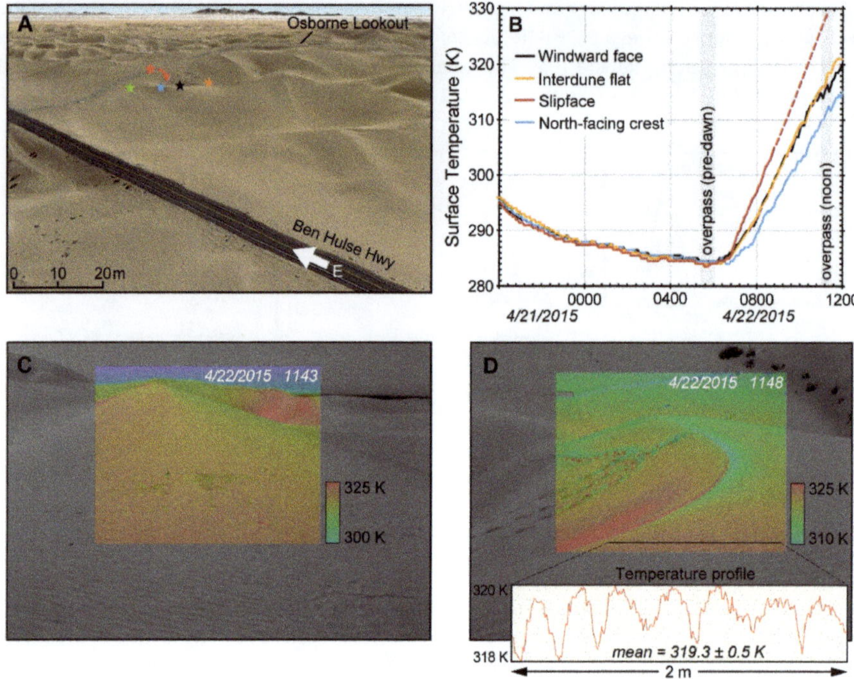

Figure 3. Ground temperature measurements. (**A**) Perspective view to the SE of the test site and the locations where time-series surface temperature measurements were obtained (image, Google Earth); (**B**) surface temperature time-series color coded according to scene elements shown in (A). Windward slope 20° to NW, slipface 30° to east-southeast (ESE). Crest measurements were from the north-facing facet of the crest; (**C**) a westward looking FLIR image taken during the day overpass time overlain on a gray-level visible context image. The dune in the center is ~5 m high. A ~20 K range in surface T is observed at sub-ASTER pixel scales (<90 m) due to surface topography; (**D**) north-looking FLIR image taken during the day overpass overlain on a gray-level visible context image. Variability of ~2 K occurs at sub-Mako pixel scales (<2 m) due to small-scale ripples.

3.3. Remotely Sensed Data

3.3.1. Hyperspectral Airborne Imaging

The Mako datasets used in the present study were collected during two overpasses on 22 April 2015. The first overpass was carried out at predawn at an elevation of 3.75 km above the surface and in an east-southeast (ESE) flight path between 0543 and 0555. The second overpass was carried out between 1125 and 1130 at the same elevation and with a WSW flight trajectory. Processing, orthorectification and analysis of the Mako data were performed using the ENVI 5.4 software package.

Mako at-sensor radiance data (R_M) from the noon overpass appear to display more high-frequency noise than that found in the predawn overpass (Figure 4). In-flight calibrations, which demonstrate similar at-sensor measurement noise levels during both overpasses, suggest that effective high-frequency atmospheric lines (e.g., CO_2, CH_4, N_2O, H_2O, and O_3) are more likely responsible for this difference between the predawn and noon radiance data. However, because H_2O is the only atmospheric component directly measured during both overpasses through the radiosonde launches (Figure 5), atmospheric corrections for the other components were based on an assumed model atmosphere, and thus cannot explicitly account for predawn and noon differences. To reduce the impact of these uncompensated high-frequency changes in atmospheric effects we co-added the adjacent native full-resolution 128-channel noon data into 43 spectral channels. Although reducing spectral resolution, this binning also reduced the noise level in the spectrally degraded noon data to match that of the predawn overpass.

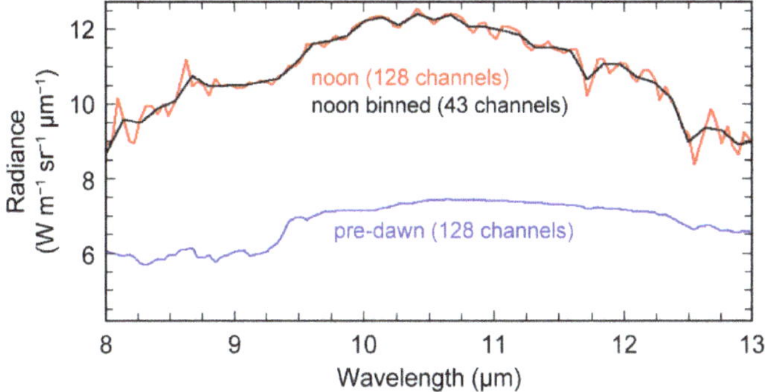

Figure 4. Mako radiance spectra. Average radiance from the 270×270 m test area. Spectral binning was applied to reduce noise levels in the noon radiance data.

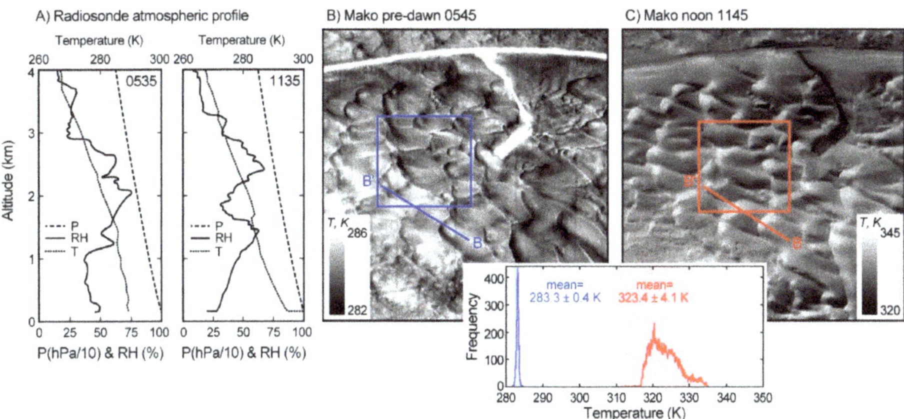

Figure 5. (**A**) Atmospheric profiles measured using radiosonde launches during the Mako overpasses; (**B,C**) Mako NEM surface temperatures images. Bottom, temperature histograms for the 270×270 m box in each image (blue, predawn and red, noon). Shaded ranges mark the T range measured on the ground at time of overpasses (Figure 3). Temperature transect B-B′ is plotted in Figure 9.

3.3.2. Mako Atmospheric Corrections and Retrieval of Surface T and ε

Two radiosonde launches from the validation site were used to characterize atmospheric conditions close in time to the two Mako overpasses (Figure 5A). The first launch was at 0535 and reached an altitude of 12 km above sea level and the second was at 1135, reaching an altitude of 9 km above sea level. Atmospheric profiles from these radiosonde data were used to drive MODTRAN-4 simulations to infer atmospheric ground-to-Mako transmissivity (τ), up-welling path radiance ($S_\uparrow$), and down-welling sky irradiance ($S_\downarrow$). Higher-level MODTRAN versions appear to have little effect at the spectral resolution of ASTER. The model parameters τ, $S_\uparrow$, and $S_\downarrow$ were, then, used to calculate surface-emitted radiation values (R), according to

$$R = R_M \tau^{-1} - S_\uparrow \tau^{-1} - \rho S_\downarrow \pi^{-1} \tag{3}$$

where $\rho = 1 - \varepsilon$ is the (unknown) surface reflectivity. For each Mako overpass this calculation involved a three-stage process as follows: (i) Application of atmospheric corrections for τ and $S_\uparrow$ to obtain land-leaving radiance values ($R_L = R_M \tau^{-1} - S_\uparrow \tau^{-1}$) for each image pixel; (ii) application of the NEM algorithm [27,28] to R_L to obtain a first approximation of ε in each of the Mako channels, assuming a maximum emissivity value (ε_{max}) of 0.985 (as determined from the laboratory spectra (Figure 2) occurs in one of the Mako channels; and (iii) correction of R_L using the value of ε found in step ii for surface-reflected $S_\downarrow$ to obtain R in all Mako channels for each image pixel.

Ultimately, surface T and ε were retrieved from the Mako surface-emitted radiance (R) values in the following two ways: (i) Re-application of the NEM algorithm (with $\varepsilon_{maximum} = 0.985$) to R to obtain final surface T estimates (Figure 5) as well as ε in all Mako channels at 2.2 m/pixel ("Mako NEM") (Figure 6), and (ii) application of the ASTER TES algorithm to Mako R values ("Mako TES") (Figure 7). For Mako TES, the Mako spectral channels were co-added to simulate the five ASTER TIR channels spectrally and the Mako pixels were co-added to obtain 90 m/pixel data to simulate ASTER spatially. T and five ASTER-like spectral channels of ε were retrieved using TES from the resampled Mako noon radiance data (Figure 7).

3.3.3. AST05

Daytime and nighttime ASTER AST05 ε images from 21 dates between 2001 and 2018 were downloaded from NASA's "EarthData" website (https://search.earthdata.nasa.gov) after visual evaluation for cloud cover and clarity (Table 1). Twelve daytime and nine nighttime acquisitions were selected to represent cloud-free conditions during all seasons of the year. The AST05 products analyzed were processed with the following three different atmospheric correction protocols: (i) The NCAR/NCEP atmospheric correction protocol ("standard correction"), (ii) the NCAR/NCEP atmospheric correction protocol using the WVS correction ("WVS correction"), and (iii) with the MODIS MOD07 [23] atmospheric correction protocol. MOD07 corrections were available only for daytime images. The mean ε values for the 3×3 pixel test area (Figures 6–8) west of the Osborne Lookout from each AST05 scene are listed in Table 1.

4. Results

4.1. Laboratory Analyses

Quartz is the dominant mineral phase at the surface of the Algodones dunes [16]. The XRD analyses we conducted for samples collected from surfaces of the four primary scene elements in the test area confirm that all have a similar mineralogical composition of >70% quartz, 5% to 20% K-feldspar, <5% plagioclase, and <5% calcite (by weight). The PSDs of the dune elements are also similar with measured median (d50) values of 240, 232, and 190 μm for the dune crest, slipface, and interdune flat, respectively (Figure 2A). Laboratory emissivity spectra reveal effectively overlapping spectra for all four scene elements (Figure 2B). All the spectra are dominated by the "quartz doublet" Reststrahlen bands at ~8.4 and 9.3 μm and the smaller quartz bands at 12.5 and 12.8 μm. Spectral measurements for grain-size splits revealed that, as expected, the depth of the Reststrahlen bands consistently decreases with decreasing grain size for all four scene elements (Figure 2C–F). Nonetheless, these prominent grain-size effects do not translate to significant spectral variability among the scene elements when measured as bulk "whole" samples. All the spectral measurements show an emissivity maximum of 0.985 at ~12.3 μm.

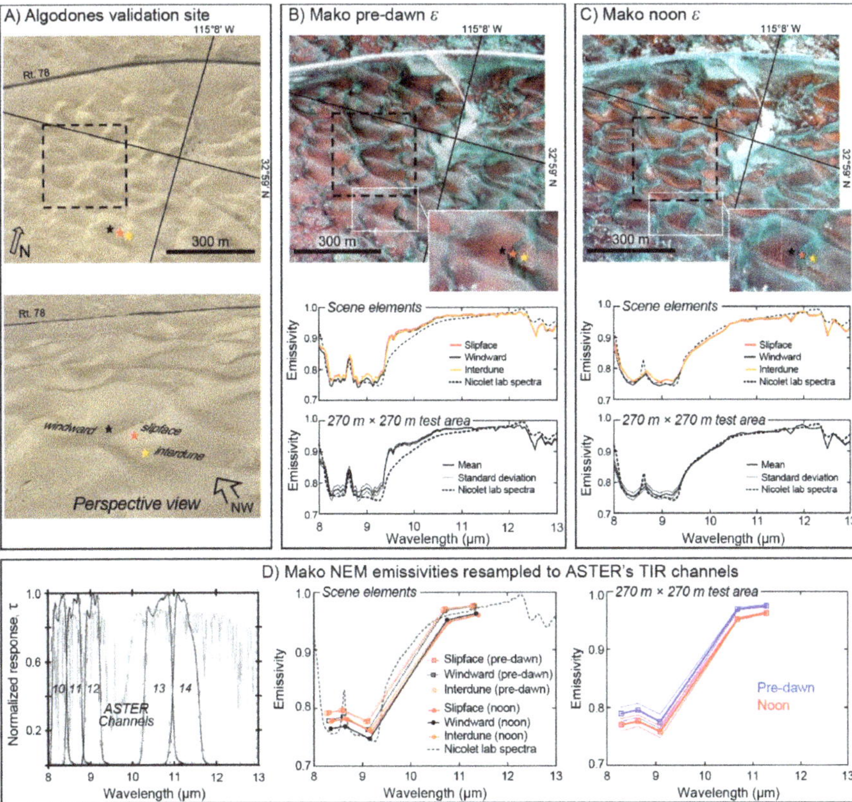

Figure 6. Mako NEM emissivity retrievals. Dashed box in A–C marks a 270 × 270 m test area equivalent to 9 ASTER pixels. (**A**) Google Earth image of the Algodones test site (top) and perspective view (bottom); Mako spectra (B,C) sample locations are color-coded accordingly; (**B**) Top, predawn 2.2 m/pixel Mako NEM emissivity image (R,G,B = 11.3, 9.1, 8.3 μm). White box marks the extent of enlarged area in the bottom right corner. Middle, Mako NEM emissivity retrievals for the primary dune element. Bottom, average Mako predawn spectra from area of the dashed box (~12,500 Mako pixels); (**C**) Top, Noon 2.2 m/pixel emissivity image (R,G,B = 11.3, 9.1, 8.3 μm). White box marks the extent of enlarged area in the bottom right corner. Middle, Mako NEM emissivity retrievals using binned radiance data (43 channels). Bottom, average Mako noon spectra from area of the dashed box (43 channels, ~12,500 Mako pixels). For B and C each "scene element" spectrum represents the mean of 32 Mako pixels. Standard deviation for all "scene element" spectra falls close to the width of the plotted line; (**D**) Left, spectral response functions for ASTER's TIR channels (black) and atmospheric transmissivity (*t*) for a "1976 US standard" MODTRAN atmosphere (grey). Middle and Right, Mako NEM emissivities from B and C spectrally resampled to ASTER's five TIR channels. Data are slightly offset along the *x*-axis for clarity.

Table 1. AST05 data used in the present study.

| ASTER Channel | | 10 | | | 11 | | | 12 | | | 13 | | | 14 | | | Remarks |
Date	Scene ID	$\varepsilon_{8.3}$ μm			$\varepsilon_{8.6}$ μm			$\varepsilon_{9.1}$ μm			$\varepsilon_{10.4}$ μm			$\varepsilon_{11.1}$ μm			
		STD	WVS	MOD07	STD	WVS	MOD07	STD	WVS	MOD07	STD	WVS	MOD07	STD	WVS	MOD07	
1/1/2016	AST_05_00301012016182807_20191101102226_31421	0.738	0.748	0.755	0.75	0.764	0.767	0.728	0.745	0.746	0.944	0.949	0.953	0.961	0.961	0.961	
1/7/2001	AST_05_00301012016182807_20191101103017_27601	0.753	0.755	0.731	0.767	0.766	0.754	0.745	0.742	0.743	0.954	0.953	0.954	0.961	0.961	0.963	day
1/16/2004	AST_05_00301162004182833_20191101103137_3958	0.730	0.732	0.749	0.753	0.752	0.767	0.738	0.739	0.748	0.953	0.954	0.948	0.962	0.962	0.961	day
3/2/2003	AST_05_00303022003182818_20191101103107_31312	0.753	0.755	0.754	0.764	0.768	0.770	0.738	0.749	0.752	0.955	0.954	0.954	0.962	0.961	0.961	day
4/9/2011	AST_05_00304092011182719_20191101102907_18505	0.769	0.755	0.769	0.775	0.768	0.775	0.752	0.749	0.752	0.956	0.954	0.956	0.96	0.961	0.961	day
5/13/2013	AST_05_00305162013182737_20191101101558_22184	0.756	0.762	0.762	0.76	0.763	0.766	0.746	0.747	0.750	0.953	0.954	0.954	0.96	0.961	0.961	day
6/17/2007	AST_05_00306172007182740_20191101103307_16982	0.741	0.748	0.747	0.75	0.754	0.753	0.73	0.734	0.732	0.948	0.948	0.948	0.961	0.961	0.961	day
7/14/2017	AST_05_00307142017182754_20191101101518_15263	0.738	0.727	0.730	0.776	0.770	0.770	0.752	0.747	0.745	0.963	0.965	0.964	0.947	0.948	0.948	day
8/10/2015	AST_05_00308102015182850_20191101102747_7226	0.743	0.735	0.736	0.754	0.748	0.748	0.735	0.730	0.727	0.96	0.957	0.958	0.962	0.962	0.962	day
9/16/2017	AST_05_00309162017182810_20191101102106_18041	0.757	0.768	0.768	0.781	0.786	0.786	0.765	0.769	0.768	0.961	0.960	0.960	0.953	0.953	0.954	day
10/5/2018	AST_05_00310052018182845_20191101102827_12174	0.723	0.756	0.756	0.744	0.774	0.774	0.726	0.757	0.758	0.953	0.958	0.958	0.962	0.961	0.961	day
12/18/2016	AST_05_00312182016182752_20191101102947_22870	0.768	0.763	0.763	0.778	0.775	0.776	0.753	0.750	0.751	0.954	0.954	0.953	0.961	0.961	0.961	day
	Mean	0.748	0.750	0.751	0.763	0.766	0.767	0.742	0.746	0.748	0.955	0.955	0.955	0.959	0.959	0.960	
	Stadard deviation	±0.015	±0.013	±0.013	±0.013	±0.011	±0.011	±0.012	±0.010	±0.011	±0.005	±0.005	±0.005	±0.005	±0.004	±0.004	
2/15/2016	AST_05_00302152016055344_20191101102747_7229	0.691	0.784		0.743	0.802		0.746	0.797		0.763	0.958		0.972	0.965		night
3/25/2015	AST_05_00303252015054743_20191101103107_31314	0.738	0.766		0.765	0.782		0.737	0.747		0.947	0.949		0.962	0.962		night, clouds
4/10/2015	AST_05_00304102015054741_20191101103558_5671	0.768			0.778			0.753			0.954			0.961			night
6/8/2019	AST_05_00306082019054731_20191101102857_15975	0.78	0.768		0.783	0.777		0.756	0.755		0.951	0.953		0.961	0.961		night
7/3/2014	AST_05_00307032014055347_20191101101929_15471	0.667	0.772		0.719	0.785		0.703	0.767		0.961	0.954		0.974	0.960		night
9/27/2013	AST_05_00309272013054723_20191101103147_4048	0.686	0.752		0.745	0.769		0.739	0.744		0.966	0.948		0.974	0.961		night
10/15/2017	AST_05_00310152017055339_20191101102046_17123	0.736	0.677		0.76	0.736		0.74	0.733		0.948	0.961		0.962	0.975		night
11/30/2013	AST_05_00311302013054719_20191101102827_12166	0.736	0.777		0.761	0.771		0.747	0.746		0.948	0.950		0.963	0.961		night
12/13/2015	AST_05_00312132015055351_20191101102316_5831	0.728	0.731		0.738	0.751		0.721	0.727		0.949	0.950		0.962	0.963		night
	Mean	0.728	0.753		0.756	0.771		0.739	0.752		0.954	0.953		0.965	0.959		
	Stadard deviation	±0.034	±0.035		±0.019	±0.020		±0.017	±0.020		±0.007	±0.005		±0.006	±0.004		

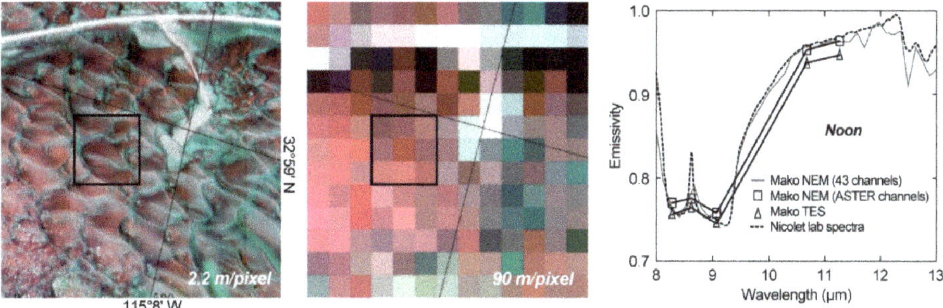

Figure 7. Simulation of ASTER TES with Mako data. Left, Mako noon radiance image (R,G,B = 11.3, 9.1, 8.3 μm) at 2.2 m/pixel. Center, image on left resampled to 90 m/pixel. The area of the 9 ASTER-like pixels examined in the right graph is within the black box. Right, mean emissivity values retrieved using the ASTER TES algorithm and surface emitted Mako radiance convolved to ASTER TIR channels and at 90 m/pixel resolution ("Mako TES"). Standard deviations fall within the symbols. Mako NEM emissivity retrievals at full Mako resolution and spectrally resampled to ASTER are also plotted.

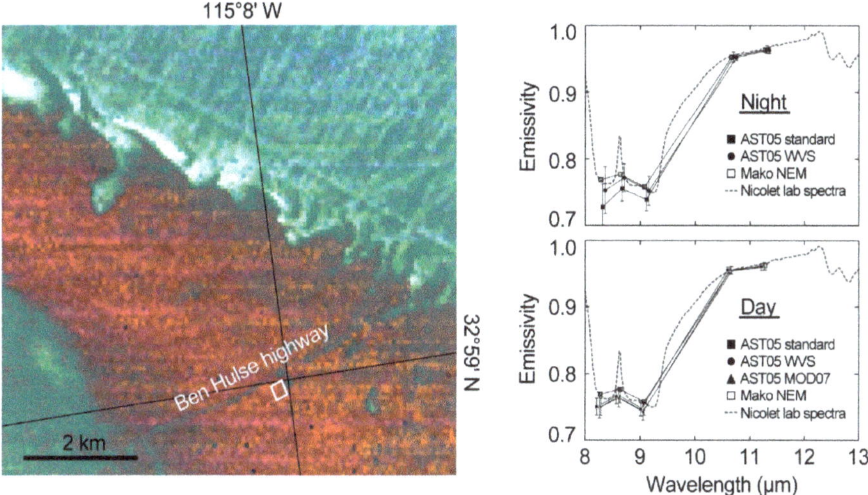

Figure 8. ASTER AST05 emissivity retrievals at Algodones. Left, AST05 90 m/pixel daytime image (R,G,B = ASTER Channels 14, 12, and 10 acquired 5 October 2018). The 270 × 270 m test site marked by white box. Right, AST05 emissivity retrievals for day and night data (Table 1). AST05 standard, WVS, and MOD07 are slightly offset along the *x*-axis for clarity.

4.2. Field-Based Temperature Measurements

Surface temperature measurements were obtained for the four dune elements during the 16 h time period between 21 and 22 April 2015 at 2000 and 1200, respectively (Figure 3). These measurements revealed tightly clustered temperatures between 284 and 285 K during the predawn Mako overpass at 0545. After sunrise, surface temperatures for all scene elements steadily increased. Because this increase in T occurred at different rates, which were effectively dictated by surface topography and surface-sun geometry, the range and variability of T across the different surface elements also steadily increased during the morning hours after sunrise. Wind-driven erosion around the iButton sensor on the directly sunlit slipface facing 30° ESE resulted in termination of T measurements for this dune element at 0900, when a range of 296–303 K was recorded among the dune elements. Extrapolation of the warming trend measured for the slipface suggests that surface temperatures during the noon overpass ranged between 311 K (measured for a shaded N-facing part of the crest) and ~330 K (extrapolated for the slipface) (Figure 3B). FLIR temperature images acquired during the day overpass support this ~20 K temperature range between the different dune elements (Figure 3C). The FLIR images also show that sub-meter T variability due to ripple morphology on the dune surfaces reached ~2 K during the noon overpass (Figure 3D). FLIR imaging also illustrates the sharp discontinuity in surface T at the crest of the dunes, where a 20 K "jump" in T can occur within 20 to 30 cm across the dune crest at noon. Sharp T transitions at the dune crests such as these cannot be resolved even in the Mako 2.2 m/pixel images.

4.3. Remotely Sensed Data

4.3.1. Atmospheric Conditions during Mako Overpasses

Near-surface air temperatures recorded by the radiosonde during the predawn overpass ranged between 288 and 289 K (Figure 5) and were 4 to 5 K higher than the predawn surface temperatures (Figure 3). Relative humidity values during the predawn overpass ranged from 49% to 38% within the lower 1 km of the atmosphere. During the noon overpass near-surface temperatures ranged between 294 and 300 K and were up to 35 K lower than surface T's on the warmest surface elements. Noontime relative humidity steadily increased upwards from ~20% to ~50% within the lower 1 km of the atmosphere.

4.3.2. Mako NEM Surface Temperatures

Predawn surface temperatures recovered from Mako within a 270 × 270 m test area near the Osborne Lookout all fell between 281.5 and 285.8 K with an average of 283.3 ± 0.4 K (Figure 4). This temperature range is in agreement with the ground measurements obtained at time of overpass for the four scene elements, which ranged between 284 and 285 K (Figure 3). Surface temperatures within the same 270 × 270 m test area recovered from the noon Mako overpass all fell between 311.2 and 335.6 K, with an average of 323.4 ± 4.1 K. These temperatures agreed with the ground measurements obtained at time of overpass for the four scene elements, which ranged between 311 and 330 K.

4.3.3. Mako NEM Emissivity Retrievals

Dune crests are narrow features that are not all resolved by Mako, and therefore are omitted as resolved scene elements in the Mako images. The mean predawn Mako NEM 128 channel spectra for the other three resolved dune elements (32 pixels for each) all display the "quartz doublet" Reststrahlen bands at ~8.4 and 9.3 μm and the smaller quartz bands at 12.5 and 12.8 μm (Figure 6). The mean ε spectra for slipfaces and interdune areas effectively overlap with each other throughout Mako's 8 to 13 μm spectral range. The mean windward spectrum effectively overlaps with the slipface and interdune spectra outside the Reststrahlen bands and is ~0.01 lower within them. Accordingly, windward slopes appear to have slightly greater spectral contrast than the other dune elements when imaged at predawn with Mako. The predawn Mako NEM spectra for all the dune elements resemble their Nicolet spectra, except between ~9.0 and 11.6 μm where the Mako spectra all plotted above the Nicolet spectrum by up to 0.04 in some wavelengths. The average Mako NEM spectrum from the 270 × 270 m test area, i.e., ~15,000 Mako 2.2 m pixels, plotted similarly as compared with the Nicolet spectra. The standard deviation of the Mako NEM spectrum for the 270 × 270 m area from the mean is less than 0.005 except within the Reststrahlen bands, where standard deviations reach 0.01.

Similar to the predawn spectra, the mean Mako noon NEM spectra for slipface and windward slopes, as well as interdune flats, all display the Reststrahlen bands and the smaller quartz bands at 12.5 and 12.8 μm (Figure 6). Slipface and interdune spectra effectively overlap with each other throughout the 8 to 13 μm spectral range. The spectrum for windward slope effectively overlaps with the spectra of the slipface and interdune areas outside the Reststrahlen bands and is up to 0.01 lower within the Reststrahlen bands. Thus, as observed in the predawn Mako data, windward slopes appear to have slightly greater spectral contrast than the other dune elements. The noon Mako NEM spectra for all dune elements effectively plot on the Nicolet ε spectrum of these dune elements except for the spectral peak near 8.6 μm, between the Reststrahlen bands, where Mako NEM spectra are ~0.04 lower. The average Mako NEM spectrum from the 270 × 270 m test area, i.e., ~15,000 Mako 2.2 m pixels, correlates similarly with the Nicolet spectra. The standard deviation of the Mako NEM spectrum for the 270 × 270 m test area from their mean is less than 0.005 except within the Reststrahlen bands, where standard deviations are as large as 0.01.

Spectrally resampled to ASTER's TIR channels predawn Mako ε for the slipface, windward, and interdune scene elements consistently plot ~0.01 above the noon ε retrievals for these scene elements (Figure 6D). In both the predawn and noon data, NEM emissivities for the windward slope in ASTER Channels 10 to 12 plot ~0.01 below the retrieved emissivities for the slipface and interdune areas. This offset in retrieved ε does not occur in ASTER Channels 13 and 14.

4.3.4. TES Simulations Using Mako Data

The robust compatibility between Mako NEM surface temperatures and emissivities with the "ground truth" measurements (Figures 5 and 6) indicates adequate atmospheric corrections and robust calibration of at-sensor radiance values to land-emitted radiation values for both overpasses. Here, we use these calibrated land-emitted radiation data to test the performance of the TES algorithm for a real geologic surface and with real TIR data that were optimally corrected for atmospheric effects (Figure 7). The mean of the noon Mako TES emissivities from the 270 × 270 m test area (nine ASTER-like pixels) all plot consistently ~0.01 below Mako NEM mean ε for the same test area.

4.3.5. Accuracy of the AST05 Products

The validated Mako ε retrievals (Figure 6) were used to test the performance of the AST05 products over the Algodones test site with the implicit assumption that the spectrum of the vegetation-free dunes does not change significantly over time (e.g., [16,32]). For ASTER Channels 13 and 14, all the AST05 emissivities for both night and day fell within ±0.005 of each other (i.e., precision) and within ±0.005 of Mako NEM (Table 1 and Figure 8). For Channels 10 to 12: (i) The means of nighttime AST05 emissivities processed with the "standard atmospheric" correction were 0.02 to 0.04 below Mako with standard deviations of up to 0.035; (ii) the means of nighttime AST05 with "WVS" correction were within 0.01 from Mako with standard deviations of up to 0.035; and (iii) the means of daytime AST05 Channels 10 to 12 emissivities processed with "standard", "WVS", and "MOD07" correction all cluster together within 0.01 below Mako with standard deviations of ~0.015.

5. Discussion

5.1. TIR Site Characterization at Sub-Mako Scales

5.1.1. Mineralogy and Grain-Size Effects

Nicolet ε spectra of the windward and slipface sections of the dunes and the interdune flats that occur between them effectively overlay each other to within 0.005 throughout the 8 to 13 μm spectral range (Figure 2). This spectral commonality reflects the relatively homogeneous quartz-dominated sand mineralogy of these different dune elements as revealed with the XRD analyses. We should not expect spectral differences over the Algodones test site due to mineralogical composition.

The Nicolet spectra demonstrate a dependency of ε on grain size. As the effective grain diameter is increased from 45 to >250 μm in splits of sieved sand the spectral contrast of the sieved samples increases (Figure 2). This is clearest for the Reststrahlen bands, which are noticeably deepened. However, this pronounced grain-size effect does not translate into significant spectral variability among the primary scene elements at Algodones because of their similar grain-size distributions (Figure 2A).

At other dune fields, in which sorting is more heterogeneous, grain-size effects could lead to greater spectral variability.

5.1.2. Anisothermal Effects

Spectral "checkerboard" mixing of scene elements having different temperatures will cause the effective emissivity of the mixed region to differ from the spectrum of the components (e.g., [35]). This is because the average of multiple Planck functions, each for a different temperature, is not itself a Planck function, but it is assumed in ε retrieval that the radiance spectrum for a blackbody is one. Model calculations for checkerboard mixing considering the T conditions observed on the ground at Algodones (Figure 3) indicate that the magnitude of the distortion of the retrieved mixed ε spectrum, even at noon, is <0.005, smaller than the predicted precision of AST05.

5.1.3. Multiple-Reflection Effects

The dunes are not topographically flat, and some scene elements are irradiating each other, leading to reduction of spectral contrast. This "cavity-radiation" effect operates for both isothermal and anisothermal surfaces. It raises effective ε selectively at wavelengths for which ε is low, and therefore the reflectivity is high. As a result, the depth of the Reststrahlen bands of dune elements that experience multiple reflection can become effectively reduced as compared with their depth in laboratory spectra of the same sand. This cavity-radiation effect will vary from pixel to pixel as is dictated by the local topographic setting and the time of day due to the local sun angle.

A simple "first reflection" model for the radiance R_i from the "interrogated" scene element i is the radiance R_e emitted from that element plus the radiance R_n from a neighboring element n that is incident upon element i and reflected to the sensor. Both terms are given by Planck's law, Equation (1), in which R is a function of local scene emissivity ε and T.

$$R_i \;=\; R_e(\varepsilon_i,\, T_i) + f_n\frac{1}{\pi}(1 - \varepsilon_i)R_n(\varepsilon_n,\, T_n) \tag{4}$$

where f_n is the fraction of the sky hemisphere subtended by element n and $(1 - \varepsilon_i)$ is the reflectivity of element i as given by Kirchhoff's law (see Danilina et al. [30] for further details). For the observed noon surface temperatures at Algodones (Figure 3) and the Nicolet spectra (Figure 2), and if f is ~5%, this model predicts that cavity radiation could account for an ε increase of ~0.05 near the Reststrahlen bands in the most extreme cases where element i is shaded and cool, and element n is warm.

5.2. Emissivity Retrievals with Mako

The hyperspectral resolution of Mako and its extended spectral range to 13 μm both facilitate the core assumption of the NEM approach that a maximum emissivity value (ε_{max}) actually occurs within one of the spectral channels. In the specific case of Mako at Algodones we assigned an assumed ε_{max} value of 0.985, which was the maximum value consistently observed in the Nicolet spectra for all scene elements. This ε_{max} occurred at ~12.8 μm, within the spectral range of Mako (Figure 2).

5.2.1. Moisture Effects in Mako Day and Night Emissivity Retrievals at Algodones

Because the mineralogical composition of the dune elements is invariant, it is generally assumed that the retrieved emissivities are too, if correctly determined. However, there is a consistent change in the retrieved Mako NEM ε between the predawn and noon acquisitions as the predawn spectra for all scene elements are up to 0.04 higher than the Nicolet spectra between ~9.0 and 11.6 μm, whereas the noon emissivities are in good agreement with the Nicolet spectra throughout the spectral range (Figure 6). The near isothermal conditions during the predawn acquisition (Figure 5) suggest that the effects of checkerboard thermal mixing or multiple scattering (cavity radiation), discussed above, are less likely to have a significant impact on the predawn emissivities. Furthermore, the spectral range of the observed increase in Mako NEM ε between ~9.0 and 11.6 µm is inconsistent with that expected from cavity radiation, which should preferentially occur in spectral regions of low ε (high ρ) such as the Reststrahlen bands near ~8.4 and 9.3 µm. Instead, the observed deviation of predawn Mako NEM emissivities from the Nicolet spectra (Figure 6B) more closely resembles the ε increase previously shown for sand dunes in the presence of soil moisture [36,37].

Moisture content in the Algodones dunes was not directly measured during the Mako overpasses. However, the significantly higher near-surface atmospheric relative humidity values measured from the radiosonde during the predawn vs. noon overpasses, i.e., ~50% vs. 20%, respectively (Figure 5), suggest that more atmospheric moisture was adsorbed on grain surfaces during the predawn overpass than at noon. Previously, soil moisture following rain events has been identified at Railroad Valley (Nevada, USA) as a source of concern for TES validation (e.g., [19]), but dune fields have been thought to have been more consistently dry. Spectrally resampled to ASTER's TIR channels, we find that this prominent day/night change in Mako emissivities at Algodones, which we attribute to increased predawn atmospherically sourced adsorbed soil moisture, translates to an overall ε increase of ~0.01 to 0.02 in all ASTER channels during the predawn acquisition with little change to spectral shape or contrast (Figure 6D). It could be that surface soil moisture should be measured as a routine part of TES validation experiments.

5.2.2. Scaling Up from Lab to Mako Emissivities

Scaling up from laboratory measurements (Figure 2) to remotely sensed measurements at 2.2 m/pixel consistently increased effective emissivities near the Reststrahlen bands at ~8.4 and 9.3 µm by ~0.01 for the slipface and interdune areas during both predawn and noon acquisitions (Figure 6). This spectral behavior is consistent with multiple reflections between the steep slipfaces and the adjacent interdune areas as compared with the lower-gradient windward faces for which f in Equation (4) is lower and cavity radiation effects do not appear to be significant. Thus, the field setting is fundamentally different than the laboratory setting in a way that influences ε, even if only at the percent level for carefully chosen sites.

5.2.3. Spectral Variability of the Algodones Site Mapped with Mako

In Figure 9, we use the Mako NEM ε ratio between ASTER-like Channel 13 and ASTER-like Channel 12, i.e., 10.7/9.1 µm, as a proxy for mapping the variability in spectral contrast among the Algodones dune elements. This ε ratio is closer to unity for spectrally flatter spectra. Mako pixels with a high fraction of vegetation cover ("vegetated pixels") display lower spectral contrast than vegetation-free pixels and are expressed as a distinct 10.7/9.1 µm ε ratio of ~1.17 in both the predawn and noon overpasses (Figure 9D). Thus, the Mako data demonstrate that the sparse desert vegetation seen in the high-resolution air photos of the Algodones Dunes is not a significant spectral element at the landscape scale because "vegetated" Mako pixels do not show up in the image histograms.

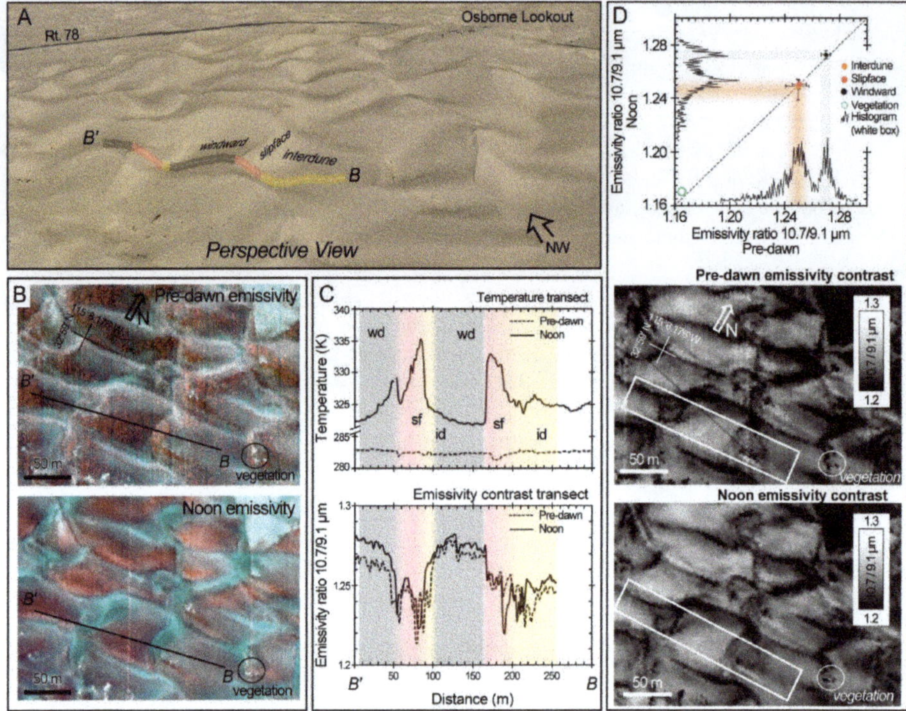

Figure 9. Surface heterogeneity mapped with Mako. (**A**) Perspective view (Google Earth) to the NW onto the Algodones test site. Transect B-B' is 300 m long and is color coded according to the three primary dune elements: interdune (id, orange), slipface (sf, red), and windward (wd, black); (**B**) predawn and daytime emissivity images (R,G,B = 11.3, 9.1, 8.3 μm) showing location of B-B'; (**C**) Top, temperature variability along the B-B' transect during the predawn and noon overpasses with sectors of the transect color-shaded according to the dune elements. Bottom, Mako 10.7/9.1 emissivity ratio (predawn and noon) along the B-B' transect with sectors of the transect color-shaded according to the scene elements; (**D**) emissivity ratio images between Mako 10.7 and 9.1 μm channels are used as a proxy for mapping the depth of the Reststrahlen band for the predawn and noon overpasses. Top, Noon vs. predawn average emissivity ratios for the scene elements (32 pixels per element). Min/max values plotted as error bars. Dashed line marks the 1:1 line. Histograms for the emissivity ratio values within the white box in the predawn and noon ratio images are plotted along their respective axes (~2500 pixels in each box). Scene elements are projected onto the histograms through their respective colors.

Windward slopes have a 10.7/9.1 μm ε ratio value of 1.27 ± 0.01 as compared with slipface and interdune areas that have distinctly lower ε ratio values of 1.25 ± 0.01 and appear to be more significantly affected by cavity radiation effects (Figure 9C). This ~2% to 4% difference in spectral contrast is similar to the predicted magnitude of cavity-radiation effects on natural geologic surfaces [31,38].

We expect that irradiation of a slope by warmer adjacent scene elements would result in greater cavity-radiation effects (reduction of spectral contrast) than if the adjacent scene elements were cooler. Yet the B-B' transect demonstrates that although predawn isothermal conditions along the transect break down to T heterogeneity of up to 15 K by noon, due to differential solar heating, the 10.7/9.1 μm ε ratio for the different dune elements does not change significantly between the predawn and noon overpasses (Figure 9C). In both predawn and noon Mako NEM ε images we find the same ~2% to 4% difference between the 10.7/9.1 μm ε ratio for the windward slopes (1.27 ± 0.01) and for the slipface and interdune elements (1.25 ± 0.01) (Figure 9D). Thus, cavity-radiation effects in the Algodones Dunes appear to be dominated by topographic roughness, and any effects of *T* heterogeneity are secondary.

Whereas Mako's 2.2 m pixels are sufficient to spatially resolve the primary scene elements (i.e., windward and slipface slopes and interdune flats), ASTER's 90 m pixels at Algodones are expected to include effective mixtures between them. Figure 10 demonstrates that mixing between windward, slipface, and interdune scene elements could fully account for the distribution of 10.7/9.1 µm ε ratio values (1.257 ± 0.016) within the 270 × 270 m test area (9 ASTER pixels) we use below to test the AST05 product. Examined at an order-of-magnitude larger area of ~2.5 × 2.5 km (Figure 10) the 10.7/9.1 µm ε ratio for the dune field remains similar with a mean of 1.251 ± 0.015 that can also be explained as a mixture of the primary dune elements described above. As this spectral homogeneity to within <1% scales up to moderate spatial resolutions, we suggest that the Algodones dune field can also serve as a validation site for emissivity retrievals from other spaceborne multispectral thermal sensors, such as ECOSTRESS (60 m/pixel) and MODIS (1 km/pixel).

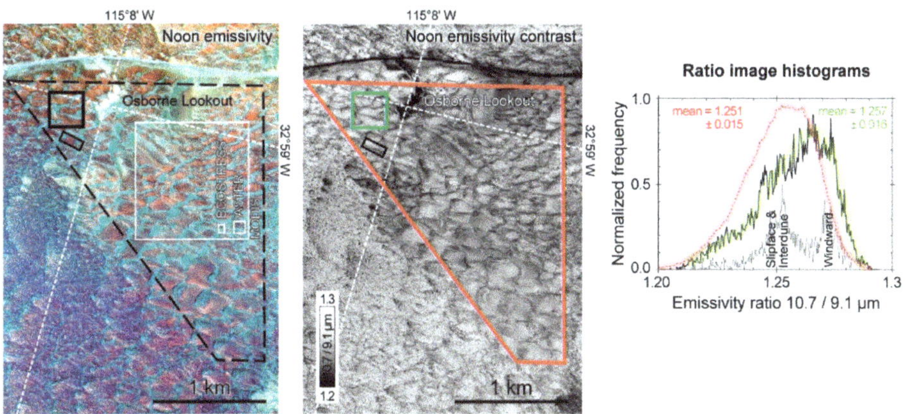

Figure 10. Emissivity heterogeneity across the Algodones dune field near the Osborne Lookout. Left, Mako 2.2 m/pixel emissivity noon image (R,G,B = 11.3, 9.1, 8.3 µm). Black box, same as histogram area in Figure 8. Dashed box, area of dashed histogram on the right. Bold black box, 3 × 3 ASTER pixel area from where AST05 emissivities were extracted (Figure 7). Pixel sizes for single MODIS, ASTER, and ECOSTRESS thermal data are marked in white. Middle, Mako 2.2 m/pixel 10.7/9.1 µm emissivity ratio noon image. Right, histograms for noon emissivity ratio values from within the small, medium, and large areas outlined and color-coded accordingly in the emissivity contrast image. Small black box ~2500 Mako pixels, green box ~15,000 Mako pixels, and red polygon ~520,000 Mako pixels.

5.3. Testing the TES Algorithm with Mako

The Mako noon data for Algodones provide an opportunity to test the performance of the TES algorithm over a natural geologic surface with high ε contrast using real TIR data optimally corrected for atmospheric effects. For this test, we applied the TES algorithm to Mako calibrated land-emitted radiance data (*R*) resampled to ASTER-like spectral and spatial resolutions (R_{AST}) to obtain "Mako TES" emissivities for ASTER's five spectral channels and at 90 m/pixel (Figure 7). While displaying a similar spectral shape, the Mako TES emissivities were consistently 0.01 below Mako NEM emissivities in all channels. This offset in Mako TES emissivities is associated with the scatter about the TES ε_{min} vs. MMD regression, Equation (2), which was predicted to be one of the inherent sources of uncertainty in TES. In addition, it appears that effective mixing between surface elements with temperatures ranging between ~310 and ~340 K during the noon acquisition (Figure 5) did not impact TES retrievals.

5.4. Validation of AST05

The mineralogy and PSD measurements, together with Nicolet spectra of the Algodones samples, allowed us to hypothesize minimal compositional and spectral variability across the validation site (Figure 2), as was previously assumed in earlier validation studies at Algodones (e.g., [16,32]). With Mako we confirmed this hypothesis and determined that the natural heterogeneity in surface T and variability in the effective ε of the primary scene elements at Algodones should not significantly affect TES retrievals at ASTER's spatial scale of 90 m/pixel, provided robust atmospheric corrections are available (Figure 7).

The average daytime AST05 emissivities, for the Algodones 9 pixel (270 × 270 m) test area, were all within the target performance of AST05 (±0.015) for all three atmospheric correction options that were tested, i.e., NCAR/NCEP, WVS, and MOD07 (Figure 8). The AST05 predawn emissivities deviated from the consistent behavior of daytime AST05. Whereas predawn Channels 13 and 14 retrievals for the two available nighttime atmospheric correction methods, i.e., NCAR/NCEP and WVS, were similar to those of the daytime retrievals (accuracy and precision of ~0.005) the repeatability in ASTER Channels 10 to 12 was lower (Figure 8). The precision of AST05 processed with NCAR/NCEP standard corrections was 0.035, 0.020, and 0.020 for channels 10, 11 and 12, respectively. The means were 0.04, 0.02 and 0.02 lower than Mako NEM for Channels 10, 11, and 12, respectively. The precision of the AST05 nighttime products processed with the WVS correction option was similar, but their accuracy was significantly better, as the means in Channels 10, 11, and 12 were 0.02, 0.005, and 0.005 lower than Mako NEM, respectively. Therefore it appears that WVS atmospheric correction is the preferred option for obtaining AST05 retrievals that meet the anticipated performance, except for Channel 10 retrievals which remain slightly below the target accuracy even with the WVS correction.

Soil moisture increased predawn Mako NEM emissivities in all the ASTER-channel wavelengths by 0.01 as compared with Mako NEM noontime emissivities for the 270 × 270 m dune area that was used to validate the AST05 retrievals (Figure 6D). Yet, it appears that this soil-moisture effect did not impact AST05 because AST05 retrievals in Channels 13 and 14 did not change and remained stable and accurate in both day as well as night acquisitions (Figure 8). We suggest that the different overpass times between ASTER night acquisitions (~2230) and the predawn Mako experiment (0535) can account for the absence of the soil moisture effect in AST05 night retrievals at Algodones. As air and surface temperatures typically decrease over the course of the night (e.g., Figure 3), the possibility of developing suitable conditions for effective adsorption of atmospheric humidity into the soil increases as the night progresses. Therefore, it appears that when spectrally significant adsorption of atmospheric humidity occurs at Algodones, it typically happens after the time of ASTER's overpass.

The differences between AST05 day and AST05 night retrievals occur mainly in Channels 10 to 12, whereas Channel 13 and 14 emissivities are stable (Figure 8). Coincidently, the spectral range of ASTER Channels 10 to 12 is where both cavity radiation as well as atmospheric effects would be expected to impact AST05 retrievals more significantly. The former because of the lower ε values of the Algodones sand at these wavelengths (Figure 2), which would account for a more prominent cavity radiation effect (Equation (4)), and the latter because of the larger atmospheric contributions at these wavelengths (Figure 6D), which have to be accurately corrected for. However, as Mako NEM retrievals reveal an invariant 2% to 4% cavity radiation effect in both the predawn and noon overpasses (Figure 9C,D) we suggest that residual atmospheric effects lead to the somewhat degraded performance of AST05 nighttime retrievals. We also find that this problem with AST05 retrievals is most pronounced in Channel 10, which has been previously identified as the ASTER TIR channel most impacted by atmospheric effects (e.g., [39]).

6. Conclusions

This paper focused on testing the performance of the ASTER TES algorithm and the AST05 product over the Algodones dunes, which provide a natural geologic surface of high ε contrast that is the type of surface that the TES algorithm was designed for. Mako hyperspectral data with 2.2 m/pixel resolution allowed us to constrain the magnitude of the inherent uncertainty associated with validation of remotely sensed thermal data acquired at 90 m/pixel against centimeter-scale laboratory spectra. Our main findings are as follows:

- Laboratory emissivity spectra for the Algodones sand revealed pronounced grain-size effects on the depth of the Reststrahlen band absorption (Figure 2). Using Mako's 2.2 m/pixel resolution to resolve the primary scene elements in the dune field we were able to demonstrate that this spectral dependency on grain size does not translate into significant emissivity variability at Mako and ASTER scales (Figures 6 and 9) because grain-size distribution is roughly uniform across the dune field. However, at other dune fields, in which sorting is more heterogeneous, grain-size effects could lead to significant spectral variability.

- Predawn (0535) and noontime (1135) Mako overpasses conducted six hours apart revealed an effective ε increase of up to ~0.04 during the predawn acquisition (Figure 6), which is most likely associated with adsorption of atmospherically sourced soil moisture. Within ASTER's TIR spectral channels, the increase in effective ε amounted to only ~0.01, which is, coincidentally, about the magnitude of the inaccuracy predicted for ASTER TES emissivity retrievals [5]. However, this soil-moisture effect did not seem to impact AST05 nighttime retrievals at Algodones. We suggest that this is because spectrally significant adsorption of atmospheric moisture at Algodones can happen later in the night after ASTER's overpass time (~2230 local). The significance of soil moisture following rainfall events was previously recognized as an important factor to consider in TIR remote sensing (cf. [19,36]). Our results highlight the similar impact of atmospherically sourced adsorbed soil moisture on remote TIR measurements.

- Spectral emissivity contrast measured with Mako for fully resolved landscape elements, such as dune slipface slopes and interdune flats, was 2% to 4% lower than the spectral emissivity contrast measured in the lab for sand samples from these landscape elements (Figures 6 and 9). This magnitude of spectral differences is similar to the predicted magnitude of "cavity-radiation" effects on natural geologic surfaces [31,38]. Surface temperature heterogeneity did not impact emissivity retrievals at Algodones.

- The Mako 2.2 m/pixel emissivity maps (Figure 10) reveal that spectral variability within the Algodones validation site when imaged at 90 m/pixel is below 1% because the dune surface at this scale is an effective mixture of the different dune elements. At Algodones, ε retrievals at 90 m/pixel can be directly compared with laboratory centimeter-scale spectra of sand collected in the field after Mako was used to confirm that spectral effects of surface heterogeneity in mineralogy, grain size, moisture, vegetation, and cavity radiation all together were below ~0.01. The Algodones site can also be used to validate other TIR sensors such as ECOSTRESS and MODIS.

- The accuracy and precision of daytime AST05 emissivity retrievals using each of the three available atmospheric correction protocols, i.e., NCAR/NCEP, WVS, and MOD07, is better than 0.01 at Algodones (Figure 8), and therefore meets the target performance of the ASTER's standard emissivity product. Nighttime AST05 emissivities in Channels 10 to 12 display lower precision (<0.03) likely due to residual atmospheric effects. WVS atmospheric correction was required to obtain accurate (<0.01) nighttime AST05 retrievals at Algodones.

Author Contributions: Conceptualization, A.M., A.R.G. and E.A.A.; Methods, all authors; validation, A.M., A.R.G., and E.A.A.; data analysis, A.M. and A.R.G.; field investigation, A.M., A.R.G., E.A.A., H.T.; writing-original draft preparation, A.M.; writing—review and editing, A.R.G., E.A.A., J.B., G.H., D.M.T. and K.M.B.; visualization, A.M., J.B. D.M.T. participated in airborne experiment design, planning, and execution; K.N.B. processed the airborne imagery and conducted a critical review of the data interpretation. All authors have read and agreed to the published version of the manuscript.

Funding: This research was partially funded from the Bear Fight Institute, NASA subcontract No. 1545008. The APC were funded from the Bear Fight Institute, NASA subcontract No. 1545008.

Acknowledgments: We thank Onn Crouvi and Navot Morag, both of the Geological Survey of Israel, for measuring the particle-size distributions and for the XRD analyses, respectively. The research described in this paper was carried out at the University of Washington (Seattle), The Bear Fight Institute (Winthrop, WA), and the Jet Propulsion

Laboratory (California Institute of Technology, Pasadena), under contracts with JPL and the National Aeronautics and Space Administration (NASA). Mako airborne hyperspectral data were acquired under the auspices of The Aerospace Corporation's Independent Research and Development program by Eric Keim. We appreciate the critical reading and formal evaluation by four reviewers that improved the manuscript.

Conflicts of Interest: The authors declare no conflict of interest.

References

1. Abrams, M.J. The Advanced Spaceborne Thermal Emission and Reflection Radiometer (ASTER): Data products for the high spatial resolution imager on NASA's Terra platform. *Int. J. Remote Sens.* **2000**, *21*, 847–859. [CrossRef]
2. Kahle, A.B.; Palluconi, F.D.; Hook, S.J.; Realmuto, V.J.; Bothwell, G. The advanced spaceborne thermal emission and reflectance radiometer (ASTER). *Int. J. Imaging Syst. Technol.* **1991**, *3*, 144–156. [CrossRef]
3. Yamaguchi, Y.; Tsu, H.; Fujisada, H. Scientific Basis of ASTER Instrument Design. In Proceedings of the SPIE 1939, Sensor Systems for the Early Earth Observing System Platforms, Orlando, FL, USA, 11–16 April 1993; pp. 150–160. [CrossRef]
4. Yamaguchi, Y.; Kahle, A.; Tsu, H.; Kawakami, T.; Pniel, M. Overview of Advanced Spaceborne Thermal Emission and Reflection Radiometer (ASTER). *IEEE Trans. Geosci. Remote Sens.* **1998**, *36*, 1062–1071. [CrossRef]
5. Gillespie, A.R.; Rokugawa, S.; Matsunaga, T.; Cothern, J.S.; Hook, S.J.; Kahle, A.B. Temperature and Emissivity Separation from Advanced Spaceborne Thermal Emission and Reflection Radiometer (ASTER) Images. *IEEE Trans. Geosci. Remote* **1998**, *36*, 1113–1126. [CrossRef]
6. Baldridge, A.M.; Hook, S.J.; Grove, C.I.; Rivera, G. The ASTER spectral library version 2.0. *Remote Sens. Environ.* **2009**, *113*, 711–715. [CrossRef]
7. Gustafson, W.T.; Gillespie, A.R.; Yamada, G. Revisions to the ASTER Temperature/Emissivity Separation Algorithm. In *Second Recent Advances in Quantitative Remote Sensing*; Sobrino, J.A., Ed.; Publicacions de la Universitat de València: Valencia, Spain, 2006; pp. 770–775. ISBN1 84-370-6533-X. ISBN2 978-84-370-6533-5.
8. Gillespie, A.R.; Abbott, E.A.; Gilson, L.; Hulley, G.; Jiménez-Muñoz, J.-C.; Sobrino, J.A. Residual errors in ASTER temperature and emissivity standard products AST08 and AST05. *Remote Sens. Environ.* **2011**, *115*, 3681–3694. [CrossRef]
9. Tonooka, H. Accurate atmospheric correction of ASTER thermal infrared imagery using the WVS method. *IEEE Trans. Geosci. Remote Sens.* **2005**, *43*, 2778–2792. [CrossRef]
10. Hook, S.J.; Vaughan, R.G.; Tonooka, H.; Schladow, S.G. Absolute radiometric in-flight validation of mid infrared and thermal infrared data from ASTER and MODIS on the Terra spacecraft using the Lake Tahoe, CA/NV, USA. *IEEE Trans. Geosci. Remote* **2007**, *45*, 1798–1807. [CrossRef]
11. Sobrino, J.A.; Jiménez-Muñoz, J.C.; Balick, L.; Gillespie, A.R.; Sabol, D.A.; Gustafson, W.T. Accuracy of ASTER Level-2 thermal-infrared Standard Products of an agricultural area in Spain. *Remote Sens. Environ.* **2007**, *106*, 146–153. [CrossRef]
12. Coll, C.; Caselles, V.; Valor, E.; Niclòs, R.; Sánchez, J.M.; Galve, J.M.; Mira, M. Temperature and emissivity separation from ASTER data for low spectral contrast surfaces. *Remote Sens. Environ.* **2007**, *110*, 162–175. [CrossRef]
13. Tonooka, H.; Palluconi, F.D. Validation of ASTER/TIR standard atmospheric correction using water surfaces. *IEEE Trans. Geosci. Remote Sens.* **2005**, *43*, 2769–2777. [CrossRef]
14. Schmugge, T.; Ogawa, K.; Jacob, F.; French, A.; Hsu, A.; Ritchie, J.C. Validation of Emissivity Estimates from ASTER Data. In Proceedings of the International Geoscience and Remote Sensing Symposium, Toulouse, France, 21–25 July 2003; pp. 1873–1875. [CrossRef]
15. Schmugge, T.; Ogawa, K. Validation of emissivity estimates from ASTER and MODIS data. In Proceedings of the 2006 IEEE International Symposium on Geoscience and Remote Sensing, Denver, CO, USA, 31 July–4 August 2006; pp. 260–262. [CrossRef]
16. Hulley, G.C.; Hook, S.J.; Baldridge, A.M. Validation of the North American ASTER Land Surface Emissivity Database (NAALSED) Version 2.0. *Remote Sens. Environ.* **2009**, *113*, 2224–2233. [CrossRef]
17. Hulley, G.C.; Hook, S.J. The North American ASTER Land Surface Emissivity Database (NAALSED) Version 2.0. *Remote Sens. Environ.* **2009**, *113*, 1967–1975. [CrossRef]

18. Hulley, G.C.; Hook, S.J.; Abbott, E.; Malakar, N.; Islam, T.; Abrams, M. The ASTER Global Emissivity Dataset (ASTER GED): Mapping Earth's emissivity at 100 m spatial scale. *Geophys. Res. Lett.* **2015**, *42*, 7966–7976. [CrossRef]

19. Sabol, D.E., Jr.; Gillespie, A.R.; Abbott, E.A.; Yamada, G. Field Validation of the ASTER Temperature-Emissivity Separation Algorithm. *Remote Sens. Environ.* **2009**, *113*, 2328–2344. [CrossRef]

20. Sirguey, P. Simple correction of multiple reflection effects in rugged terrain. *Int. J. Remote Sens.* **2009**, *30*, 1075–1081. [CrossRef]

21. Hall, J.L.; Boucher, R.H.; Buckland, K.N.; Gutierrez, D.J.; Keim, E.R.; Tratt, D.M.; Warren, D.W. Mako airborne thermal infrared imaging spectrometer—Performance update. In Imaging Spectrometry XXI. *Int. Soc. Opt. Photonics* **2016**, *9976*, 997604. [CrossRef]

22. Buckland, K.N.; Young, S.J.; Keim, E.R.; Johnson, B.R.; Johnson, P.D.; Tratt, D.M. Tracking and quantification of gaseous chemical plumes from anthropogenic emission sources within the Los Angeles Basin. *Remote Sens. Environ.* **2017**, *201*, 275–296. [CrossRef]

23. Seemann, S.W.; Borbas, E.E.; Li, J.; Menzel, W.P.; Gumley, L.E. MODIS atmospheric profile retrieval algorithm theoretical basis document, version 6. *Coop. Inst. Meteorol. Satell. Stud.* **2006**, *6*, 37.

24. Hulley, G. A Water Vapor Scaling (Wvs) Method for Improving Atmospheric Correction of Thermal Infrared (Tir) Data. In *Thermal Infrared Remote Sensing: Sensors, Methods, Applications*; Künzer, C., Dech, S., Eds.; Springer: Berlin/Heidelberg, Germany, 2013; Volume 17, pp. 253–265. [CrossRef]

25. Palluconi, F.D.; Hoover, G.; Alley, R.; Jentoft-Nilsen, M.; Thompson, T. An atmospheric correction method for ASTER thermal radiometry over land. *Algorithm Theor. Basis Doc.* **1999**. Available online: http://www.aster.jspacesystems.or.jp/t/jp/documnts/pdf/2b01t.pdf (accessed on 29 February 2020).

26. Nicodemus, F.E. Directional reflectance and emissivity of an opaque surface. *Appl. Opt.* **1965**, *4*, 767–773. [CrossRef]

27. Gillespie, A.R. Lithologic Mapping of Silicate Rocks Using TIMS. In Proceedings of the TIMS Data Users' Workshop, NASA Stennis Space Center, MS, USA, 18–19 June 1985; JPL Publication 86-38. pp. 29–44.

28. Realmuto, V.J. Separating the Effects of Temperature and Emissivity: Emissivity Spectrum Normalization. In Proceedings of the 2nd TIMS Workshop, Pasadena, CA, USA, 7–8 July 1990; JPL Publication 90-55. pp. 31–35.

29. Adams, J.B.; Gillespie, A.R. *Remote Sensing of Landscapes with Spectral Images*; Cambridge University Press: Cambridge, UK, 2006.

30. Danilina, I.; Gillespie, A.R.; Balick, L.K.; Mushkin, A.; O'Neal, M.A. Performance of a thermal-infrared radiosity and heat-diffusion model for estimating sub-pixel radiant temperatures over the course of a day. *Remote Sens. Environ.* **2012**, *124*, 492–501. [CrossRef]

31. Danilina, I.; Gillespie, A.R.; Balick, L.K.; Mushkin, A.; Smith, M.R.; Blumberg, D. Compensation for sub-pixel roughness effects in thermal-infrared images. Special Issue: Third International Symposium on Recent Advances in Quantitative Remote Sensing. *Int. J. Remote Sens.* **2013**, *34*, 3425–3436. [CrossRef]

32. Hulley, G.; Baldridge, A. Validation of Thermal Infrared (TIR) Emissivity Spectra Using Pseudo-Invariant Sand Dune Sites. In *Thermal Infrared Remote Sensing: Sensors, Methods, Applications*; Künzer, C., Dech, S., Eds.; Springer: Berlin/Heidelberg, Germany, 2013; Volume 17, pp. 515–527. [CrossRef]

33. Crouvi, O.; Amit, R.; Enzel, Y.; Porat, N.; Sandler, A. Sand dunes as a major proximal dust source for late Pleistocene loess in the Negev Desert, Israel. *Quat. Res.* **2008**, *70*, 275–282. [CrossRef]

34. Korb, A.R.; Salisbury, J.W.; D'Aria, D.M. Thermal-infrared remote sensing and Kirchhoff's law: 2. Field measurements. *J. Geophys. Res. Solid Earth* **1999**, *104*, 15339–15350. [CrossRef]

35. Gillespie, A.R. Spectral mixture analysis of multispectral thermal infrared images. *Remote Sens. Environ.* **1992**, *42*, 137–145. [CrossRef]

36. Hulley, G.C.; Hook, S.J.; Baldridge, A.M. Investigating the effects of soil moisture on thermal infrared land surface temperature and emissivity using satellite retrievals and laboratory measurements. *Remote Sens. Environ.* **2010**, *114*, 1480–1493. [CrossRef]

37. Masiello, G.; Serio, C.; Venafra, S.; DeFeis, I.; Borbas, E.E. Diurnal variation in Sahara desert sand emissivity during the dry season from IASI observations. *J. Geophys. Res. Atmos.* **2014**, *119*, 1626–1638. [CrossRef]

Remote Sens. **2020**, *12*, 815

38. Danilina, I.; Gillespie, A.; Balick, L.; Mushkin, A.; Smith, M.; O'Neal, M. Subpixel Roughness Effects in Spectral Thermal Infrared Emissivity Images. In Proceedings of the 2009 First Workshop on Hyperspectral Image and Signal Processing: Evolution in Remote Sensing, Grenoble, France, 26–28 August 2009.
39. Hulley, G.C.; Hughes, C.G.; Hook, S.J. Quantifying uncertainties in land surface temperature and emissivity retrievals from ASTER and MODIS thermal infrared data. *J. Geophys. Res. Atmos.* **2012**, *117*, D23113. [CrossRef]

Technical Note

Technical Methodology for ASTER Global Water Body Data Base

Hiroyuki Fujisada [1,*], Minoru Urai [1,2] and Akira Iwasaki [1,3]

[1] Sensor Information Laboratory Corp, 2-23-36 Shihaugaoka, Tsukubamirai, Ibaraki 300-2359, Japan; urai-minoru@aist.go.jp (M.U.); aiwasaki@sal.rcast.u-tokyo.ac.jp (A.I.)

[2] Geological Survey of Japan, National Institute of Advanced Industrial Science and Technology (AIST), Tsukuba Central 7, 1-1-1 Higashi, Tsukuba, Ibaraki 305-8567, Japan

[3] Research Center for Advanced Science and Technology, University of Tokyo, 4-6-1, Komaba, Meguro, Tokyo 153-8904, Japan

* Correspondence: fujisada@silc.co.jp; Tel.: +81-297-21-7244

Received: 3 October 2018; Accepted: 16 November 2018; Published: 22 November 2018

Abstract: A waterbody detection technique is an essential part of a digital elevation model (DEM) generation to delineate land–water boundaries and set flattened elevations. This paper describes the technical methodology for improving the initial tile-based waterbody data that are created during production of the Advanced Spaceborne Thermal Emission and Reflection radiometer (ASTER) GDEM, because without improvement such tile-based waterbodies data are not suitable for incorporating into the new ASTER GDEM Version 3. Waterbodies are classified into three categories: sea, lake, and river. For sea-waterbodies, the effect of sea ice is removed to better delineate sea shorelines in high latitude areas: sea ice prevents accurate delineation of sea shorelines. For lake-waterbodies, the major part of the processing is to set the unique elevation value for each lake using a mosaic image that covers the entire lake area. Rivers present a unique challenge, because their elevations gradually step down from upstream to downstream. Initially, visual inspection is required to separate rivers from lakes. A stepwise elevation assignment, with a step of one meter, is carried out by manual or automated methods, depending on the situation. The ASTER global water database (GWBD) product consists of a global set of 1° latitude-by-1° longitude tiles containing water body attribute and elevation data files in geographic latitude and longitude coordinates and with one arc second posting. Each tile contains 3601-by-3601 data points. All improved waterbody elevation data are incorporated into the ASTER GDEM to reflect the improved results.

Keywords: ASTER instrument; stereo; digital elevation model; global database; optical sensor; water body detection

1. Introduction

The Advanced Spaceborne Thermal Emission and Reflection radiometer (ASTER) is an advanced multispectral imaging sensor that was launched on board the Terra spacecraft in December, 1999 [1,2]. ASTER has an along-track stereoscopic viewing capability in its visible and near-infrared (VNIR) bands at 15-m spatial resolution with a base-to-height ratio of 0.6. Because of ASTER's excellent satellite ephemeris and instrument parameters [3–6], this along-track stereoscopic viewing capability makes it possible to generate excellent digital elevation model (DEM) data products from ASTER data without referring to ground control points (GCPs) for individual scenes [5].

After nearly a decade of ASTER data acquisition, sufficient cloud-free data had been acquired such that it was possible to create a global DEM from ASTER data (ASTER GDEM). Versions 1 and 2 of the ASTER GDEM, based on 1.2 and 1.5 million scene-based ASTER DEMs, respectively, were released jointly by the Ministry of Economy, Trade, and Industry (METI) of Japan and the U.S. National

Aeronautics and Space Administration (NASA) in 2009 and 2011 [6]. ASTER GDEM Version 3, which was derived from about 1.9 million scene-based DEMs, will be released to the public sometime in 2018.

Waterbody detection is an essential part of DEM generation, because image matching is not directly possible for waterbodies. For ASTER GDEM Version 2, waterbody detection included a methodology for separating land water bodies from the rest of the land surface and then assigning them a flattened elevation value [5,6]. The methodology applied was valid only for waterbodies contained in the same 1° latitude-by-1° longitude tile area. Lakes that cross tile boundaries and are situated in two adjacent 1° latitude-by-1° longitude tiles may have slightly different elevations in the two adjacent tiles. Another shortcoming of the ASTER GDEM Version 2 approach to waterbody detection and correction was that river elevations did not uniformly step-down from upstream to downstream. No global water data base was released to the public with ASTER GDEM Version 2.

In recent years, many attempts have been made to create global waterbody databases because of their importance in studying global biogeochemical cycles [7–14]. Such databases still have shortcomings related to nonglobal coverage, spatial resolution, and public availability. Although the Shuttle Radar Topography Mission (SRTM) Waterbody Data product (SWBD) satisfies spatial resolution and public availability requirements, the coverage is not global. Rather, data from that mission were collected only between 56° south 60° north latitudes. ASTER data and the ASTER GDEM cover land surface areas between 83° south 83° north latitudes, an important attribute in the generation of a global waterbody database. This paper describes the methodology applied in the production and improvement of a global water database (GWBD) from ASTER data (ASTER GWBD).

In spite of its shortcomings, the SWBD was still useful in the creation of the ASTER GWBD. The SWBD's ESRI Shapefile format was converted to a raster format for comparison with ASTER GWBD. Another dataset useful in creating the ASTER GWBD was the GeoCover2000 [15], which was produced from Landsat 7 data. The original GeoCover2000 dataset, covering the Earth with 14.25 m spatial resolution and UTM coordinates, was converted to the same spatial resolution and coordinates as the ASTER GWBD i.e., to geographic latitude/longitude coordinates with 1 arcsecond postings, and 1° latitude-by-1° longitude tile size. This conversion facilitates accurate comparison with the ASTER GWBD.

ASTER GWBD generation consists of two parts: separation of waterbodies from land areas and classification of detected waterbodies into three categories: sea, river, and lake. The separation process was carried out during scene-based DEM generation using an algorithm described in our previous paper [5]. However, many aspects of ASTER GWBD generation and enhancement required manual intervention, including visual feature identification. Such work was accomplished using our support tool which utilizes 'region of interest' (roi) and 'masking' functions of 'ENVI' image analysis software by Harris Geospatial Solutions.

As mentioned previously, the tile-based water body data are generated from the scene-based waterbody data simultaneously with ASTER GDEM generation, but they were not publicly released as ASTER GWBD with ASTER GDEM Version 2 because of the imperfections previously noted. The new ASTER GWBD was developed in conjunction with ASTER GDEM Version 3 to incorporate the improved water body data into ASTER GDEM Version 3. Important improvements were made to the ASTER GWBD:

(1) Waterbodies are classified into three categories: sea, lake, and river waterbodies based on their features.
(2) Sea-waterbodies have zero elevation.
(3) Lake-waterbodies have flattened (uniform) elevations.
(4) River-waterbody elevations step down monotonically from upstream to downstream.

This paper describes how these improvements to the ASTER GWBD were accomplished. The new ASTER GWBD product consists of a global set of 1° latitude-by-1° longitude tiles that contain water body attribute and elevation data files in geographic latitude and longitude coordinates and with one arc-second postings. Consequently, each tile contains 3601-by-3601 data points, including one common column and one common row with its neighboring tiles. Section 2 describes the basic configuration the GWBD product. Section 3 describes the processing algorithm for sea-waterbody. The major part of the algorithm is zero elevation setting and sea ice removal. Section 4 describes the processing algorithm for lake-waterbodies. The major part of the algorithm is the unique elevation value regardless of the size. Section 5 describes the processing algorithm for river-waterbodies. The major of the algorithm is step down elevation from upstream to downstream. Section 6 describes the processing algorithm how to incorporate the improved waterbody elevation data into GDEM to reflect the improved results. In addition to ASTER GDEM V3, the improved ASTER GWBD also will be released to public sometime in 2018.

2. Basic Configuration

Figure 1 shows the ASTER GWBD folder structure, and Table 1 shows data format. Each tile is composed of an attribute file and a DEM file. The attribute file distinguish a type of waterbody; sea-waterbody (attribute 1), river-waterbody (attribute 2), and lake-waterbody (attribute 3). The DEM file shows zero elevation for sea-waterbody, one unique (flattened) elevation for lake-waterbody, and stepwise elevation with a step of one meter for river elevation.

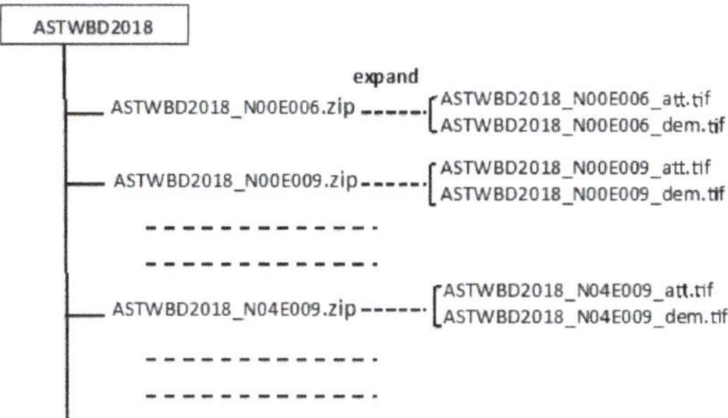

Figure 1. Advanced Spaceborne Thermal Emission and Reflection radiometer (ASTER) global water database (GWBD) folder structure.

Table 1. Data format.

Tile Size	3601 × 3601 (1 degree by 1 degree)
Posting	1 arc-second
Geographic coordinates	Geographic latitude and longitude
Output format	Geotiff, 8 bits for attribute Attribute DN values: 1 for sea, 2 for river, 3 for lake, and 0 for land Geotiff, signed 16 bits, and 1 m/DN for dem files
Special DN values	Referenced to the WGS84/EGM96 geoid
Coverage	North 83 degree to south 83 degree

3. Sea-Waterbody

In order to set the elevation of sea-waterbodies to zero, they first must be separated from inland lakes and rivers. This separation was carried out for scene-based DEM generation using the global sea-waterbody database that was created using GTOPO30 [5,16]. If the sea-waterbody area is larger than 80% of the sea-waterbody GTOPO30 database, this area is identified as a sea-waterbody. The 80% criterion was adopted to compensate for the inaccuracy of the database. The land–sea interface (sea shoreline) is determined during ASTER GDEM generation by calculating the ratio of the number of stacked-sea-waterbody data to the total number of stacked-pixel data. If the ratio is larger than 0.5, the pixel is assigned as a part of the sea waterbody. Otherwise, the pixel is considered to be land. The 50% criterion was adopted in consideration of tidal effects to present an average delineation [6].

In high latitude areas, another obstacle to accurate delineation of the sea shoreline is the presence of sea ice, whose effects must be removed if sea shorelines are to be accurately delineated in the ASTER GWBD and GDEM. Target areas for sea ice removal were selected using the global coarse mosaic image that was generated from original ASTER GDEM data. A sea ice removal process was carried out for the following high latitude target areas.

(1) Latitudes of 60 degrees north and further north areas
(2) Latitudes of 60 degrees south and further south areas
(3) Extreme south of Greenland
(4) Hudson Bay
(5) James Bay
(6) Ungava Bay
(7) Sea of Okhotsk
(8) Bering Sea
(9) Patagonia area

Figure 2 shows the algorithm flow for sea ice removal. It is difficult to delineate sea ice that occurs near sea shorelines using DEM data alone, because sea ice elevations frequently are similar to land elevations near the sea shoreline. Most sea ice exhibits elevations lower than 30 m, but land topography near sea shorelines often does not exceed 30 m, thus the possibility for confusion. Consequently, sea ice removal for the ASTER GWDB utilized ancillary data wherever useful data exist. Two such useful datasets were (1) the Canadian Digital Elevation Data (CDED) [17], which covers all of Canada with postings every 3 arc-seconds for latitude and every 3, 6, or 12 arc-seconds for longitude, depending on latitude and (2) Alaska Digital Elevation Data [18] that covers the State of Alaska with postings every 2 arc-seconds of latitude and longitude. The processing steps employed in sea ice removal are summarized as follows.

(1) Generate a 2° latitude-by-2° longitude tile mosaic from unimproved Version 3 ASTER GDEM data.
(2) If the mosaic area includes any sea shoreline, continue the processing.
(3) If ancillary reference data exist in the mosaic area, delineate by comparing the mosaic data with the reference data by visual identification under the support tool. If reference data are not available, delineate by using brightness contrast under the support tool. In this case, the baseline technique is to designate all areas less than 30 m as sea ice, and then exclude land areas from the sea ice areas by comparing to Google Earth and GeoCover images and/or by visual judgments, as necessary and possible.
(4) The improved sea-waterbody data were incorporated into the ASTER GWBD.
(5) Repeat step (1) to step (4) for all sea ice removal target areas.

Figure 3 presents typical results from the sea ice removal process. Several examples are shown of original and corrected DEM images. Most, but not all, of the gray scale image areas shown in each first image of Figure 3 represent sea ice with elevations lower than 30 m. Some, however, are land areas that had to be identified and retained by manual intervention.

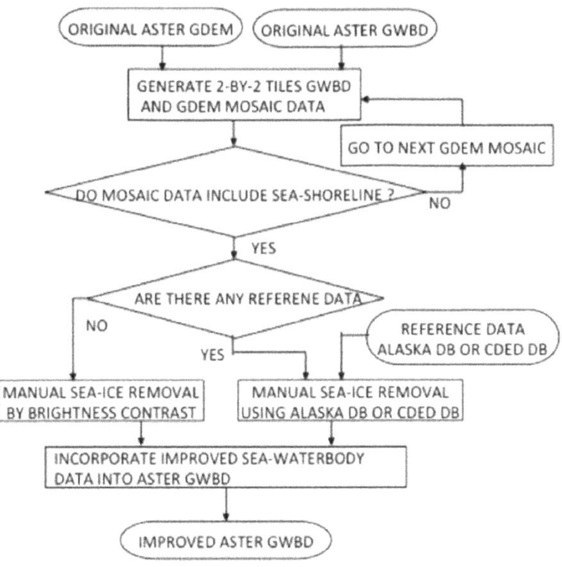

Figure 2. Algorithm flow for sea ice removal.

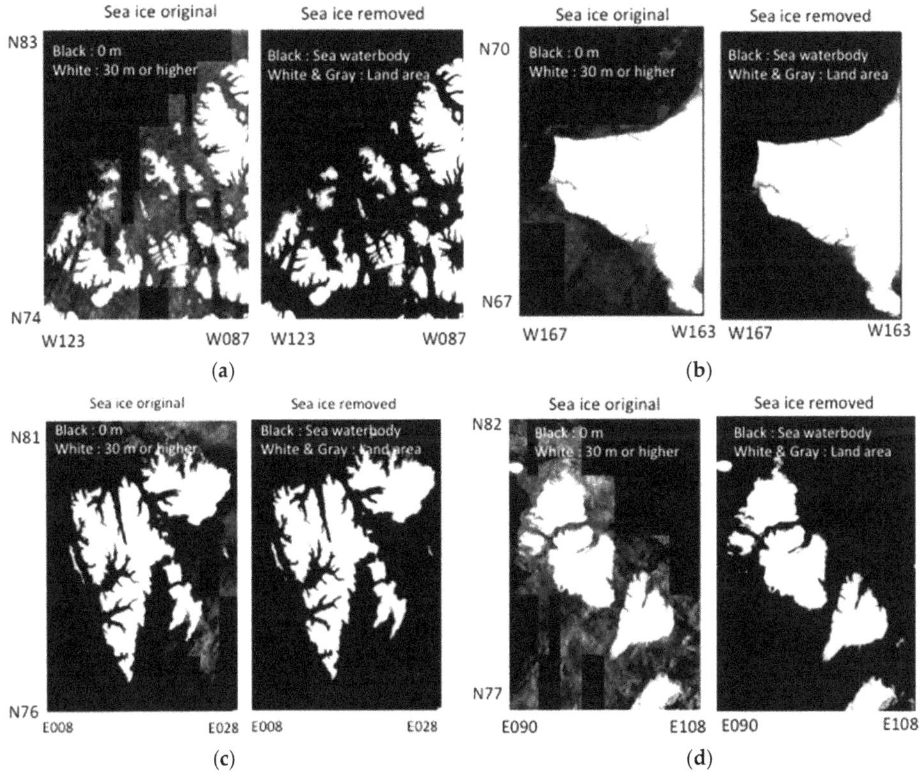

Figure 3. *Cont.*

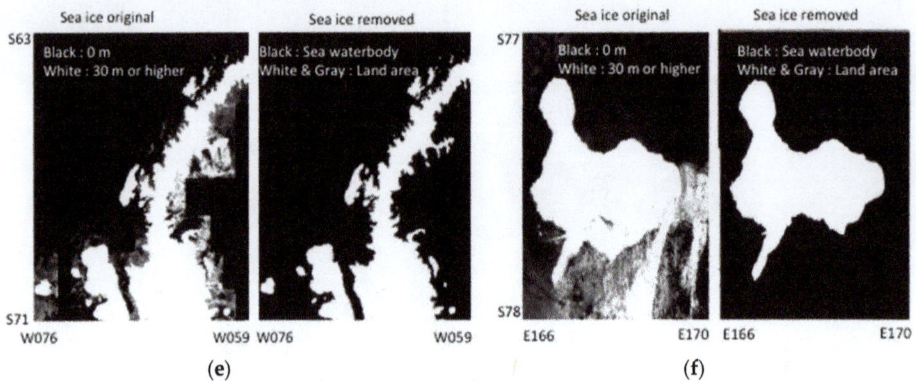

Figure 3. Typical examples of original and corrected DEM images before and after sea ice removal. (**a**) Queen Elizabeth Island; (**b**) Alaska facing Chukchi Sea; (**c**) Svalbard; (**d**) Severnaya Zemlya Islands; (**e**) Antarctic Peninsula; (**f**) Ross Island.

4. Lake-Waterbody

The main goal of lake-waterbody anomaly correction is to set the unique elevation value for each lake regardless of size. The lake grouping is needed because processing algorithm depends on the size of waterbodies. Lake-waterbodies are classified into three groups based on their size. Group1 lakes have sizes larger than a scene-based DEM with 61.5 km in cross-track and 63 km in along-track directions. Table 2 lists seven lakes that belong to group1. Group2 lakes are lakes that are larger than a 2° latitude-by-2° longitude tiles mosaic image of ASTER GWBD and that do not belong to group1 lakes. Table 3b lists lakes belonging to group2. Group3 lakes are all lakes that do not belong to group1 or group2. Group3 lakes are expressed within a 2° latitude-by-2° longitude tiles mosaic image of ASTER GWBD data. The processing algorithm applied to any given lake depends on the waterbody group of that lake.

4.1. Processing Algorithm for Group1 Lake Waterbodies

Group1 lakes all lie between 60°N latitude and 56°S latitude, so errors in the input ASTER GWBD covering these seven lakes can be corrected by replacing ASTER GWBD attributes with SWBD attributes. Figure 4 shows the algorithm flow for group1 lake anomaly correction. The process is carried out as follows.

(1)   Input original ASTER GWBD and SWBD.

(2)   Select one of the seven group1 lakes for correction.

(3)   Generate mosaic image data from both the input ASTER GWBD and the SWBD that cover the total area of the selected group1 lake.

(4)   Copy the selected group1 lake SWBD attributes to the corresponding ASTER GWBD attribute nuarea.

(5)   Assign the nominal elevation value in Table 2 to the selected group1 lake.

(6)   Decompose the ASTER GWBD mosaic image data into individual tiles.

(7)   Incorporate the improved tiles into original input ASTER GWBD.

(8)   If uncorrected group1 lakes remain, repeat Step (2) to Step (7).

Table 2. Group1 lakes.

Name	Location	ASTER GWBD Elevation (m)	Area (km^2)	SWBD Elevation (m)	Area (km^2)
Superior	USA, Canada	179	81,482	179	81,330
Michigan, Huron	USA, Canada	175	118,659	175	118,490
Erie	USA, Canada	172	25,957	172	25,906
Ontario	USA, Canada	73	19,359	73	19,304
Victoria	Kenya, Tanzania, Uganda	1133	67,540	1134	67,455
Bikal	Russia	456	32,212	449	32,021
Caspian	Russia etc.	−28	397,547	−29	377,244

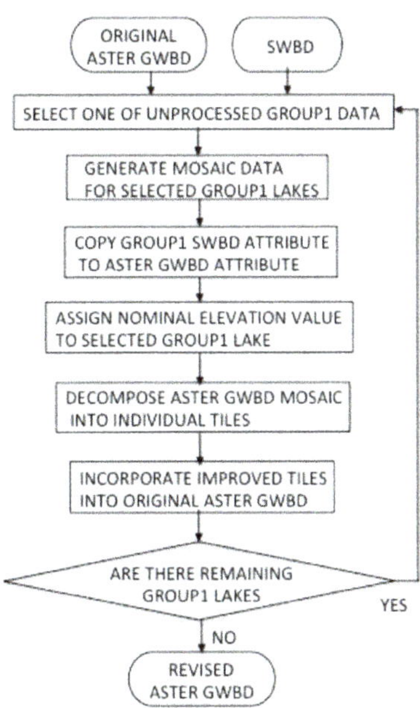

Figure 4. Algorithm flow for group1 lake anomaly correction.

4.2. Processing Algorithm for Group2 Lake Waterbodies

Group2 lake elevations were calculated using unimproved ASTER GDEM V3 data. The corrected elevation values for group2 lakes are reported in Table 3b and compared with corresponding SWBD lake elevation values, where available. Figure 5 shows the algorithm flow for correcting group2 lake elevations. The process is carried out as follows.

(1) Input original ASTER GWBD and ASTER GDEM.
(2) Select one of the group2 lakes for correction.
(3) Generate mosaic image data from both the unimproved input ASTER GWBD and unimproved ASTER GDEM that cover the total area of the selected group2 lake.
(4) Calculate the elevation value of the selected lake surface by averaging perimeter elevations from the input ASTER GDEM data, using only the 10% of perimeter values that fall between 45% and 55% of the perimeter elevations in ascending order from the bottom to top, since the perimeter elevations have the random errors, and the center value is close to real value without the random errors. The parameter 10% is empirically selected.
(5) Decompose ASTER GWBD mosaic image data into individual tiles.
(6) Incorporate the improved tiles into original input ASTER GWBD.
(7) If uncorrected group2 lakes remain, repeat Step (2) to Step (6).

Table 3a. Group2 lakes.

Name	Location	ASTER GWBD		SWBD	
		Elevation (m)	Area (km^2)	Elevation (m)	Area (km^2)
Great Bear	Canada	154	30,586	-	-
Great Slave	Canada	157	26,965	-	-
Winnipeg	Canada	215	24,396	215	24,370
Manitoba	Canada	237	4553	245	4578
Winnipegosis	Canada	250	5089	251	5157
Cedar	Canada	250	2553	253	2573
Nipigon	Canada	259	4464	258	4480
Athabasca	Canada	209	7697	207	7689
Reindeer	Canada	335	5258	335	5469
Nettiling	Canada	24	4744	-	-
Amadjuak	Canada	99	2813	-	-
Churchill, Peter·Pond	Canada	413	1626	415	1796
Kinbasket	Canada	745	361	729	324
Great Salt	USA	1276	4434	1282	2770
Nicaragua	Nicaragua	34	7896	31	7868
Titicaca	Bolivia, Peru	3811	7654	3815	7549
Vanern	Sweden	41	5475	44	5459
Vattern	Sweden	95	1866	88	1870
Peipsi	Estonia, Russia	25	3497	28	3514
Onega	Russia	33	9784	-	-

Table 3b. Group2 lakes.

Name	Location	ASTER GWBD		SWBD	
		Elevation (m)	Area (km^2)	Elevation (m)	Area (km^2)
Ladoga	Russia	3	17,688	4	-
Aral Sea North	Russia etc.	37	3147	39	3031
Aral Sea South	Russia etc.	26	18,333	29	23,743
Yssyk-kol	Kyrgyuzstan	1603	6224	1601	6217
Balkhash	Kazakhstan	337	17,341	338	17,053
Volta	Ghana	75	6329	75	6140
Turkana	Kenya, Ethiopia	348	7490	361	7512
Tanganyika	Tanzania etc.	773	32,971	767	62,971
Malawi	Tanzaniaf, Malawi Mozambique	476	29,688	476	29,653
Kariba	Zambia, Zimbabwe	482	5287	487	5349
Bratsk1	Russia	293	4600	291	4831
Bratsk2	Russia	392	1798	391	1850
Krasnoyarskoye	Russia	236	1920	223	1678
Vilyuy	Russia	234	2072	-	-
Svetogorskaye	Russia	83	497	-	-
Hantalka	Russia	48	2235	-	-
Khantaiskoe	Russia	57	1007	-	-
Keta	Russia	77	461	-	-
Dyupkun	Russia	102	240	-	-
Taymyr	Russia	5	4829	-	-

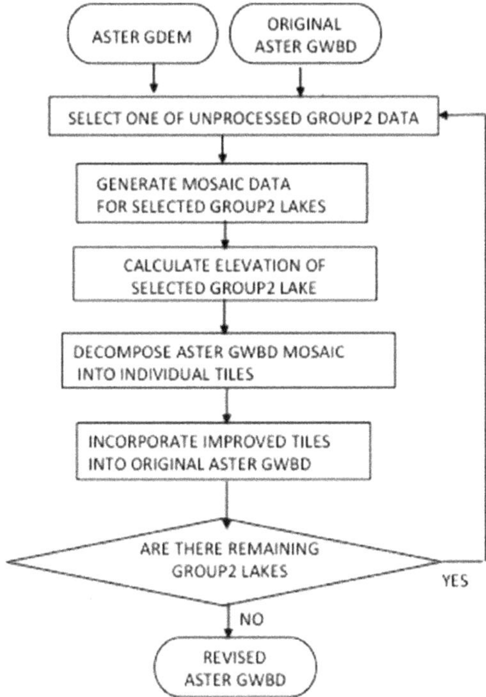

Figure 5. Algorithm flow for group2 lake elevation anomaly correction.

4.3. Processing Algorithm for Group3 Lake Waterbodies

As mentioned previously, group3 lakes are expressed in 2° latitude-by-2° longitude tiles mosaics of ASTER GWBD and GDEM data created from 1° latitude-by-1° longitude tiles. Almost all lakes belong to group3. Most lakes within the original 1° latitude-by-1° longitude tiles were corrected during ASTER GDEM generation. However, some lakes may extend across adjacent tiles, regardless of the size. Thus, the lake elevations may be slightly different for the same lake connected through a tile boundary, such that they still exhibit plural elevations in the 2° latitude-by-2° longitude tiles mosaics. This type of anomaly must be corrected, so these lakes have one unique (flattened) elevation. Figure 6 shows the algorithm flow for group3 lakes anomaly correction. The process is carried out as follows.

(1) Input the unimproved Version 3 ASTER GWBD and ASTER GDEM.

(2) Generate a 2° latitude-by-2° longitude tiles mosaic image data from both the input ASTER GWBD and the ASTER GDEM.

(3) If any lake in the mosaic area has plural elevations, go to next step to calculate the unique elevation for each abnormal lake. Otherwise, return to previous Step (2) to generate the next mosaic image data.

(4) Calculate the unique elevation value for each abnormal lake with plural elevations by averaging the perimeter elevation data. For averaging, use only the data between 45% and 55% from the bottom in ascending order by the same reason as the group2 lakes.

(5) Decompose ASTER GWBD mosaic image data into individual tiles.

(6) Incorporate the improved tiles into the original input ASTER GWBD.

(7) Repeat Step (2) to Step (6) for all land areas between latitudes 83°N and 83°S.

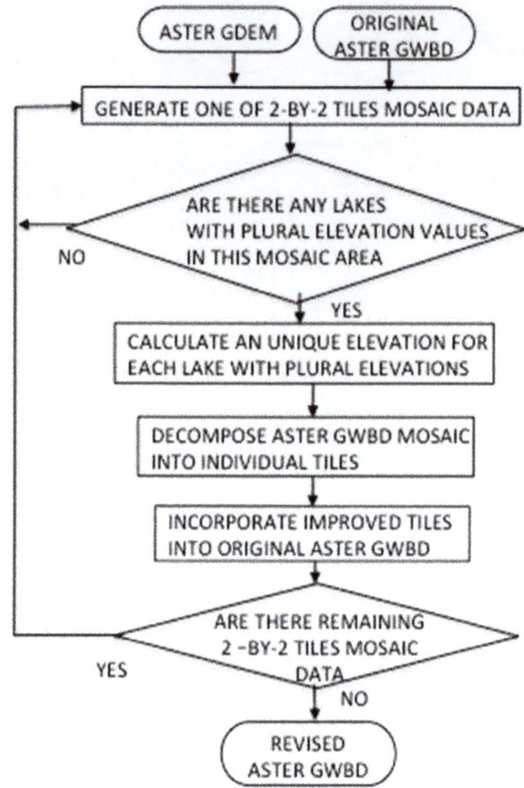

Figure 6. Algorithm flow for group3 lake anomaly correction.

5. River-Waterbody

River elevations are not constant, but rather gradually become lower from upstream to downstream. In order to assign rivers proper declining elevations, they first must be separated from lakes. Unfortunately, there is no automated way to achieve such separation. Consequently, in the initial stage of waterbody detection and assignment of elevation, all inland waterbodies are treated as lakes with a constant elevation value for each tile [5,6]. The separation must be carried out by visual identification for each tile using the support tool. After separating rivers from lakes, river elevations are assigned stepwise elevations with a step of one meter. The stepwise elevation assignment is carried out by a manual or automated method, depending on the situation using the support program. In case of waterfall, one meter step is changed to proper gap elevation. The locations of the waterfalls can be easily identified from Google Earth image.

5.1. Basic Algorithm Flow for River Stepwise Elevation

Figure 7 shows the basic algorithm flow for determining river stepwise elevations. The process involves selecting a series of reference points along the course of a river and assigning those reference points unique elevations based on perimeter elevation values or based on SWBD data, where available. Reference point elevations must always step down (decrease) from upstream to downstream, and they must always be lower than their perimeter elevations. The process is carried out as follows.

(1) Input original ASTER GWBD and ASTER GDEM.

(2) Generate one of 2° latitude-by-2° longitude tiles mosaic image data from both the unimproved input ASTER GWBD and the ASTER GDEM.

(3) Manually separate rivers from lakes, if any rivers exist.

(4) Designate all existing rivers as nominal attribute 2.

(5) Select river reference points from which stepwise elevations between adjacent reference points will be calculated by manual or automated editing, depending on the situation. Set reference point elevations based on perimeter data or SWBD data.

(6) Temporarily designate river reference points as attribute 5.

(7) Apply manual or automated editing to calculate stepwise elevations between adjacent reference points, using the support program.

(8) Decompose ASTER GWBD mosaic image data into individual tiles.

(9) Incorporate improved tiles into original input ASTER GWBD.

(10) Change temporary attribute 5 to formal river attribute 2.

(11) Repeat Step (2) to Step (10) for all land areas between latitudes 83°N and 83°S.

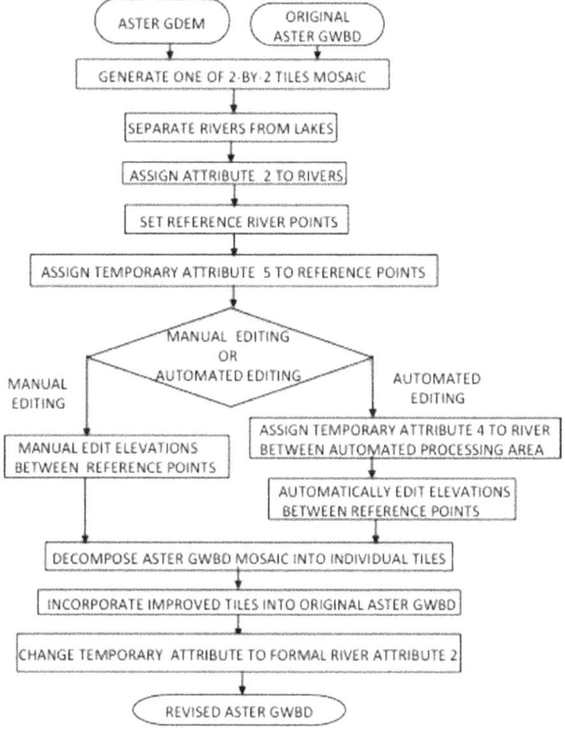

Figure 7. Basic algorithm flow for river stepwise elevation editing.

5.2. Manual Stepwise Elevation Editing

Figure 8 shows one example of manual stepwise elevation editing process. The process is very simple, and is usually used in cases where the elevation difference between adjacent reference points is equal to or less than 16 m, although this criterion can be flexibly changed depending on the situation. The process is carried out as follows.

(1) Original river image for stepwise elevation editing is first masked for each elevation value as shown in Figure 8b using the support tool. The elevation difference is roughly allocated in equal distance between adjacent reference points.

(2) The attribute image area for each elevation value is selected by logical and operation for each masked area and attribute 2 area.

Figure 8c shows the final color density slice elevation image.

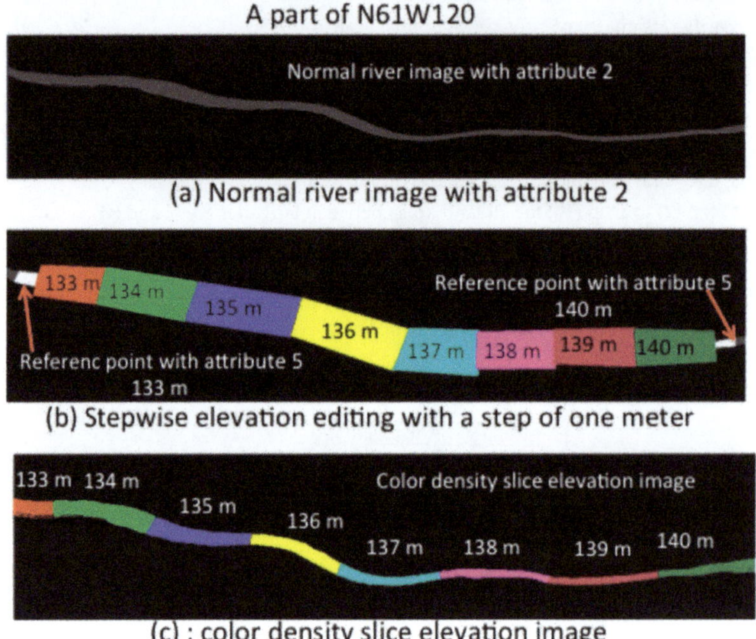

Figure 8. One example of manual editing process for river stepwise elevation.

5.3. Automated Stepwise Elevation Editing

Figure 9 shows one example of the automated stepwise elevation editing process, which utilizes double-line cell strings to define stepwise river segments with decreasing elevations from upstream to downstream. Elevations for the stepwise river segments are calculated from the perimeter river bank elevations, as defined in the ASTER GDEM. Borderlines between connecting river segments are automatically defined and then adjusted to achieve the shortest path length between opposing shorelines. The process is carried out as follows.

(1) Assign the river area between adjacent reference points as temporary attribute 4 to designate the automated stepwise elevation editing area, including the start point (Figure 9a).

(2) Create double-line cell strings for the attribute 4 river segment from perimeter lines along the left and right river banks (Figure 9b). The perimeter string for the island in the river is excluded from the double-line cell strings, as shown in Figure 9b.

(3) Number each cell-string boundary line adjacent to the river from one reference point with the lowest elevation to the other reference points with the highest elevation. The starting minimum number is zero. The total numbers on the left bank side and right bank side may be slightly different, because the two banks of the river do not have exactly the same shoreline distance between reference points.

(4)   Allocate consecutive cell string numbers for each one-meter step to calculate the borders at both bank sides and define the stepwise river segments. Tentative borderlines on the river are connecting lines between two positions ($\overline{AB}$ and $\overline{CD}$ lines) at left and right banks as shown in Figure 10.

(5)   Rotate all border lines to have the shortest path lengths, since the borderlines with the shortest lengths are virtually at right angles to the river flow directions.

(6)   Assign the corresponding elevation values to all river sections between adjacent borders.

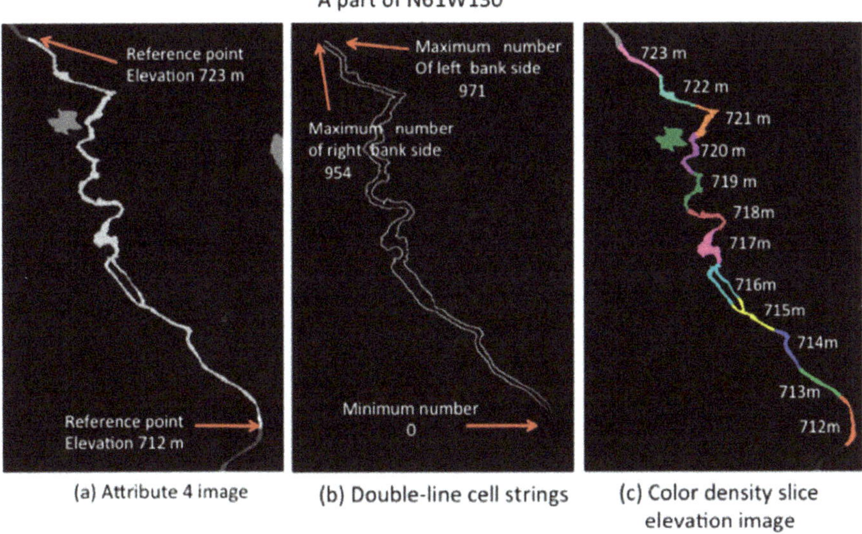

Figure 9. One example of automated editing process for river stepwise elevation.

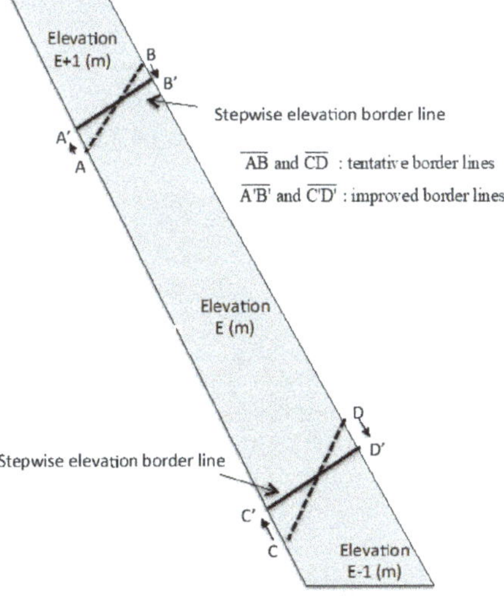

Figure 10. Stepwise river elevation border line rotation to find shortest length.

Figure 9c shows the final color density slice image of the newly defined stepwise river segments.

6. Incorporation of Waterbody Data into GDEM

ASTER GWBD folders include two files: an attribute file and a DEM file, as shown in Figure 1. The improved waterbody elevation data, based on previous sections, are described in the DEM files, and must be incorporated into ASTER GDEM to reflect the improved results. At the same time, it is essential to keep the consistency between waterbodies and their perimeter elevation values. The perimeter elevations must be higher than the waterbody elevation. Figure 11 shows the algorithm flow to incorporate ASTER GWBD data into ASTER GDEM. The process is carried out as follows.

(1) Input original Version 3 ASTER GDEM and the now improved ASTER GWBD.
(2) Select one ASTER GWBD tile and the corresponding original ASTER GDEM tile.
(3) Copy the waterbody elevation data into the original ASTER GDEM tile.
(4) Edit the land elevations along sea shorelines such that those are equal to or higher than one meter.
(5) Edit the perimeter elevations for all lakes in the tile such that those are at least one meter higher than the lake elevation values.
(6) Edit the perimeter elevations for all one-meter steps of rivers in the tile such that those are at least one meter higher than the elevations of the one-meter step area.
(7) Repeat Step (2) to Step (6) for all ASTER GWBD tiles.

Typical examples of the final improved results are shown in Figures 12–14. These shaded-relief images clearly show that all sea and lake waterbodies are completely flattened regardless of their sizes. For river-waterbodies, each stepwise elevation area is flattened, as shown in Figure 14.

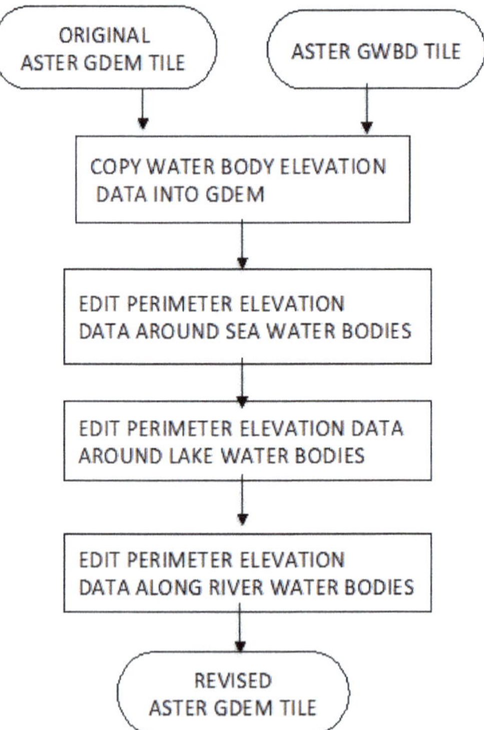

Figure 11. Algorithm flow to incorporate ASTER GWBD.

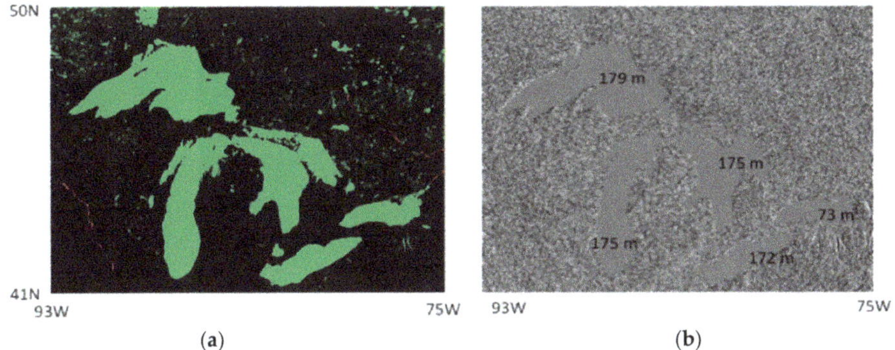

Figure 12. Final improved images of Great Lakes. (**a**) Attribute image of Great Lakes green denotes lake, red denotes river; (**b**) Shaded-relief elevation image of Great Lakes.

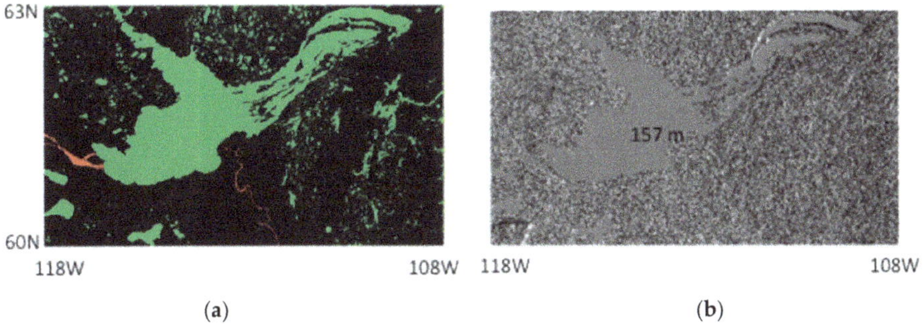

Figure 13. Final improved images of Great Slave Lake. (**a**) Attribute image of Great Slave Lake green denotes lake, red denotes river; (**b**) Shaded-relief elevation image of Great Slave Lake.

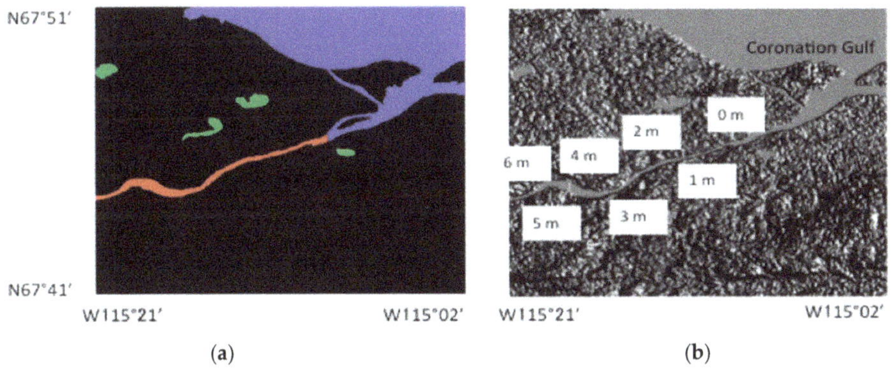

Figure 14. Final improved images of Coronation Gulf area. (**a**) Attribute image of Coronation Gulf area, blue denotes sea, red denotes river, green denotes lake; (**b**) Shaded-relief elevation image of Coronation Gulf area.

7. Summary

A waterbody detection technique is an essential part of DEM generation to delineate land–water boundaries and to set flattened elevations. This paper described the technical methodology for improving the initial tile-based waterbody data that are created during generation of the ASTER

GDEM, but which are not suitable for incorporating into new ASTER GDEM Version 3. Waterbodies are classified into three categories: sea, lake, and river.

Sea-waterbodies were separated from inland waterbodies, and their elevations were set to zero. The effects of sea ice were removed to better delineate sea shorelines in high latitude areas, because sea ice prevents accurate delineation of sea shorelines. This process was enhanced by reference to ancillary data, specifically Google Earth and GeoCover images.

Lake waterbodies are classified into three groups based on size. Group1 lakes are much larger than scene DEMs, which thus do not include enough land area to define the lake or calculate its elevation. For Group1 lakes the corresponding SWBD attribute image was used to define the ASTER GWBD area, and the nominal elevation value was used to assign the lake elevation. Group2 lakes have a size larger than a 2° latitude-by-2° longitude tiles mosaic image of ASTER GDEM data and do not belong to group1 lakes. Group3 lakes are all other lakes. For group2 and group3 lakes, the elevation for each lake was calculated from the perimeter elevation data using the mosaic image that covers entire area of the lake.

River elevations are not constant but gradually decline from upstream to downstream. Rivers were separated from lakes by visual inspection, because there is no automated way to discriminate between rivers and lakes. A stepwise elevation assignment was carried out for rivers using manual or automated methods, depending on the situation under support program.

All improved waterbody elevation data were incorporated into the ASTER GDEM Version 3 to reflect the improved results. At the same time, the waterbody perimeter elevations were edited such that those were at least one meter higher than the waterbody elevation.

Author Contributions: Conceptualization, H.F.; methodology, H.F., M.U. and A.I.; investigation, H.F., M.U. and A.I.; writing—original draft preparation, H.F.; writing—review and editing, H.F., M.U. and A.I.

Funding: This research received no external funding.

Acknowledgments: The authors would like to acknowledge Japan Space Systems for supplying the ASTER data. The authors would also like to thank the ASTER Science Team members, specifically Level-1 and DEM Working Group members for their useful discussion. They would also like to thank G. Bryan Bailey (U.S. Geological Survey, retired) for his suggestions and comments that improved the quality of the manuscript significantly.

Conflicts of Interest: The authors declare no conflict of interest.

References

1. Fujisada, H.; Sakuma, F.; Ono, A.; Kudo, M. Design and preflight performance of ASTER instrument protoflight model. *IEEE Trans. Geosci. Remote Sens.* **1999**, *36*, 1152–1160. [CrossRef]
2. Fujisada, H. ASTER Level-1 data processing algorithm. *IEEE Trans. Geosci. Remote Sens.* **1998**, *36*, 1101–1112. [CrossRef]
3. Fujisada, H.; Bailey, G.B.; Kelly, G.G.; Hara, S.; Abrams, M.J. ASTER DEM performance. *IEEE Trans. Geosci. Remote Sens.* **2005**, *43*, 2707–2714. [CrossRef]
4. Fujisada, H.; Iwasaki, A.; Hara, S. ASTER stereo system performance. *Proc. SPIE* **2001**, *4540*, 39–49.
5. Fujisada, H.; Urai, M.; Iwasaki, A. Advanced methodology for ASTER DEM generation. *IEEE Trans. Geosci. Remote Sens.* **2011**, *49*, 5080–5091. [CrossRef]
6. Fujisada, H.; Urai, M.; Iwasaki, A. Technical methodology for ASTER global DEM. *IEEE Trans. Geosci. Remote Sens.* **2012**, *50*, 3725–3736. [CrossRef]
7. Carroll, M.L.; Townshend, J.R.; DiMiceli, C.M.; Noojipady, P.; Sohlberg, R.A. A new global raster water mask at 250 m resolution. *Int. J. Digit. Earth* **2009**, *2*, 291–308. [CrossRef]
8. Feng, M.; Sexton, J.O.; Channan, S.; Townshend, J.R. A global, high-resolution (30-m) inland water body dataset for 2000: First results of a topographic–spectral classification algorithm. *Int. J. Digit. Earth* **2015**. [CrossRef]
9. Verpoorter, C.; Kutser, T.; Seekell, D.A.; Tranvik, L.J. A global inventory of lakes based on high-resolution satellite imagery. *Geophys. Res. Lett.* **2014**, *41*, 6396–6402. [CrossRef]

10. Verpoorter, C.; Kutser, T.; Tranvik, L. Automated mapping of water bodies using Landsat multispectral data. *Limnol. Oceanogr. Methods* **2012**, *10*, 1037–1050. [CrossRef]

11. Pekel, J.; Cottam, A.; Gorelick, N.; Belward, A.S. High-resolution mapping of global surface water and its longterm changes. *Nature* **2016**, *540*, 418–422. [CrossRef] [PubMed]

12. SRTM Data Editing Rules, Version 2.0. 2003. Available online: https://dds.cr.usgs.gov/srtm/version2_1/Documentation/SRTM_edit_rules.pdf (accessed on 2 November 2018).

13. SRTM Water Body Data Product Specific Guidance, Version 2.0. 2003. Available online: https://dds.cr.usgs.gov/srtm/version2_1/SWBD/SWBD_Documentation/SWDB_Product_Specific_Guidance.pdf (accessed on 2 November 2018).

14. Slater, J.A.; Garvey, G.; Johnston, C.; Haase, J.; Heady, B.; Kroenung, G.; Little, J. The SRTM Data Finishing Process and Products. *Photogramm. Eng. Remote Sens.* **2006**, *72*, 237–247. [CrossRef]

15. LandsatGeocoverMosaics. Available online: http://glcf.umd.edu/data/mosaic/ (accessed on 2 November 2018).

16. GTOPO30. Available online: https://lta.cr.usgs.gov/GTOPO30 (accessed on 2 November 2018).

17. Canada Digital Elevation Data. Available online: https://uwaterloo.ca/library/geospatial/collections/canadian-geospatial-data-resources/canada/canada-digital-elevation-data-cded (accessed on 2 November 2018).

18. Alaska Digital Elevation Models. Available online: https://catalog.data.gov/dataset/5-meter-alaska-digital-elevation-models-dems-usgs-national-map78c4c (accessed on 2 November 2018).

MDPI
St. Alban-Anlage 66
4052 Basel
Switzerland
Tel. +41 61 683 77 34
Fax +41 61 302 89 18
www.mdpi.com

Remote Sensing Editorial Office
E-mail: remotesensing@mdpi.com
www.mdpi.com/journal/remotesensing